Lecture Notes in Networks and Systems

Volume 980

Series Editor

Janusz Kacprzyk, Systems Research Institute, Polish Academy of Sciences, Warsaw, Poland

Advisory Editors

Fernando Gomide, Department of Computer Engineering and Automation—DCA, School of Electrical and Computer Engineering—FEEC, University of Campinas—UNICAMP, São Paulo, Brazil

Okyay Kaynak, Department of Electrical and Electronic Engineering, Bogazici University, Istanbul, Türkiye

Derong Liu, Department of Electrical and Computer Engineering, University of Illinois at Chicago, Chicago, USA
Institute of Automation, Chinese Academy of Sciences, Beijing, China

Witold Pedrycz, Department of Electrical and Computer Engineering, University of Alberta, Alberta, Canada
Systems Research Institute, Polish Academy of Sciences, Warsaw, Poland

Marios M. Polycarpou, Department of Electrical and Computer Engineering, KIOS Research Center for Intelligent Systems and Networks, University of Cyprus, Nicosia, Cyprus

Imre J. Rudas, Óbuda University, Budapest, Hungary

Jun Wang, Department of Computer Science, City University of Hong Kong, Kowloon, Hong Kong

AF811982

The series “Lecture Notes in Networks and Systems” publishes the latest developments in Networks and Systems—quickly, informally and with high quality. Original research reported in proceedings and post-proceedings represents the core of LNNS.

Volumes published in LNNS embrace all aspects and subfields of, as well as new challenges in, Networks and Systems.

The series contains proceedings and edited volumes in systems and networks, spanning the areas of Cyber-Physical Systems, Autonomous Systems, Sensor Networks, Control Systems, Energy Systems, Automotive Systems, Biological Systems, Vehicular Networking and Connected Vehicles, Aerospace Systems, Automation, Manufacturing, Smart Grids, Nonlinear Systems, Power Systems, Robotics, Social Systems, Economic Systems and other. Of particular value to both the contributors and the readership are the short publication timeframe and the world-wide distribution and exposure which enable both a wide and rapid dissemination of research output.

The series covers the theory, applications, and perspectives on the state of the art and future developments relevant to systems and networks, decision making, control, complex processes and related areas, as embedded in the fields of interdisciplinary and applied sciences, engineering, computer science, physics, economics, social, and life sciences, as well as the paradigms and methodologies behind them.

Indexed by SCOPUS, EI Compendex, INSPEC, WTI Frankfurt eG, zbMATH, SCImago.

All books published in the series are submitted for consideration in Web of Science.

For proposals from Asia please contact Aninda Bose (aninda.bose@springer.com).

Nabendu Chaki · Agostino Cortesi ·
Rituparna Chaki · Khalid Saeed
Editors

Applied Computing for Software and Smart Systems

Proceedings of ACSS 2024

Springer

Editors
Nabendu Chaki
Department of Computer Science
and Engineering
University of Calcutta
Kolkata, West Bengal, India

Rituparna Chaki
A. K. Choudhury School of Information
Technology
University of Calcutta
Kolkata, West Bengal, India

Agostino Cortesi
Department of Informatics and Statistics
Ca'Foscari Univeristy
Venice, Italy

Khalid Saeed
Faculty of Computer Science
Bialystok University of Technology
Białystok, Poland

ISSN 2367-3370 ISSN 2367-3389 (electronic)
Lecture Notes in Networks and Systems
ISBN 978-981-97-9761-5 ISBN 978-981-97-9762-2 (eBook)
https://doi.org/10.1007/978-981-97-9762-2

© The Editor(s) (if applicable) and The Author(s), under exclusive license to Springer Nature
Singapore Pte Ltd. 2025

This work is subject to copyright. All rights are solely and exclusively licensed by the Publisher, whether the whole or part of the material is concerned, specifically the rights of translation, reprinting, reuse of illustrations, recitation, broadcasting, reproduction on microfilms or in any other physical way, and transmission or information storage and retrieval, electronic adaptation, computer software, or by similar or dissimilar methodology now known or hereafter developed.
The use of general descriptive names, registered names, trademarks, service marks, etc. in this publication does not imply, even in the absence of a specific statement, that such names are exempt from the relevant protective laws and regulations and therefore free for general use.
The publisher, the authors and the editors are safe to assume that the advice and information in this book are believed to be true and accurate at the date of publication. Neither the publisher nor the authors or the editors give a warranty, expressed or implied, with respect to the material contained herein or for any errors or omissions that may have been made. The publisher remains neutral with regard to jurisdictional claims in published maps and institutional affiliations.

This Springer imprint is published by the registered company Springer Nature Singapore Pte Ltd.
The registered company address is: 152 Beach Road, #21-01/04 Gateway East, Singapore 189721, Singapore

If disposing of this product, please recycle the paper.

Preface

On behalf of Team ACSS, we would like to introduce the proceedings of the eleventh edition of the International Symposium on Applied Computing for Software and Smart Systems (ACSS). The proceedings consist of five chapters covering the domains of Algorithm, Software Engineering, Distributed Systems, Security, Machine Learning and Image Analysis. ACSS began its journey in the year 2014 with the objective of providing a discussion platform for the researchers working on their Ph.Ds. Over the years, it transformed into a full-fledged conference. ACSS uses a double-blind peer review of all to select papers from those submitted to the conference. The conference is known for its highly engaging discussions during the presentations as well as after the presentations. This has been the main reason for the conference in attracting many potential students towards a career in research.

The Symposium was organized by the University of Calcutta in collaboration with Ca'Foscari University of Venice, Italy and Bialystok University of Technology, Poland. The goal of ACSS has been to promote a culture of healthy exchange of research findings among computer scientists across the world.

We, the editors, are extremely thankful to the members of the review committee for taking time off their busy schedules to review each paper. The diverse expertise of the reviewers has made the reviews gateways to improved research articles.

The dissemination initiatives from Springer have drawn many high-quality submissions from scholars primarily, but not exclusively from India. The reviewers mainly considered the technical quality and the originality of each paper, specifically considering the clarity of presentation. The entire process of paper submission, review and acceptance process was done online. After carefully considering the reviews, the Program Committee selected only 18 papers for publication out of a total of 82 submissions.

We take this opportunity to express our sincere gratitude towards the members of the Program and Organizing Committees, whose sincere efforts before and during the symposium have resulted in strong technical program and in effective discussions. We acknowledge the support of the projects SERICS (PE00000014—CUP H73C2200089001) and iNEST (ECS00000043—CUP H43C22000540006) funded by PNRR NextGeneration EU, and we thank Springer Nature for sponsoring the best paper award. We are grateful to Mr. Aninda Bose and his colleagues in Springer Nature for their towards publishing this post-symposium book in the series "Lecture Notes in Networks and Systems". We would also like to thank ACM for the continuous support towards the success of the symposium. Last, but not the least, we thank all the authors without whom the Symposium would not have reached up to this standard.

On behalf of the editorial team of ACSS 2024 we sincerely hope this volume will be beneficial to all its readers and motivate them towards better research works.

Kolkata, India — Rituparna Chaki
Venice, Italy — Agostino Cortesi
Kolkata, India — Nabendu Chaki
Białystok, Poland — Khalid Saeed

Contents

Algorithms

A Graph-Oriented Context-Aware Activity-Based Behavioural Model for Geriatric Care 3
Moumita Ghosh and Sankhayan Choudhury

Streamlining Talent Acquisition by Gaining Insights from Resumes 17
Het Patel, Aayush Patel, Devarsh Patel, Dhruv Shah, Sarthak Tailor, and Novarun Deb

CCD Array Spectrometer-Based FBG Bragg Wavelength Detection Using ANN Algorithm 29
Dipten Kumar, Nandini Basumallick, and Debotosh Bhattacharjee

Performance Assessment of Sequential Models for Solar Radiation Forecasting Over Varying Forecast Horizon 49
Asif Iqbal Middya, Sarbani Roy, and Agnish Paul

Distributed Systems

Advancing Smartphone Sensor-Based Keystroke Dynamics for Implicit and Active Authentication: Addressing Challenges and Enhancing Usability Control 63
Soumen Roy, Utpal Roy, Devadatta Sinha, and Rajat Kumar Pal

A Prototype for Peer-to-Peer Transient Microservices with Wi-Fi Direct 97
Deepak Gohil, Dharmik Jethva, Krish Patel, Ujjwal Singh, Ved Vaghela, and Novarun Deb

Image Analysis

Comparison of the Effectiveness of Face Recognition Algorithms in Terms of Photo Distortion Level 115
Daniel Grabowski and Khalid Saeed

Multi-class Histology Image Analysis Using Handcrafted Texture and Colour Features for Breast Abnormality Classification 129
Anindita Mohanta, Sourav Dey Roy, Niharika Nath, and Mrinal Kanti Bhowmik

FvFc-Net: Forged Video Frame Classification Network 145
Santanu Das, Sourav Dey Roy, Priya Saha, and Mrinal Kanti Bhowmik

Semi-automated Ground Truth Generation System for Bangla Offline Handwritten Text Documents 161
Rejoy Chakraborty, Payel Rakshit, Chayan Halder, and Kaushik Roy

SWFEM: Sparse Weighted Fine-Tuned Ensemble Model for Leaf Disease Detection 173
Rakesh Paul, Susobhan Ghosh, Mridul Ghosh, Sk Md Obaidullah, and Kaushik Roy

Machine Learning

A Comparative Study of Hospital Length of Stay Prediction of Indoor Patients for Admission, Post Admission, and Discharge Stage Data Using Machine Learning Algorithms 187
Somnath Ghosh and Debotosh Bhattacharjee

BEN-RS-ANN: An Innovative Approach for Revealing Emotion from Bengali Text with Exposure to Polysemy Resolution 207
Taniya Seal, Anuran Bhattacharya, Shatarupa Das, Saheli Patra, Debapratim Das Dawn, Abhinandan Khan, Sanjit Kumar Setua, and Rajat Kumar Pal

Flood Detection in UAV Images of Urban Area Using Machine Learning and Deep Learning Techniques 225
Tirthankar Ghosh, Arati Paul, and Nabendu Chaki

Supervised Classification Approach for Precise Cell Type Identification Improves Single Cell Data Analysis 241
Adrija Das, Gourab Das, Amlan Chakrabarti, and Zhumur Ghosh

Quality-Driven Web Service Selection: Machine Learning Based Approach 257
Dewashish Kumar and Anirban Sarkar

Software Engineering and Security Machine Learning

Contextual Correlation Inference in Multi-fleet Robotic Systems 275
Raunak Bag, Nabendu Chaki, and Agostino Cortesi

Risk Assessment in Agile Software Development 295
Oindrilla Ghosh, Rituparna Chaki, Nabendu Chaki, and Agostino Cortesi

Adaptive Cyber Defence: Leveraging GANs for Simulating and Mitigating Advanced Network Attacks in IoT Environments 309
P. Kiran Rao, Subarna Chatterjee, P. Suman Prakash, and K. Seshadri Ramana

Author Index 323

About the Editors

Nabendu Chaki is Professor and Head of the Department Computer Science and Engineering, University of Calcutta, Kolkata, India. He is Editor-in-Chief of the Springer Nature book series on Services and Business Process Reengineering. Besides editing about 40 conference proceedings with Springer, Dr. Chaki has authored 9 text and research books with CRC Press, Springer Nature, etc. He has published more than 250 Scopus indexed research articles in journals and international conferences with h-index of 17. Professor Chaki has served as Visiting Professor in different places including US Naval Postgraduate School, California, and in different Universities in Italy and Poland. He had been Founder Chair of ACM Professional Chapter in Kolkata and served in that capacity for 3 years since January 2014.

Agostino Cortesi Ph.D., is Full Professor of computer science at Ca' Foscari University, Venice, Italy. He has extensive experience in the area of static analysis and software verification techniques. His main research interests concern programming languages theory, software engineering, and static analysis techniques, with particular emphasis on security applications. He serves as Co-Editor-in-Chief of the book series "Services and Business Process Reengineering" published by Springer Nature. He has been Adviser of several doctoral and postdoctoral students from Italy and abroad (e.g., India, Cuba, and then USA) and has published more than 200 papers in high-level international journals and proceedings of international conferences. According to Scopus, his h-index is 21, with more than 1.600 citations. Currently, he holds Chairs of "Software Engineering", "Program Analysis and Verification", "Computer Networks and Information Systems", and "Data Programming".

Rituparna Chaki is Full Professor in the A. K. Choudhury School of Information Technology, University of Calcutta, India, since June 2015. She joined academia as Faculty Member in the West Bengal University of Technology in 2005. Before that she has served under Government of India in maintaining industrial production database. Besides, she has served as Visiting Professor in the AGH University of

Science and Technology, Cracow, Poland, since 2013 for consecutive years. Rituparna did her Ph.D. from Jadavpur University in 2002. She has been associated in organizing many conferences in India and abroad as Program Chair, OC Chair, or Member of Technical Program Committee. She has published more than 60 research papers in reputed journals and peer-reviewed conference proceedings. Her research interest is primarily in ad hoc networking and its security. She is Professional Member of IEEE and ACM. Rituparna had been Secretary for ACMW-India and is Chapter Chair for ACM Professional Chapter, Kolkata.

Khalid Saeed is Full Professor of computer science at Bialystok University of Technology and Half-Time Visiting Professor at Universidad de La Costa, Barranquilla, Colombia. He was with Warsaw University of Technology in 2014–2019 and with AGH Krakow in 2008–2014. He received his B.Sc. degree from Baghdad University in 1976 and M.Sc. and Ph.D. (distinguished) degrees from Wroclaw University of Technology in Poland in 1978 and 1981, respectively. He received his D.Sc. degree (habilitation) in computer science from the Polish Academy of Sciences in Warsaw in 2007. He was nominated by the President of Poland for the title of Professor in 2014. He supervised more than 15 Ph.D. and 160 M.Sc. theses. He gave more than 50 invited lectures and keynotes in different universities in Europe, China, India, South Korea, Serbia, Germany, Japan, and Canada, on biometric image processing and analysis.

Algorithms

A Graph-Oriented Context-Aware Activity-Based Behavioural Model for Geriatric Care

Moumita Ghosh and Sankhayan Choudhury

Abstract In lieu of activity recognition, the broader objective of a geriatric health-care monitoring system is to identify cognitive impairments. The prerequisite of this identification is to model the behavioural pattern of the concerned person in terms of activity sequence. The behavioural pattern may change with the change in context parameters. Thus, a context-sensitive formal representation of behaviour is highly needed for apprehending the deviation within behavioural patterns. In this domain, most of the works are focused on activity discovery or activity recognition; minimal efforts are there on behavioural study. Moreover, the existing models handle simple transactional queries efficiently but for analytical queries, the situation is managed at run-time through the data retrieval from the lower-level recognition model. Instead of mining the entire dataset for each such query, an independent higher-level formal model for the representation of behaviour becomes effective. The caregiver can directly interact with the higher-level model for getting the output in less response time. In this paper, a graph-based behavioural analysis model is formally proposed on top of the underlying recognition model for achieving the above-said objective. The change in activity pattern for a specific context is carefully addressed within the proposed model. The model has been illustrated through a benchmark dataset and a rigorous experiment has been done on benchmark datasets ARUBA, TULUM, and KYOTO to measure the efficiency in terms of response time.

Keywords Behavioural model · Activity recognition · Geriatric care · Graph-based model

Supported by organization x.

M. Ghosh (✉)
Haldia Institute of Technology, Icare Complex, Haldia, West Bengal, India
e-mail: ghoshmoumita06@gmail.com

M. Ghosh · S. Choudhury
University of Calcutta, JD Block, SaltLake, Kolkata, India

© The Author(s), under exclusive license to Springer Nature Singapore Pte Ltd. 2025
N. Chaki et al. (eds.), *Applied Computing for Software and Smart Systems*, Lecture Notes in Networks and Systems 980, https://doi.org/10.1007/978-981-97-9762-2_1

1 Introduction

The growing ageing population face a progressive decline in physical and cognitive skills. Cognitive impairment is one of the major factors that contribute significantly to the loss of the ability to perform Activities in Daily Life (ADL) [15]. Thus close monitoring of daily activities is one of the ways to apprehend the degree of cognitive impairment that helps to take corrective measures. This person-centric approaches in healthcare domains are visible like [16]. In this work also, the attempt is to model the activity pattern of a person for apprehending the chance of occurring Mild Cognitive impairments(MCI).

The widely used ambient assisted living system is based on continuous [17] monitoring of a person using ambient and wearable sensors. However, for apprehending the cognitive impairments, i.e. deviation from his (her) own behavioural practice, the prerequisite is to propose a way for expressing the regular behaviour of the concerned person in terms of activity sequence. Here, the term behaviour implies the collection of possible activity sequences and a specific activity sequence is a collection of consecutive activities executed by the concerned person.

Most of the works in this domain are focused on activity recognition [6] whereas the works related to behavioural modelling are less. For example, instead of finding the simple activity chart, the caregiver may like to execute analytical queries like "Identify the habits of the person concerned" or "Is there any change in habit with the change in context parameters?" In general, most of the existing works are handling these situations by mining the data at run-time using a recognition model instead of providing an independent model to express the behavioural pattern of the person. In this paper, we propose a graph-based formal model for capturing the behaviour of a person concerned. As a result, the user (caregiver) can execute the above-said queries involving semantic interpretation on the proposed higher-level model in a much easier way with less execution time.

Today's assisted living systems are targeting to remind people with cognitive disorders to accomplish daily activities [5]. The paper [9, 12] presented a scenario about activity recognition and reminder systems combined with environmental and wearable sensors for a dementia patient. These attempts can be more fruitful provided the activity pattern can be expressed in a formal way in terms of a model. The subsequent analysis on the generated model may provide an insight into the cognitive state of the person.

The focused objective of this work is to propose a context-sensitive behaviour model on the top of the activity recognition model for better handling of assisted living solutions. For execution of the complex analytical queries, repeated mining from the dataset through the recognition model is not needed; rather the higher-level model will be capable for handling such queries in less response time. The overall system may be visualized as a three-layer architecture. These are the Physical layer, Activity recognition layer, and Behavioural description layer. The Physical layer contains the raw sensor data, the Activity recognition layer uses various data-dependent models for activity recognition, and the proposed Behavioural Description layer describes the behavioural pattern of the concerned person through activity sequences. The development of the topmost layer is only within the scope of this work.

The structure of the paper is as follows. Section 2 describes the existing methodologies for activity recognition and also mentions the motivation behind the current work. In Sect. 3, the proposed Graph-Oriented Behavioural Analysis Model (GOBAM) is presented and illustrated with necessary examples. Section 4 mentions the execution of GOBAM considering benchmark datasets and highlights the inferences that come out through the experiment. Section 5 concludes the article by mentioning the possible future scopes.

2 Related Work

Several assisted living systems are developed for monitoring geriatric people. In general, there are two types of approaches for activity monitoring: statistical and relational approaches. Statistical approaches mainly focus on creating models from a dataset with adequate data using probabilistic reasoning [11]; whereas relational approaches are built based on specific rules. The statistical methods are better at handling the randomness and uncertainty of human activity data. The probabilistic approaches such as the hidden Markov model [1]are also widely used approaches in this category. These models are unable to explain the reason behind the anomaly as they learn directly from sensor data. These are intelligible as no such rules are used for anomaly detection. The models are weak in capturing any external information like contextual values, common sense knowledge, or domain knowledge. On the contrary, relational models are based on clauses and logic [4]. It is a better alternative to capture additional information like context parameter values, information provided by the caregiver, or recommendations from domain experts. These models are easy to understand and offer a clear understanding of anomaly detection.

The above works are focused on activity recognition assuming the simple transaction type queries of the caregiver. These attempts are good enough to answer the queries like "Prepare the daily chart of the activities" or "What is the average duration of sleeping within a day?" But for the execution of queries like "Is there any habit changes due to changes in context parameter?" or "Identify the habits (activity sequence performed by the person in general) in winter nights," the existing approaches are not suitable. These approaches are mining the results from the low-level sensory data through the recognition model for the response of each such query. As a result, the response time becomes significant. Thus instead of the above-said approach, the alternative proposal is to propose a higher-level model for capturing the behavioural pattern (in terms of activity sequences) that can be accessed for answering the said queries within less response time.

Some attempts are visible where the behaviour is considered as a distinct entity [8, 13, 14]. The work presented in these papers uses the term behaviour in different ways; The paper [14] defines the behaviour as a sequence of frequently fired sensors' pattern, but no context parameter is used. The paper [8] defines behaviour as activity sequence or as different features of different home activities in various context parameter settings. But, till now, there has been less attempt to propose a distinct formal model for expressing the behavioural pattern considering specific contexts. An effort [10] is there to develop a personalized behavioural model in terms of activity sequence in a given context.

Graph-based model could be a better alternative for this higher-level representation that captures more semantics for expressing the behaviour of the concerned person. Inherently graphs are more suitable for depicting the existing complex relations within the artifacts. Thus, we have proposed a graph-based behavioural model where an activity is represented as a vertex and the relation between activities (i, j) indicates that activity i precedes activity j. Moreover, the context parameters are also expressed within the model as a different type of vertices and may change the association property for a specific context. Such expression of behavioural pattern is more understandable and the existing graph algorithms may be applied on the model for extracting the information to satisfy the user's requirement from a higher level instead of consulting the underlying crude-level data.

The overall contributions of this proposal are listed below.

- A Graph-Oriented Behavioural Analytical Model (GOBAM) is proposed that occurs on top of the lower-level activity recognition model to formally express the behaviour of a geriatric person in terms of activity sequence.
- Context awareness is carefully addressed within the model for making it compatible with real-life scenarios.
- Complex queries can be executed directly on the proposed model to offer the result in less response time.

3 Proposed Solution: Graph-Oriented Behaviour Analysing Model (GOBAM)

Let us describe the problem scenario first. A geriatric person is living alone in an apartment. She is capable of performing the daily basic activities like sleeping, Bed to Toilet, Eating, Meal Preparation, etc. The frequency of the activities on respective days may be different. The sensors are placed in strategic locations within the apartment to capture data for monitoring the activities of the said person. Sensors can be ambient or wearable. The ambient sensors trigger while a motion is detected and continuously send signals for the time intervals. For a specific activity, a single or more than one sensor may be involved. It is also assumed that the location of the sensors is known beforehand and the activities are done in a sequential manner, i.e. more than one activity is not executed simultaneously. Thus a repository, i.e. a dataset is available that contains labelled sensor data regarding the activity of the person. The challenge is to develop a Graph-Oriented Behavioural Analysis Model (GOBAM) through the analysis of sensor data for expressing the behaviour of the said person in terms of activity sequences.

Table 1 Raw data as triggered in ARUBA dataset

Date	Time	Sensor id	Value	Annotation
04-11-2010	08:11:10 AM	M018	ON	Meal_Preparation begin
04-11-2010	08:11:15 AM	M019	ON	
04-11-2010	08:11:16 AM	M015	ON	
04-11-2010	08:11:17 AM	M018	OFF	

The benchmark dataset ARUBA [2] is used as baseline data for the necessary illustration of GOBAM in this section. The snippet of the ARUBA dataset is shown in Table 1. Each tuple in the table is considered as an event. Collection of events forms a sequence and an activity is represented by a group of consecutive sequences. As for example, in the first event the sensor M018 was triggered and as per annotation the meal preparation activity started. In the next part of this section, the basic artifacts needed for GOBAM are defined and illustrated through examples based on the ARUBA dataset.

3.1 Definitions

Definition 1 Activity(A): An activity is a performed action caused by triggering a set of sensors in a specific sequence. The definition of an activity is clarified by a tuple of five qualifiers as represented in Eq. 1.

$$A_i = ([X_i], [Loc_i], [Start_i], [Dur_i], [Freq_i]) \quad (1)$$

The qualifier $[X_i]$ indicates a set of top k triggering sensors $[x_1, x_2, x_3]$ for the activity A_i considering k is 3 and each sensor is expressed as x_i. The location set $[Loc_i]$ is the possible location of the activity A_i. $[Start_i]$ is the average starting time of the activity considering all occurrences of the activity per time shift. The entire day is considered as a collection of six-time shifts such as morning (5:00–8:59 AM), mid-morning (9:00–11:59 AM), afternoon (12:00 PM–4:59 PM), evening (5:00–9:00 PM), night (9:00 PM–12:59 AM), and late-night (1:00 AM–4:59 AM). So this set contains more than one value provided the activity occurs in various time shifts. $[Dur_i]$ is the average duration of the activity considering all occurrences of the activities in various time shifts. The frequency $[Freq_i]$ indicates the median value from the number of occurrences of the activities in each day.

Example 1 A benchmark dataset ARUBA [2] is used for illustrating the above-said concept of activity. The ARUBA dataset consists of eleven primary activities. These are Sleeping (SL), Bed to Toilet (BTT), Meal Preparation (MP), Eating (ET), Housekeeping (HK), Wash Dishes (WD), Relax (RX), Work (WK), Enter Home (EH),

Leave Home (LH), and Respirate(RS). As per our definition, the activity Sleeping is expressed as follows. Sleeping = {[M002, M003, M007], Bed Room, [11 pm], 240 minutes, 1}. The top 3 triggering sensors (considering k = 3) are [M002, M003, M007]; the location is Bed Room, The qualifier $[Freq_i]$ indicates that the activity Sleeping has occurred in a specific time slot within a day. The value is computed by taking the median of the number of occurrences of the activity per day from the dataset. The sleeping always occurred in the night time shift and the average starting time for the specific time shift of the activity is 11 pm. The average duration of the activity Sleeping is 240 minutes considering all days within the dataset.

Definition 2 A habit (H) is defined as a frequently generated ordered list of activities. Habit is expressed as the following Eq. 2; whereas A_k^{quali} signifies an activity A_K with subsequent qualifiers.

$$H_x = [A_1^{[quali1]}, A_2^{[quali2]}, A_3^{[quali3]}, ..., A_n^{[qualin]}] \tag{2}$$

A habit can be expressed for a specific context. Context is one of the most crucial concerns for apprehending the behaviour of a person. Practically, the habit may change with the change in context. The equation for the context-aware habit description is shown in Eq. 3;

$$H_x^{[context_i]} = [A_1^{[quali1]}, A_2^{[quali2]}, A_3^{[quali3]}, ..., A_n^{[qualin]}] \tag{3}$$

Example 2 Again with reference to the dataset ARUBA [2], the habit is illustrated through examples. A specific activity sequence pattern, occurring more than a fixed threshold for an individual, creates habit. It could be context-unaware or context-specific. For example, two habits for an individual generated from the ARUBA dataset H_1 and H_2 are expressed in detail in Eqs. 4 and 5. In the above-said habit, in H_1, context values are not considered. But the dataset reflects that the pattern or the frequency of the pattern is changing with the change in context values. For example, Meal Preparation takes place with a higher frequency in winter after the activity Sleeping but the frequency of occurrence of Meal Preparation after Sleeping is significantly low in summer. Thus context-sensitive habit declaration is more realistic and examples are described in Eq. 5. Context parameters time shift and season are considered in these examples and accordingly, two habits of the person concerned are mentioned.

$$\begin{aligned} H_1 &= [Sleeping^{[quali1]}, MealPreparation^{[quali2]}, Relax^{[quali3]}] \\ quali1 &= [[M002, M003, M007], BedRoom, [11pm], 240minutes, 1] \\ quali2 &= [[M018, M019, M016], Kitchen, [5am, 8am, 12noon, 4pm, 7pm, 10pm], 33minutes, 8] \\ quali3 &= [[M007, M009, M016], Dining, [8am, 4pm, 8pm], 50minutes, 5] \end{aligned} \tag{4}$$

$$\begin{aligned} &H_2^{shift=Morningandseason=Winter} \\ &= [Sleeping^{[quali1]}, MealPreparation^{[quali2]}, Relax^{[quali3]}] \end{aligned} \tag{5}$$

Definition 3 A Behaviour (B) is a collection of habits and it is expressed as 6

$$B = [H_1^{[context_1]}, H_2^{[context_2]}, H_3^{[context_3]}, .., H_n^{[context_n]}] \tag{6}$$

Example 3 As an example from the ARUBA dataset [2] [Sleeping, Meal Preparation Eating], [Meal Preparation, Relax, Eating] together may be said as behaviour.

$$B_1 = [H_3, H_4] \tag{7}$$

3.2 Context-Unaware Model

A Graph-Oriented Behaviour Analysing Model (GOBAM) is proposed that captures the behavioural pattern of an individual in terms of activity sequences. Initially, the model is presented without considering context parameters; Later an enhanced context-aware GOBAM is proposed in Sect. 3.3. The GOBAM is conceptualized as a graph G; where activities are vertices and execution order between vertices A_i and A_j is represented through an edge. $G(V, E, W)$ where V denotes set of activities.
$V = \{A_1, A_2, A_3.., A_n\}$
$E = \{< e_{ij} > i, j \in V\}$.
E denotes the set of directed edges between a pair of vertices A_i and A_j.
$W = \{[w_{ij}], i, j \in Vand < i, j > \in E\}$.
W, The weight between a pair of vertices A_i and A_j is represented by a 2-tuple. The first indicates a sub-sequence and the remaining value is the frequency of the sub-sequence [7]. The sub-sequence, represented through a solid edge, is treated as post-sequence, i.e. the sequence occurs after the sequence $[A_i, A_j]$ but the dotted edge indicates pre-sequence, i.e. the sequence occurs before $[A_i, A_j]$. For example, in Fig. 1 the weight between MP and ET is represented by a solid edge with tuple value [RX[3]]. It symbolizes that the occurrence of activity Relax (RX) after the activity sub-sequence (Meal Preparation(MP), Eating (ET)) is 3.

Illustration In this section, the context-unaware GOBAM is illustrated through an example. Assume that Table 2 reflects the activity sequences of the resident as extracted from a hypothetical dataset. Based on the extracted activity sequence, the equivalent GOBAM is represented in Fig. 1. As per the description of the GOBAM, activities like Sleeping (SL), Bed to Toilet (BTT), Meal Preparation (MP), Eating (ET), and Relax (RX) are represented as vertices of the graph. There are two types of edges for representing the activity patterns; solid edges and dashed edges. Solid edges between (A_i, A_j) symbolize the post-sequence after the sub-sequence (A_i, A_j). The dashed edges symbolize the pre-sequence before the sub-sequence (A_i, A_j). The weight between a pair of vertices A_i and A_j is represented by a 2-tuple. The first indicates a sub-sequence and the remaining value is the frequency of the sub-sequence. The sub-sequence, represented through a solid edge, is treated as post-sequence, i.e. the sequence occurs after the sequence $[A_i, A_j]$ but the dashed edge indicates pre-

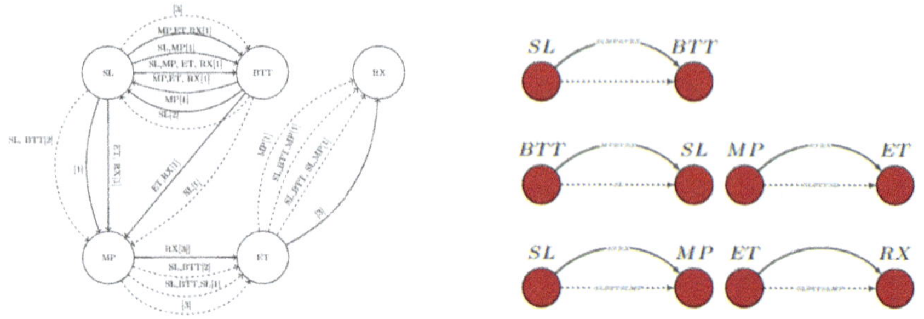

(a) Graph-based representation of the four activity sequences represented in table 2

(b) Transformation of an activity sequence into a Graph

Fig. 1 Activity sequences represented as graph

Table 2 A set of activity sequences

SL no	Activity sequences
1.	[SL, BTT, MP, ET, RX]
2.	[SL, BTT, SL, MP, ET, RX]
3.	[SL, BTT, SL, MP]
4.	[MP, ET, RX]

sequence, i.e. the sequence occurs before $[A_i, A_j]$. For example, in Fig. 1 there are four edges between Meal Preparation (MP) and Eating (ET). The weight between MP and ET is represented by a solid edge with a tuple value [RX[3]]. It symbolizes that the occurrence of activity Relax (RX) after the activity sub-sequence (Meal Preparation (MP), Eating (ET)) is 3. On the contrary, the weight ([SL, BTT[2]]) assigned to a dashed edge symbolizes that [SL, BTT] occurs two times as pre-sequence before [MP, BTT] and it can be verified also from Table 2. The generation of the graph-oriented model GOBAM is described based on the dataset containing sensory data about the daily activities of a person described in Algorithm 1.

3.3 *Context-Aware Model*

Context is one of the most crucial concerns for apprehending the behaviour of a person. In a real-life situation, the behavioural pattern may change with the change in context. As a result, to make the GOBAM more realistic, context-aware GOBAM is presented in this Sect. 3.3. In this presentation, context information is fused with the context-unaware model for a better understanding of behaviour. Various context parameters like time shift, weather, season, etc. are used in general. The context parameters, used for a specific case, are considered as another type of vertices within

Algorithm 1 CUGC: Context Unaware Graph Creation.

Require: Dataset (D) contains the labeled sensor data grabbed by all the sensors within the apartment(as depicted in table 1.)
Ensure: Graph model GOBAM in the form of the adjacency matrix.
Initialize a $GM = M * M$ matrix where M is the number of activities present in the dataset.
Create the activity set Act_set with subsequent qualifiers; $Act\_set = [A_1, A_2, ..A_n]$
An activity sequence list(Act_seq_list) is created containing n number of activity sequences (Act_seq) where n is the number of days for which the data is available in the dataset. ▷ *An Act_seq is the consecutive activity list for a day.*
for each consecutive activity pair $(A_i, A_j) \in Act\_seq[A_{i-1}, A_i, A_j, A_{j+1}]$ **do**
  if A_i is the first activity within the activity sequence **then**
    A dashed edge is created between $A_i \rightarrow A_j$
    $weight(W_{i,j}) = [Null, w]$
    Assign $GM_{i,j} = W_{i,j}$
    ▷ *w is the total occurrences of the presequence*$[A_{i-1}, A_i, A_j]$
  else if $A_{i-1} \rightarrow A_i$ **then**
    a dashed edge is created between $A_i \rightarrow A_j$
    $weight(W_{i,j}) = [A_{i-1}, 1]$
    $GM_{i,j} = W_{i,j}$
    ▷ *w is the total occurrences of the presequence*$[A_{i-1}, A_i, A_j]$
  else if A_j is the last activity in the activity sequence. **then**
    a solid edge is created between $A_i \rightarrow A_j$
    $weight(W_{i,j}) = [Null, w]$
    $GM_{i,j} = W_{i,j}$
    ▷ *w is the total occurrences of the post-sequence*$[A_i, A_j]$
  else if $A_j \rightarrow A_{j+1}$ **then**
    A solid edge is created between $A_i \rightarrow A_j$
    $weight(W_{i,j}) = [A_{j+1}, w]$
    $GM_{i,j} = W_{i,j}$
    ▷ *w is the total occurrences of the post-sequence*$[A_i, A_j, A_{j+1}]$
  end if
end for

GOBAM compared to activity vertices and are represented as squares. A specific coloured edge from a context vertex C_i to activity vertex A_i with weight w represents that according to the given dataset, w is the probability of occurrence of the activity A_i for a given context value C_i. Moreover, the edge (association) between a pair of activities with the same colour indicates the execution order of the activities with specific pre-sequence and/or post-sequence for the given context value.

Context-aware GOBAM is described as $G(V, E, W)$ where $V = [V_1, V_2]$; V_1 and V_2 are set of context nodes and Activity nodes, respectively. $E = \{< e_1, e_2 > i, j \in V_2\}$; e1 is the edge between the context node and to activity node whereas e2 indicates the edge between two consecutive occurred activities. $W = \{[W_1, W_2]\}$. W_1, is the weight between the context vertex C_i and the activity vertex A_i; whereas the weight W_2 signifies the weight between a pair of activity vertices A_i and A_j.

Consider Table 3 where each tuple contains the execution order of two activities in a given context with occurrence probability and corresponding pre-sequence, post-sequence pair. The equivalent context-aware GOBAM model is shown as Fig. 2a. In this diagram, the context nodes are represented as squares, and activity nodes are represented as circles. The diagram reflects that the activity of Sleeping occurs mostly at night time shift(say context C1) with a probability 0.7 whereas it also occurs at the morning but with a probability 0.3. Moreover, it also reflects that for context C1 [i.e. time shift = night]; the next activity of Sleeping (SL) is "Bed to Toilet (BTT) with post-sequence [MP, ET, RX] with frequency 1. Similarly, the

Table 3 Activity sequences with context parameters

SL no.	Present activity	Probability	Context parameters	Next activity	Pre-sequence	Post-sequences
1.	SL	0.7	Night	BTT	Null	[MP, ET, RX]
2.	SL	0.6	Morning	MP	[SL, BTT]	[ET, RX]
3.	SL	0.3	[Winter, Morning, Weekday]	MP	[SL, BTT]	Null
4.	MP	0.3	Morning	LH	[SL, BTT]	[LH, EH, ET]
5.	MP	0.4	Afternoon	ET	[RX, LH, EH]	[ET]

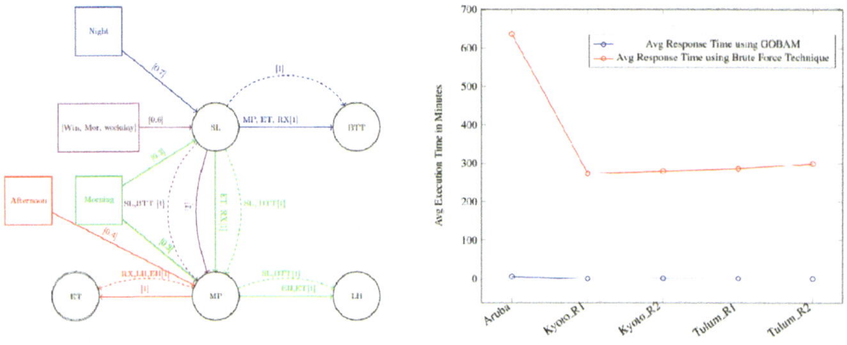

(a) Context Aware Behavioral Model

(b) Compare among average response time for different datasets for Context-Aware Model

Fig. 2 Context-aware behavioural model and average response time

dashed blue line between SL and BTT for same context (marked as blue) indicates that the pre-sequence is null corresponding to the activity sequence [SL, BTT] in the same context C1.

3.4 *Information Extraction from GOBAM*

In this application domain of "Geriatric care," the caregivers may be interested to analyse the behavioural pattern of the person for apprehending mild cognitive impairments. This typical requirement cannot be satisfied by simple transaction queries. In contrast, it demands the execution of more analytical queries like "What are the habits of the person?" or "Is there any change in the habit for a specific context?"

We have listed a short list of such queries like "List the activities exceeding a reference range as chosen by the caregiver" or "Find the possible next n activities for a given activity in a specific context." All these queries can be executed based on the proposed model GOBAM without consulting the underlying data.

The first query is to determine the habits of the person concerned. In the GOBAM as depicted in Fig. 1, consider all the edges between the activities "MP" and "ET." Each edge forms a sequence. The number of activities present in a sequence is termed support. For example, the four sequences are formed from four edges. One of the sequences is [MP, ET,3] which implies it's a sequence of support 2 with frequency value 3. The others are [SL, BTT, SL, MP, ET,1], [SL, BTT, MP, ET,2], and [MP, ET, RX,3]. As per the given definition, habit is a frequently occurred activity sequence, i.e. the sequence with a frequency more than the selected threshold value. Let the threshold for this example be 1; as a result, the sequence [SL, BTT, SL, MP, ET] is going to be deducted from the already formed four sequences. The next task is to check whether a sequence of support n exists within a sequence of greater support or not. Accordingly, after getting a match, the sequence of lower support is deducted. This process is continued, and finally, the existing sequences are formed as a habit. The same procedure is executed between each pair of vertices, i.e. activities, and accordingly, a set of habits is formed. To answer the second query, a graph edit distance is used. Several habits are generated under different context parameter settings. A graph edit distance-based approach is used to determine how the habit changes under two different contextual situations. The graph edit distance is the number of edges/nodes change needed to make two graphs isomorphic. A higher numeric value means two habits are very much dissimilar and a lower value signifies two nearly similar habits.

The caregiver may be interested to know whether the resident is performing the activities within a normal range or not. This information is highly important as the activity frequency could be a marker to find cognitive impairment. All this information like the average frequency of an activity, the average duration of an activity, and the average start time of any activity can be easily provided using the qualifier value concerned to a specific activity within GOBAM. The possible next n activities can be stated also following a given activity in a specific context.

4 Experiment and Inferences

In this section, we have demonstrated the experiments for validating the necessity of the proposed model GOBAM. We use three publicly available datasets from the CASAS smart home. The ARUBA [2] dataset consists of nine months of data of a single resident. The ARUBA dataset consists of eleven primary activities. These are Sleeping (SL), Bed to Toilet (BTT), Meal Preparation (MP), Eating (ET), Housekeeping (HK), Wash Dishes (WD), Relax (RX), Work (WK), Enter Home (EH), Leave Home (LH), and Respirate (RS). The TULUM dataset [3]represents sensor

Table 4 Context-unaware and context-aware activity sequences for ARUBA dataset

SL No.	Context-unaware frequent habits	Habits are generated in Winter Evening
1.	[Wash Dishes, Relax]	[Meal Preparation, Eating, Wash Dishes]
2.	[Sleeping, Bed to Toilet, Sleeping, Meal Preparation, Relax]	[Eating, Wash Dishes, Relax]
3.	[Meal Preparation, Relax, Meal Preparation, Eating]	[Meal Preparation, Relax]
4.	[Leave Home, Enter Home, Meal Preparation, Relax, Sleeping]	[Meal Preparation, Work, Meal preparation, Relax]

Table 5 Dissimilarity index for two habits under different contextual parameter setting for ARUBA dataset

Context parameter for first habit			Context parameter for second habit			Dissimilarity score
Weekday	Season	Shift	Weekday	Season	Shift	
Monday	Summer	Morning	Monday	Summer	Afternoon	15
Tuesday	Winter	Morning	Tuesday	Summer	Morning	8
Sunday	Winter	Night	Monday	Winter	Night	2
Sunday	Summer	Evening	Sunday	Late afternoon	Winter	7

events collected in the WSU smart apartment testbed from April to July of 2009. The proposal is further executed on the KYOTO dataset [3].

Table 4 listed context-unaware and context-aware habits of ARUBA dataset. In this experiment, the sequences are considered as habit provided the sequence repeats more than fifty percent of the days for which the data is available within the dataset. ARUBA contains 180 days of data; thus each habit is a sequence that repeats for more than 90 days.

The outcome as listed in Table 5 reflects that the change in behavioural pattern is significant (edit distance = 15) if the context changes from (Monday, Summer, Morning) to (Monday, Summer, Afternoon). But no such significant change happened (Edit distance = 2) considering the change in context from (Sunday, Winter, Night) to (Monday, Winter, Night). We have also trapped the response time for finding the habits using GOBAM and the traditional brute force approach that mines the data from lower level. Significantly the average response time is reduced compared to the brute force approach which in turn establishes the necessity of the proposed higher-level behavioural representation through GOBAM. The response time is measured for habit generation in various contexts (100 times for each dataset considering random context) for the benchmark datasets ARUBA, TULUM, and KYOTO, and the result is depicted in Fig. 2b.

5 Conclusion

In this paper we proposed a graph-based contextualized behaviour analysis model through activity sequences. The model captures the typical behavioural pattern of a person concerned in various contexts. It can be used as a crucial tool to apprehend the mild cognitive impairments as the deviation can be identified through the comparison among the GOBAM snapshots of various time instants. To understand the habits becomes a key prerequisite for apprehending deviation, and that can be derived smoothly using GOBAM in lesser response time. We have planned to extend the work towards dementia patient care. The behavioural pattern of dementia patients can be generated while sufficient data of various dementia patients are available. The plan is to find the commonality within the generated graphs through which we may infer a general activity pattern for a patient having dementia.

References

1. Atallah L, Yang GZ (2009) The use of pervasive sensing for behaviour profiling-a survey. Pervasive Mob Comput 5(5):447–464
2. Cook DJ (2010) Learning setting-generalized activity models for smart spaces. IEEE Intell Syst 2010(99):1
3. Cook DJ, Schmitter-Edgecombe M (2009) Assessing the quality of activities in a smart environment. Methods Inf Med 48(05):480–485
4. Dey AK (2009) Modeling and intelligibility in ambient environments. J Ambient Intell Smart Environ 1(1):57–62
5. Evans JJ, Emslie H, Wilson BA (1998) External cueing systems in the rehabilitation of executive impairments of action. J Int Neuropsychol Soc 4(4):399–408
6. Forkan ARM, Khalil I, Tari Z, Foufou S, Bouras A (2015) A context-aware approach for long-term behavioural change detection and abnormality prediction in ambient assisted living. Pattern Recogn 48(3):628–641
7. Halim Z, Ali O, Khan MG (2019) On the efficient representation of datasets as graphs to mine maximal frequent itemsets. IEEE Trans Knowl Data Eng 33(4):1674–1691
8. Hoque E, Dickerson RF, Preum SM, Hanson M, Barth A, Stankovic JA (2015) Holmes: a comprehensive anomaly detection system for daily in-home activities. In: International conference on distributed computing in sensor systems. IEEE, pp 40–51
9. Kautz H, Arnstein L, Borriello G, Etzioni O, Fox D (2002) An overview of the assisted cognition project. In: AAAI-2002 workshop on automation as caregiver: the role of intelligent technology in elder care, No 2002, p 6065
10. Lago P, Jiménez-Guarín C, Roncancio C (2017) Contextualized behavior patterns for change reasoning in ambient assisted living: a formal model. Expert Syst 34(2):e12163
11. Monekosso DN, Remagnino P (2010) Behavior analysis for assisted living. IEEE Trans Autom Sci Eng 7(4):879–886
12. Osmani V, Balasubramaniam S, Botvich D (2007) Self-organising object networks using context zones for distributed activity recognition. In: Proceedings of the ICST 2nd international conference on Body area networks, pp 1–9
13. Rashidi P, Cook DJ, Holder LB, Schmitter-Edgecombe M (2010) Discovering activities to recognize and track in a smart environment. IEEE Trans Knowl Data Eng 23(4):527–539
14. Saives J, Pianon C, Faraut G (2015) Activity discovery and detection of behavioral deviations of an inhabitant from binary sensors. IEEE Trans Autom Sci Eng 12(4):1211–1224

15. Sands LP, Yaffe K, Lui LY, Stewart A, Eng C, Covinsky K (2002) The effects of acute illness on ADL decline over 1 year in frail older adults with and without cognitive impairment. J Gerontol A Biol Sci Med Sci 57(7):M449–M454
16. Sarkar P, Chaki R, Cortesi A (2022) A patient-centric nurse scheduling algorithm. SN Comput Sci 3:1–16
17. Thapliyal H, Nath RK, Mohanty SP (2017) Smart home environment for mild cognitive impairment population: solutions to improve care and quality of life. IEEE Consum Electron Mag 7(1):68–76

Streamlining Talent Acquisition by Gaining Insights from Resumes

Het Patel, Aayush Patel, Devarsh Patel, Dhruv Shah, Sarthak Tailor, and Novarun Deb

Abstract This report investigates an advanced Resume Analyzer that employs Natural Language Processing (NLP) to streamline the candidate screening process. Utilizing sophisticated algorithms, the system extracts pivotal information from resumes, such as skills and experience, and employs machine learning model for precise job relevance categorization. The architectural framework encompasses essential stages, including data preprocessing, feature extraction, and model training. Evaluation results underscore the system's accuracy and scalability, with detailed case studies illustrating its practical applicability. The Resume Analyzer is positioned to revolutionize conventional recruitment processes, presenting a tool that enhances objectivity and facilitates more informed hiring decisions in today's competitive job market.

Keywords Resume analyzer · Automated categorization · Natural language processing · Job matching

H. Patel · A. Patel · D. Patel · D. Shah (✉) · S. Tailor · N. Deb
Indian Institute of Information Technology, Vadodara, India
e-mail: 202151142@iiitvadodara.ac.in

H. Patel
e-mail: 202151108@iiitvadodara.ac.in
URL: https://iiitvadodara.ac.in/

A. Patel
e-mail: 202151106@iiitvadodara.ac.in

D. Patel
e-mail: 202151107@iiitvadodara.ac.in

S. Tailor
e-mail: 202151169@iiitvadodara.ac.in

N. Deb
e-mail: novarun_deb@iiitvadodara.ac.in

© The Author(s), under exclusive license to Springer Nature Singapore Pte Ltd. 2025

N. Chaki et al. (eds.), *Applied Computing for Software and Smart Systems*, Lecture Notes in Networks and Systems 980, https://doi.org/10.1007/978-981-97-9762-2_2

1 Introduction

In the evolving landscape of talent acquisition, the process of efficiently screening and categorizing resumes has become increasingly crucial for organizations seeking to make informed hiring decisions. This report delves into the development and implementation of a sophisticated Resume Analyzer, a novel tool designed to enhance the efficiency and objectivity of candidate screening through the integration of Natural Language Processing (NLP) and machine learning techniques. Leveraging Python programming language and several key libraries, including NLTK, Scikit-learn, and Pandas, the project undertakes a comprehensive exploration of resume data to extract pertinent details, such as skills and experience. The methodology encompasses text preprocessing, employing techniques such as tokenization, stemming, and vectorization, to transform raw textual data into a format suitable for machine learning analysis.

The project employs a RandomForestClassifier within a robust pipeline, integrating text preprocessing and vectorization seamlessly. The model is trained on a dataset comprising diverse resumes and their associated categories, paving the way for accurate job relevance categorization. The system's efficacy is evaluated using metrics such as accuracy, precision, recall, and F1-score, showcasing its proficiency in classifying resumes according to predefined job categories. The exploration further extends to a practical application, wherein the Resume Analyzer is employed to assess the relevance of resumes to a provided job description, showcasing its adaptability to real-world scenarios.

Additionally, the report introduces a novel approach to enhancing the relevance scoring of resumes through cosine similarity, demonstrating the tool's capability to prioritize the most pertinent profiles. The integration of this scoring mechanism represents a critical innovation in the resume analysis process, enabling organizations to identify and shortlist candidates with the highest degree of alignment to specific job requirements.

This introduction sets the stage for a comprehensive exploration of the Resume Analyzer, elucidating its development, functionality, and practical applications. The ensuing sections will delve into the intricacies of the project, including the methodology, results, and implications for the recruitment landscape. Through this exploration, we aim to provide a detailed understanding of the project's significance, its contributions to the field of talent acquisition, and its potential to revolutionize traditional recruitment processes.

2 Literature Survey

- Resume Parser Using Natural Language Processing Techniques The paper focuses on parsing the resume from /pdf/doc/rtf format into plain document and tokenized data entities. It then extract skillsets from the Resumes, Ranking algorithms are used, and best resumes are catered according to employer's provided skillset constraints. The results are finally provided in the form of Pie Chart and Bar Graph [1].

- Resume Analyzer and Suggestion Making Using Natural Language Processing: This paper introduces an information retrieval prototype that can extract relevant information from resumes and provide suggestions to the recruiters and the applicants. The prototype can also generate a summary of the resume and a score based on the skills and experience of the applicant [2].
- A Machine Learning Approach for Automation of Resume Recommendation: This paper proposes a resume recommendation system using Machine Learning-based classification techniques and existing similarity measures used to find the most relevant resumes. The paper shows classification using four different models, Random Forest, Multinomial Naive Bayes, Logistic Regression, and Linear Support Vector Machine, and the accuracy scores of these models are compared for the task. It is a Machine Learning solution whose complete framework consists of preprocessing, similarity comparison, classification, and ranking leading to results [3].
- Resume Screening Classification Using Artificial Intelligence and Natural Language Processing: The paper has leveraged the AI and ML understanding and applied a custom algorithm for matching and predicting best job applications. It first applies NLP and Artificial Intelligence to extract features and then gives scores to resumes using matching algorithms. The application is flexible and customizable for different organizations. The drawback is that the paper focuses and gives accuracy on boolean output, either the resume is selected or is rejected, and no ranking is given [4].
- AI Resume Analyzer: This paper aims at enhancing the job-seeking process by using advanced technologies such as NLP and machine learning. It has a systematic approach, from user registration to job recommendation, giving a good security of data. The project's workflow involves resume parsing, user categorization, and then dynamic job recommendations according to extracted individual skills and experiences. Notably, it stands out for its adaptability, ensuring tailored job recommendations that evolve with the dynamic demands of the job market. It offers a scalable solution for both job seekers and employers [5].

3 Comparative Analysis

In this section, we'll compare how we conducted our research to other methodologies used as a part of other existing studies. By comparing methodologies, we aim to improve our research and contribute to the broader conversation in our field.

3.1 Methodologies in Feature Extraction

There are multiple methodologies to execute the feature extraction, which is done after the preprocessing of the text of the resumes, namely Bag-of-Words (BoW), Word2Vec, and Term Frequency, Inverse Document Frequency (TF-IDF). Bag-Of-Words (BoW) is a simple, yet widely used method for vectorizing text data, which involves creating a dictionary of unique words and counting the number of times each unique word appears in a unique document.

Word2Vec, a neural network-driven method, acquires continuous word embeddings to grasp the semantic connections among words. It has its limitations: it cannot handle out-of-vocabulary (OOV) words well, and cannot capture different meanings of a word based on context.

TF-IDF works by giving more weights to the words which are important for distinguishing between different documents and reducing the accuracy of clustering or retrieval.

In our case, the use of such methods is to extract the features of the resumes in which the majority of required skills, education, and requirements are almost the same for applicants, and resumes are screened based on the same unique skills or some distinguished projects which make it different from other applications. We have chosen TF-IDF over BoW for feature extraction, since, by giving higher weightage to rare words, it helps in capturing the meaning of a document. Also many times, some words in the resume could be out of vocabulary or have been used in different contexts in the corpus, which is why TF-IDF is preferred over Word2Vec.

3.2 Methodologies in Classification Methods

There are multiple methodologies to implement the ML classification, which is done after the feature extraction of the resumes, namely Logistic Regression, Random Forest, Linear Support Vector Classification, and Multinomial Naive Bayes.

Logistic Regression uses a logistic function to model a binary dependent variable. The resume data is passed to LR as well and the accuracy is recorded for a new resume.

The Linear Support Vector Classifier (LCVC), also known as a Support Vector Machine (SVM), is a supervised classification algorithm that works by making a separating hyperplane. In simpler terms, in a two-dimensional setting, a hyperplane refers to a line that divides a plane into two distinct regions, each corresponding to a different class.

Multinomial Naive Bayes (MNB), a type of Naive Bayes classifier, is a straightforward "probabilistic classifier" relying on Bayes' theorem while making strong assumptions regarding the independence of features.

We chose Random Forest Classifier over other types for the following reasons:

- Random Forest classifiers excel in handling non-linear relationships in resumes, outperforming linear classifiers like Logistic Regression.
- They are less prone to overfitting, even with complex feature spaces, enhancing their suitability for resume analysis.
- Random Forest models provide feature importance scores, aiding recruiters in identifying influential skills and experiences.
- They generally perform well across metrics like accuracy, F1-score, precision, and recall, optimizing the hiring process.
- Random Forests efficiently handle large datasets and high-dimensional feature spaces, ensuring scalability.

3.3 Methodologies Used in Similarity Measures

There are multiple methodologies for measurement of similarity like Cosine Similarity, Jaccard Similarity, Euclidean Distance, and Pearson Correlation Coefficient.

Euclidean Distance: It measures the straight-line distance between two points in a multidimensional space. It is intuitive and easy to compute but sensitive to the scale and magnitude of features.

Jaccard Similarity: It calculates the similarity between two sets by measuring the intersection over the union of their elements. It's commonly used for comparing the similarity of binary or categorical data but doesn't consider the magnitude of features.

Pearson Correlation Coefficient: It assesses the linear correlation between two variables by measuring their covariance divided by the product of their standard deviations. It's useful for measuring the strength and direction of the linear relationship between continuous variables but assumes a linear relationship and is sensitive to outliers.

We chose cosine similarity over other available similarity measures because

- In resume analysis, our resume dataset exhibits class imbalance, where certain job categories or qualifications are more prevalent than others. Cosine similarity is robust to class imbalance and can effectively handle variations in the distribution of features across different resume categories.
- Cosine similarity is suitable for text analysis, where emphasis is placed on semantic meaning rather than occurrence of words or their frequency.
- In resume analysis, we are extracting various features such as skills, experiences, and education, which result in a high-dimensional feature space. Cosine similarity allows you to compare resumes based on the relevance of these features without being affected by their scale.

4 Problem Statement

In the contemporary job market, the ever-increasing volume of resumes poses a significant challenge for human resource professionals and recruiters in effectively categorizing and analyzing applicant profiles. The manual screening of resumes is time-consuming and prone to biases. To address this challenge, our project aims to develop an automated resume categorization system leveraging natural language processing and machine learning techniques.

4.1 Objectives

1. **Automated Categorization:** Create a system that can automatically categorize resumes into predefined job categories based on their content.
2. **Efficient Preprocessing**: Develop a robust preprocessing pipeline to standardize textual data, including lowercasing, punctuation removal, tokenization, stopword removal, and stemming, ensuring a consistent and clean dataset.
3. **Effective Vectorization:** Implement TF-IDF vectorization to transform preprocessed resumes into numerical representations, capturing the unique importance of words within each document.
4. **Machine Learning Classification:** Utilize the RandomForestClassifier to train a model capable of accurately categorizing resumes into predefined job categories based on their TF-IDF vectors.
5. **Comprehensive Evaluation**: Assess the model's performance using key metrics such as accuracy, confusion matrix, classification report, and macro-averaged precision, recall, and F1-score to ensure its effectiveness in automating the categorization process.

4.2 Significance

The successful implementation of this automated resume categorization system will significantly streamline the recruitment process, allowing for more efficient and unbiased initial screening of job applicants. By leveraging machine learning, we aim to enhance the accuracy and speed of resume analysis, ultimately contributing to a more effective and equitable hiring process.

4.3 Dataset

3446 resumes make up our dataset, and each one is labeled with the category to which it most closely fits. There are twenty-four classes that these resumes fall into. Because of the large number of UTF-8 characters and information that has been suppressed for privacy, the dataset is rather straightforward but noisy. Additionally, several records had to be deleted since they were formatted incorrectly. There are four columns in the dataset:

1. ID, which has an ID for every row.
2. Resume_str, which is the resume's text.
3. Resume_html, which is the resume in HTML format.
4. Category, which refers to the field or area of employment where the resume is located.

5 Proposed Pipeline

There are three steps in our pipeline. Our workflow involves Data Preprocessing, Text Vectorization, and Classification, followed by model evaluation (Fig. 1).

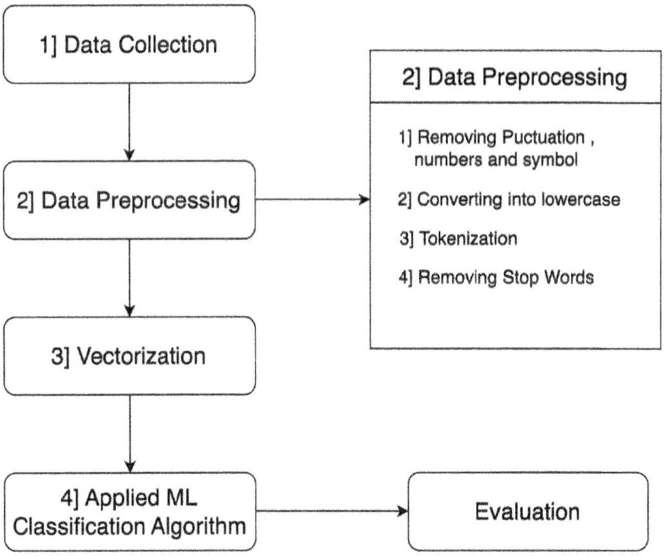

Fig. 1 Proposed pipeline for the model

5.1 Preprocessing

Objective: In our resume analysis project, the initial step involves preprocessing the textual content of resumes. The purpose of this phase is to standardize and clean the data, ensuring a consistent format that is conducive to subsequent analysis. Our approach, as reflected in the provided code, encompasses several key steps:

- **Lowercasing:** All text is converted to lowercase, ensuring uniformity in the representation of words.
- **Punctuation Removal**: Extraneous punctuation marks are eliminated to reduce noise in the data and maintain focus on meaningful words.
- **Tokenization:** The resumes are broken down into individual words or tokens, facilitating a more granular analysis.
- **Stopword Removal**: Common stopwords, which contribute little to the overall meaning, are removed to enhance the relevance of the remaining words.
- **Stemming**: Words are reduced to their root form using stemming, promoting consistency, and capturing core meanings effectively.

Reference to Code: The `preprocess_text` function encapsulates these preprocessing steps, ensuring that each resume undergoes a standardized transformation. Additionally, the `ColumnExtractor` class facilitates the extraction of the relevant column (`Resume_str`) for processing.

5.2 Vectorization

Objective: Following preprocessing, the next phase involves vectorization—converting textual data into a numerical format suitable for machine learning models. In our proposal, we employ Term Frequency-Inverse Document Frequency (TF-IDF) vectorization, a technique that evaluates the importance of words based on their frequency within individual resumes and rarity across the entire dataset.

Reference to Code: The vectorization pipeline is created using scikit-learn's `Pipeline` class. This pipeline seamlessly combines the preprocessing steps with TF-IDF vectorization. The `TfidfVectorizer` is configured to utilize the previously defined `preprocess_text` function, ensuring a consistent and meaningful transformation.

5.3 Classification

Objective: The core of our proposal lies in the classification phase, where we leverage machine learning to categorize resumes into predefined job categories. For this purpose, we have chosen the RandomForestClassifier, an ensemble learning method that taps into the collective intelligence of multiple decision trees.

Reference to Code: Our classification pipeline, encapsulated in a scikit-learn `Pipeline` object, integrates the preprocessing, vectorization, and classification steps seamlessly. The `RandomForestClassifier` is trained on the labeled dataset (`train_df`), enabling it to discern patterns and relationships between TF-IDF vectors and corresponding resume categories. Subsequently, the model is used to predict categories for the test set (`test_df`).

5.4 Scoring

Objective: Evaluation is a critical aspect of our proposal, and we employ various metrics to assess the performance of our trained model. The scoring phase provides insights into how well the model is performing in categorizing resumes, guiding any necessary adjustments or improvements.

Reference to Code: The code calculates and prints key metrics after making predictions on the test set. These metrics include accuracy, confusion matrix, classification report, and macro-averaged precision, recall, and F1-score. This comprehensive evaluation ensures a thorough understanding of the model's strengths and areas for potential refinement.

In summary, our proposed approach encompasses a systematic pipeline, addressing preprocessing, vectorization, classification, and thorough evaluation to ensure the effectiveness of our resume analysis system. This method aligns with best practices in natural language processing and machine learning, offering a robust and scalable solution for categorizing resumes accurately (Fig. 2).

6 Results and Discussion

6.1 Model Training and Evaluation

The implementation of the Resume Analyzer project yielded notable results during model training and evaluation. The RandomForestClassifier demonstrated robust training on a diverse dataset, showcasing its ability to categorize resumes accurately. The evaluation metrics, including accuracy, precision, recall, and F1-score, underpin the model's proficiency in classifying resumes into predefined job categories (see Fig. 3).

```
Accuracy: 0.61
Confusion Matrix:
[[26  0  0  0  0  0  0  0  0  0  0  0  0  0  0  0  3  0  0  0  0  0  0  0]
 [ 0 20  0  0  0  0  0  1  0  0  0  0  0  0  0  0  0  0  4  1  0  1  1  2]
 [ 1  0  1  0  0  0  0  0  0  0  0  0  0  0  0  1  1  0  0  0  1  1  0  2]
 [ 0  0  0  9  1  0  0  2  0  0  0  1  0  1  0  2  0  0  0  0  0  0  2  2]
 [ 0  0  0  1  2  0  0  0  0  1  1  0  0  1  0  2  1  0  0  1  1  0  1  6]
 [ 0  1  1  0  1  0  1  0  0  0  0  0  0  0  0  0  1  0  0  0  1  0  0  0]
 [ 0  0  0  0  0  0 17  0  0  0  0  0  0  0  0  1  0  0  0  0  3  0  0  0]
 [ 1  0  0  0  1  0  0 14  0  0  0  0  0  0  0  1  3  0  0  0  3  0  0  0]
 [ 0  0  0  0  0  0  0  0  0  0  0  0  0  0  0  0  0  0  1  0  1  0  0  0]
 [ 0  0  0  0  0  0  0  1  0 13  0  0  0  0  1  1  1  0  2  1  2  3  2  0]
 [ 0  0  0  0  0  0  2  0  0  1 17  0  0  2  0  0  0  0  1  0  0  0  0  1]
 [ 0  1  0  0  1  0  0  0  0  2  0 26  0  2  0  0  0  1  0  0  0  0  1  0]
 [ 0  0  0  0  0  0  0  1  0  1  0  1  4  0  1  1  1  0  2  1  6  1  0  0]
 [ 0  0  0  1  0  0  0  0  0  0  0  0  0 15  0  2  0  0  0  0  0  1  0  0]
 [ 0  0  0  0  0  0  0  0  0  1  0  1  0  2 13  0  0  0  0  0  2  3  3  0]
 [ 0  0  0  0  0  0  1  0  0  0  0  2  1  0  0 14  0  0  0  0  3  0  0  0]
 [10  0  0  0  0  0  0  2  0  0  0  0  0  0  0  0  7  0  0  0  0  0  0  0]
 [ 0  0  0  0  0  0  2  0  0  2  0  0  0  0  0  0  0  9  1  0  0  0  4  1]
 [ 0  2  0  0  0  0  0  1  0  0  0  1  0  0  1  0  0  0 11  2  2  0  0  0]
 [ 0  1  0  0  0  0  0  0  0  0  1  0  0  0  0  0  0  0  0 16  0  0  0  0]
 [ 0  0  0  0  0  0  0  0  0  0  0  0  0  0  0  1  0  0  0  0 24  0  0  1]
 [ 0  0  0  0  0  0  0  0  0  2  0  0  0  0  0  0  0  0  0  1  1 11  1  1]
 [ 0  2  0  3  0  0  0  0  0  2  2  0  1  1  0  0  0  1  1  0  1  1 14  0]
 [ 0  0  0  0  0  0  0  0  0  0  1  0  0  1  0  1  0  0  0  0  0  1  0 18]]
```

Fig. 2 Accuracy and confusion matrix of the model

6.2 Relevance Scoring and Candidate Prioritization

The innovative addition of relevance scoring through cosine similarity represents a pivotal advancement in the project. This mechanism enables the prioritization of resumes based on their alignment with a provided job description. For example see Fig. 4. The resulting relevance scores provide detailed insights into the relevance of individual profiles, offering a practical solution for efficiently shortlisting candidates

7 Conclusion

In conclusion, the Resume Analyzer project, fortified by the robust integration of Natural Language Processing (NLP) and machine learning techniques, stands as a significant advancement in automating the resume categorization process. The systematic pipeline, comprising preprocessing, vectorization, classification, and evaluation, has demonstrated effectiveness in enhancing the efficiency and objectivity of candidate screening.

The project's contribution is highlighted by its ability to accurately categorize resumes, providing a valuable tool for recruiters facing the challenge of handling vast volumes of applicant profiles. The introduction of relevance scoring through cosine similarity further refines the system, enabling recruiters to prioritize candidates based on their alignment with specific job requirements.

```
Classification Report:
                         precision    recall  f1-score   support

            ACCOUNTANT       0.68      0.90      0.78        29
              ADVOCATE       0.74      0.67      0.70        30
           AGRICULTURE       0.50      0.12      0.20         8
               APPAREL       0.64      0.45      0.53        20
                  ARTS       0.33      0.11      0.17        18
            AUTOMOBILE       1.00      0.00      0.00         6
              AVIATION       0.74      0.81      0.77        21
               BANKING       0.64      0.61      0.62        23
                   BPO       1.00      0.00      0.00         2
  BUSINESS-DEVELOPMENT       0.52      0.48      0.50        27
                  CHEF       0.77      0.71      0.74        24
          CONSTRUCTION       0.81      0.76      0.79        34
            CONSULTANT       0.67      0.20      0.31        20
              DESIGNER       0.60      0.79      0.68        19
         DIGITAL-MEDIA       0.81      0.52      0.63        25
           ENGINEERING       0.52      0.67      0.58        21
               FINANCE       0.39      0.37      0.38        19
               FITNESS       0.82      0.47      0.60        19
            HEALTHCARE       0.48      0.55      0.51        20
                    HR       0.70      0.89      0.78        18
INFORMATION-TECHNOLOGY       0.47      0.92      0.62        26
      PUBLIC-RELATIONS       0.48      0.65      0.55        17
                 SALES       0.48      0.48      0.48        29
               TEACHER       0.53      0.82      0.64        22

              accuracy                           0.61       497
             macro avg       0.64      0.54      0.52       497
          weighted avg       0.62      0.61      0.59       497

Precision: 0.64
Recall: 0.54
F1 Score: 0.52
```

Fig. 3 Classification report, precision, recall, and F1-scores of the model

```
In [31]: print(df[['Resume_str', 'Relevance Score']].sort_values(by='Relevance Score', ascending=False))

                                       Resume_str  Relevance Score
283           INFORMATION TECHNOLOGY CONSULTANT    ...         0.871372
236          INFORMATION TECHNOLOGY SPECIALIST/SYS...         0.867652
238           DIRECTOR OF INFORMATION TECHNOLOGY   ...         0.866165
253           DIRECTOR OF INFORMATION TECHNOLOGY   ...         0.864151
337           IT MANAGER              Highlights   ...         0.861474
...                                               ...              ...
1231          CONSULTANT            Summary     Over 2...         0.000000
1232          IT CONSULTANT              Professional ...         0.000000
1233          CONSULTANT             Executive Profil...         0.000000
1234          CONSULTANT          Executive Profile    ...         0.000000
3445          STOREKEEPER II         Professional Sum...         0.000000

[3446 rows x 2 columns]
```

Fig. 4 Relevance scores of resumes obtained, in descending order

Looking ahead, the Resume Analyzer holds the potential to revolutionize traditional recruitment processes, offering a scalable and adaptable solution for organizations seeking to make more informed and unbiased hiring decisions. As the project continues to evolve, future enhancements and refinements will further solidify its position as an innovative tool in the dynamic landscape of talent acquisition.

Acknowledgements We extend our heartfelt thanks to IIIT Vadodara for supplying the essential infrastructure and assistance throughout this design project. A special acknowledgment goes to Dr. Novarun Deb, our mentor, for his invaluable guidance and motivation. His expertise and consistent support significantly aided in the triumphant conclusion of this project. We are grateful for the educational experiences and the cooperative atmosphere nurtured by both the institution and our mentor.

References

1. Bhor S, Gupta V, Nair V, Shinde H, Kulkarni MS (2021) Resume parser using natural language processing techniques. Int J Res Eng Sci (IJRES) 9(6):01–06
2. Surekha A, Teja NSSS, Chowdary KV, Poddar K, Iliyaz S (2023) Resume analyzer and suggestion making using NLP. J Emerg Technol Innov Res 10(4)
3. Roy PK, Chowdhary SS, Bhatia R (2020) A machine learning approach for automation of resume recommendation system. Procedia Comput Sci 167:2318–2327
4. Sinha AK, Akhtar Md AK, Kumar M, Upadhyay S (2023) Resume screening classification using artificial intelligence and natural language processing. Eur Chem Bull:14473–14480
5. Chavan A, Tatewar N, Naicker P, Deore S (2023) AI resume analyzer. Int J Creat Res Thoughts (IJCRT) 11(12):e546–e548

CCD Array Spectrometer-Based FBG Bragg Wavelength Detection Using ANN Algorithm

Dipten Kumar, Nandini Basumallick, and Debotosh Bhattacharjee

Abstract An artificial neural network (ANN)-based algorithm for detecting Bragg wavelengths is proposed in this paper for interrogating fiber Bragg grating (FBG) sensors using a CCD array spectrometer. The proposed algorithm addresses the actual non-ideal behaviors of components, including optical intensity variation over time. It also accommodates the fact that FBG sensor nominal Bragg wavelengths often change during field deployment due to variations in pre-strain or ambient conditions. Finally, the proposed ANN model is compared with traditional methods of determining FBG Bragg wavelengths, such as curve fitting and centroid finding.

Keywords FBG interrogation · ANN · Bragg wavelength · CCD spectrometer

1 Introduction

In recent years, optical sensors based on fiber Bragg Grating technology have gained popularity, especially in applications where conventional electrical sensors are either impractical or unusable. Fiber Bragg Gratings are structured defects inscribed in optical fibers, reflecting a specific wavelength while allowing others to pass through. Each Fiber Bragg Grating (FBG) sensor is designed to reflect a particular wavelength (called Bragg wavelength) of light under normal conditions. However, when subjected to external strain or a temperature change, the Bragg wavelength shifts from its nominal value. The strain and temperature applied to the FBG sensor can be determined by referencing a suitable calibration table against the shift in Bragg wavelength value. This process of determining the instantaneous wavelength value of the Bragg wavelength from an FBG sensor is known as the interrogation of the sensor. When the FBG sensor is subjected to strain or temperature, the profile (i.e.

D. Kumar (✉) · D. Bhattacharjee
Jadavpur University, Kolkata 700032, India
e-mail: diptenkumar.ju@gmail.com

D. Kumar · N. Basumallick
CSIR-CGCRI, Kolkata 700032, India

© The Author(s), under exclusive license to Springer Nature Singapore Pte Ltd. 2025

N. Chaki et al. (eds.), *Applied Computing for Software and Smart Systems*, Lecture Notes in Networks and Systems 980, https://doi.org/10.1007/978-981-97-9762-2_3

shape) of the FBG reflection spectrum serving as a sensor sensing element does not change or hardly changes, as observed in [8]. However, there is a translation of the spectrum profile towards a higher or lower wavelength side resulting in a positive or negative shift of the Bragg wavelength of the FBG, respectively, compared to its nominal wavelength.

Various methods for fiber Bragg grating (FBG) interrogation exist, but two trendy ones in recent years are spectrometer-based and tunable filter-based techniques [6]. The spectrometer-based technique has been discussed in [1, 4, 8]. This method directs the reflected light from an FBG onto a one-dimensional CCD array through a diffraction grating. The wavelength of the reflected light dictates which set of pixels on the CCD array is illuminated and to what extent. By analyzing the optical spectrum distribution across the pixels, the Bragg wavelength of the light can be determined. The tunable filter-based approach is described in [2, 5, 7]. In this method, a laser or monochromatic light with a narrow bandwidth is swept across the entire range of wavelength spectrum. The Bragg wavelength is then determined from the time-stamp of the samples of the reflected optical signal from the FBG sensor as recorded by a photodetector.

A standard FBG spectrum typically spans a few hundred pico-metres, while the necessary resolution of an interrogator is typically within five pico-metres. Therefore, a central wavelength finding algorithm is essential to identify the corresponding Bragg wavelength of an FBG sensor spectrum peak. In the case of a CCD array-based interrogator, this involves determining a sub-pixel value. Each CCD element is called a pixel, and its position number (an integer) in the array is its pixel number. The central wavelength (referred to as Bragg wavelength) of the FBG spectrum is derived from the sub-pixel. In this context, "sub-pixel" denotes that the pixel number is expressed as real (or floating-point/fractional) continuous values rather than integers.

Several algorithms exist for determining sub-pixel values from the pixel values of a CCD array-based interrogator. Two commonly used traditional algorithms for this purpose are curve-fitting and centroid-finding algorithms. Among curve fitting algorithms, Gaussian and polynomial fit algorithms are more frequently employed than others. Alves et al. [1] implements centroid (central detection algorithm) and polynomial fitting to find the sub-pixel for a CCD pixel array. In [4], the implementation of the centroid, Gaussian fit, and polynomial (quadratic) fit algorithms have been discussed for the sub-pixel determination of a CCD pixel array. In [5, 7], the implementation of centroid and curve fitting algorithms is discussed for a tunable filter-based interrogator aimed at determining the central wavelength of the FBG peak.

Of these two popular algorithms, the iterative curve fitting type is not suitable for real-time determination of the Bragg wavelength using an FPGA. As pointed out in [1], "*The fit functions are recursive, and they need many iterations to achieve a good result. So, it becomes cumbersome to use this algorithm in applications where sharper time restrictions are an issue to be addressed*".

In contrast, the centroid-finding algorithm is very suitable for implementation on an FPGA, simplifying the realization of real-time Bragg wavelength detection. The centroid algorithm yields satisfactory results when the sampling is dense, such as one sample per pico-metres. However, its accuracy diminishes significantly when the sampling is sparse (like one sample per 156 pico-metres) and when the spectral peak is distorted. Often, in a FBG spectrum peak profile, the values lower than a threshold (referred to as 'Minimum Acquisition Limit' in [1]) needs to be rejected as a result of which the spectral peak shape gets distorted. Both [4, 7] highlight that the centroid algorithm's accuracy is lower than other algorithms used for interrogation.

More recently, algorithms based on neural networks have been suggested for Bragg wavelength determination. Unlike curve fitting algorithms, neural network-based algorithms are more suitable for real-time computation of Bragg wavelength because they do not require iteration during the prediction stage. All iterations occur solely in the training stage for weight and bias calculation during model building, which is completed prior to the deployment of the interrogator in real-time field applications. As mentioned in [3], an FPGA can implement a neural network-based architecture effectively. Further, it is noted in [5] that an adequately constructed ANN can act as a '*Universal Function Approximator*'.

Consequently, a neural network-based algorithm is suitable for detecting Bragg wavelength for CCD-based and tunable filter-based interrogators in real-time/time critical applications. In [8], Timur Agliullin et al. proposed a neural network-based algorithm for determining the Bragg wavelength of an FBG sensor for a CCD array-based interrogator. In that work, the model is trained by the relation 'pixel amplitude (i.e., optical signal intensity captured on each pixel) of CCD sensor vs. absolute wavelength data of FBG spectrum.' This training is done for some fixed wavelength range (3 nm in their work). However, this method has two significant disadvantages. First, there are cases where the FBG spectrum, due to some precondition (like pre-stain etc.) shifts significantly and goes out of the wavelength range window used for training (3 nm in the discussed case). The ANN model developed for the FBG sensor that determines the central wavelength from the pixel data can no longer be used in such a thing happens. Second, if the wavelength range window for training increases, the number of training samples will increase proportionately. Therefore, the training window cannot be increased arbitrarily; otherwise, the implementation becomes very expensive.

In the context of the preceding discussion, this research paper introduces innovative Artificial Neural Network-based algorithms for CCD array-based FBG Interrogators. The proposed algorithm possesses the following features:

1. Being a neural network-based algorithm, it can be implemented in an FPGA (like the centroid algorithm), making it suitable for real-time prediction of the Bragg wavelength. However, in terms of accuracy, this algorithm outperforms the centroid finding algorithm for Bragg wavelength determination in CCD array-based FBG Interrogators.
2. This algorithm and the trained model remain effective even if the nominal Bragg wavelength of the FBG sensor shifts arbitrarily due to preconditions such as pre-strain or initial ambient temperature changes.

3. The proposed algorithm's training data size is independent of the interrogator's wavelength range of operation.

The remaining part of this paper, as in Sect. 2, discusses the setup and methodology of FBG interrogation. Section 3 describes our ANN training algorithm, prediction algorithm, and results for CCD array-based Interrogation. Section 4 compares our proposed algorithm with traditional techniques like curve-fitting and centroid finding.

2 Set-Up and Methodology of FBG Interrogation

This section describes the setup and methodology of CCD array-based interrogation.

In CCD array-based interrogation, an optical circulator directs a broadband light onto an FBG sensor. The reflected light from the FBG sensor is then directed onto a one-dimensional array of CCD pixels through a diffraction grating. The diffraction grating spatially distributes the incoming reflected light from the FBG sensor so that, based on the wavelength spectrum of this light (dependent on the FBG sensor spectrum), it determines which pixels are illuminated and to what extent. The intensity of light incident on a given pixel corresponds to the charge build-up, resulting in a higher voltage when the pixel getting higher intensity optical signal is read out.

The pixel voltages serve as samples of the FBG sensor spectrum profile. Each pixel's voltage value is digitized and tabulated along with the pixel number. Subsequently, the exact FBG sensor spectrum central wavelength is determined from these values corresponding to each pixel. This process is akin to predicting the FBG spectrum from a few spatial samples and then determining its central wavelength. All the algorithms discussed in this paper for this interrogator, including curve-fitting, centroid, and the proposed neural network-based method, aim to predict the central wavelength from the voltage distribution in the pixels (Fig. 1).

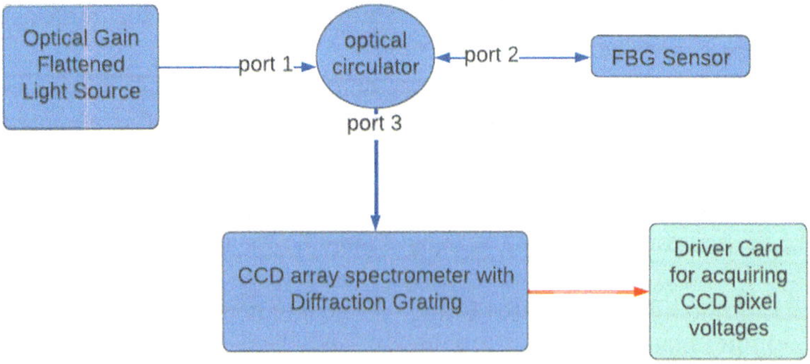

Fig. 1 Block diagram of CCD-based interrogation

3 Neural Network for CCD Array Based Interrogator

3.1 ANN Training Algorithm for CCD Array-Based Interrogator

In a CCD array-based interrogation system for a specific FBG sensor, the actual voltage of each CCD pixel may fluctuate depending on the intensity of the reflected light from the FBG sensor. This intensity is influenced by factors such as the source light's incident intensity on the FBG sensor, signal attenuation during its passage through the optical circuit, and the duration of exposure of the reflected light from the FBG sensor on the CCD pixels of the array. Variation in these parameters may lead to variation in CCD pixel voltage. However, as long as the reflectivity profile of the FBG sensor spectrum remains constant, the relative voltage of the pixels remains unaffected. The neural network training algorithm should be capable of handling all these variations.

This paper proposes a two-layer neural network-based algorithm for predicting the FBG sensor Bragg wavelength in a CCD array-based interrogation system. This prediction is based on the voltage values of the CCD pixels when the reflected optical signal from an FBG sensor is incident on the CCD array. The number of nodes in each layer depends on parameters like the 'wavelength per pixel' of the CCD array, the bandwidth of the FBG sensor spectrum, and the required interrogation resolution. In our case for illustration, we consider a 512-pixel CCD interrogator with a 'wavelength per pixel' of 156 pm having a wavelength range spanning over 79,872 pm from 1510.064 nm to 1589.936 nm. The FBG spectrum simulated in our case has a symmetric normalized profile 1500 pm wide with FWHM (Full Width Half Maximum) of 250 pm. Our target resolution is 1 pm.

This problem of predicting the FBG sensor central wavelength (or Bragg wavelength) from the pixel voltages of a CCD array-based sensor is framed as a classification problem. Given that the spectrum of the FBG sensor is 1500 pm wide and the 'wavelength per pixel' is 156 pm, a maximum of ceil (1500/156) = 10 pixels will be required in our input layer. However, to maintain an odd number of nodes for symmetry reasons and allow for flexibility with broader spectrums of FBG sensors in the input layer, we opt for 15 nodes. Since 'wavelength per pixel' is 156 pm and the required resolution of the interrogation is 1 pm, the number of nodes in the output layer is 156.

Now, when there are adjacent pixels in the CCD array with voltages surpassing a certain threshold, that implies some FBG optical signal in that corresponding region of the spectrum. Each pixel in the CCD array corresponds to a specific wavelength in the optical spectrum, information typically provided by the device manufacturer. However, with 79,872 wavelengths in increments of 1 pm and only 512 pixels, not all wavelengths have a corresponding assigned pixel number. Using the voltage values of the adjacent pixels, the ANN will discern patterns and determine the central wavelength of the FBG sensor. Relative to the pixel with the highest voltage value, the ANN outputs an integer value ranging from −77 to +78.

If the pixel number with the maximum voltage is P_{max} (which lies between 0 and 511), its corresponding wavelength is W_{max} (which lies between 1510064 pm and 1,589,936 pm), and the output of the ANN is X (which lies between −77 and +78), then the predicted Central Wavelength of the FBG sensor is $W_{max} + X$ pm.

In the proposed training method, a set of training data is generated for each FBG sensor. At first, the profile (or shape) of the FBG is plotted densely (i.e., one sample at each pm). This can be done with a high-resolution Optical Spectrum Analyzer. Then, this densely plotted profile shape can be sampled arbitrarily at all discrete points, maintaining the constraint that any two consecutive discrete samples are fixed wavelengths apart, the same as the difference in corresponding wavelength between two consecutive pixels. From the dense plot, we know where precisely the central wavelength lies. So, the wavelength difference between the 'corresponding wavelength of sampled point with maximum value' and the 'central wavelength of the FBG spectrum' is also known.

This is illustrated with an example below. Consider the blue line in Fig. 2, which is the normalized (i.e., max value is 1) spectrum of a given FBG sensor. The part of the FBG sensor spectrum shown in the figure is 1500 pm wide. This spectrum is incident on a 512-pixel CCD array with a wavelength range of ~80 nm. So, the wavelength difference between two consecutive pixels is ~80 nm/512 = 156 pm. The red dot marks are the discrete points where this spectrum is sampled. In this case, the discrete samples are 156pm apart, as it is the wavelength difference between the corresponding wavelengths of two consecutive pixels. This FBG spectrum is sampled arbitrarily at all possible points with the only constraint that the two consecutive discrete sample points are permanently fixed distance apart (156 pm in our case). As the consecutive samples are always 156 pm apart, there will be only 156 distinct sample sets, after which, due to the periodicity of sampling points, we get a repetition of the same sample sets.

In Fig. 2, we can see that the highest-valued sample is at $x = 691$. (x is the x-axis of the plot representing the relative wavelength of the spectrum in pm.) As the central wavelength of the FBG is at $x = 750$, it is +59 pm higher than the highest-valued sample.

Therefore, our training data includes all possible sample sets, in which each pattern set (i.e., a set of discrete samples from the dense plot of the spectrum) is accompanied by its respective target value. The target value is determined as the FBG sensor spectrum's central wavelength minus the corresponding wavelength value of the pixel with the highest sampled value in the pattern set. To create the training data, the sample amplitudes within each pattern set are normalized relative to the maximum amplitude in the set (pattern normalization). Additionally, the amplitudes are centered in a manner such that the maximum amplitude is shifted to the middle position of the pattern-set (pattern-centering). By pattern normalization, we mean that the value of the highest-valued sample in the pattern set is scaled to 1. Other sample values in the pattern set are scaled accordingly so that their relative magnitudes among themselves and concerning the highest value sample are not altered. Similarly, by pattern-centering, we mean that the samples in the pattern set are arranged in such a manner that the sample with the highest value is the middle sample of the set. So, after

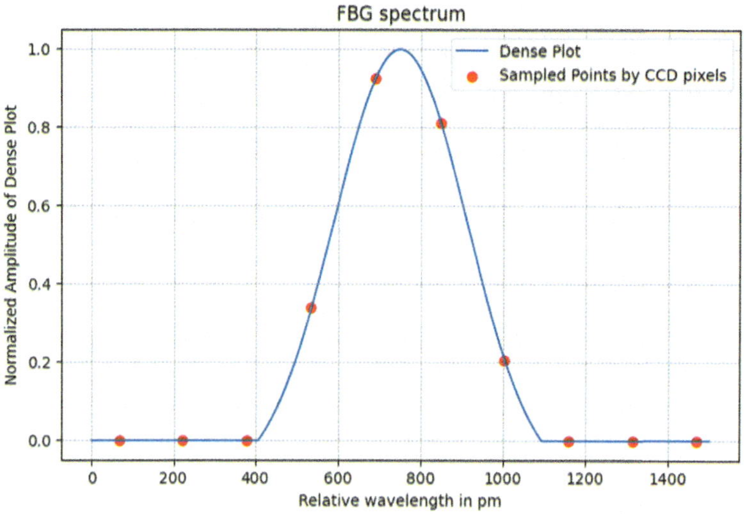

Fig. 2 FBG sensor spectrum (blue line) sampled by CCD pixels at discrete points (red dots)

pattern-normalization and pattern-centering of the pattern set, the central value of the set (which is at index 7 as the set contains 15 elements) contains value 1 (the highest value in the set). This central value is replaced by the [centroid value of the pattern set −6] and fed into ANN as input nodes. The target output value computation has already been discussed earlier. So, the steps of the training algorithm are illustrated as follows.

Steps of Training Algorithm. Using a training algorithm, training data is formed. The training data is a matrix in which each row is a training sample. Each training sample has two components: the pattern set and a target value. All elements, excluding the last one, compose the pattern set in each row, while the last element represents the target value.

Consequently, in the training matrix, the last column represents the target set, containing the training samples' target values. In the following steps, the first five steps describe how the pattern set of the training data is formed. The 6th step describes how the corresponding target value for each pattern set is computed. The training data for a given FBG sensor spectrum is done in the following steps:

1. First, the normalized profile of the FBG sensor spectrum is plotted. This is done in two steps. First, the reflection intensity value of the sensor is densely tabulated in a row for every wavelength in steps of 1 pm. Then, this FBG spectrum is normalized by dividing all the values in the row by the maximum value. So, after normalization, the peak value of the FBG spectrum is 1. This spectrum is plotted with the x-axis representing the wavelength axis of the spectrum. This wavelength, denoted by the x-axis, is not the actual wavelength of the spectrum but a relative wavelength with respect to the lowest wavelength in the FBG sensor spectrum. The unit of the x-axis is pico-metre (pm). In the above-illustrated

example (Fig. 2 blue line), in a single row, we have 1500 values as the FBG sensor spectrum is plotted for 1500 pm, and the spectrum is symmetric and peaks at x = 750 nm.

2. After the FBG spectrum profile is plotted, next in our illustrated case, it is sampled at all possible points so that consecutive samples are 156 pm apart. In the illustrated case, 156 sample sets (called pattern sets) are possible if we sample at discrete pm. These sets are amplitudes at x = [0, 156, 312, 468, 624, 780, 936,1092,1248, 1404], amplitudes at x = [1, 157, 313, 469, 625, 781, 937,1093,1249, 1405],amplitudes at x = [2, 158, 314, 470, 626, 782, 938,1094,1250, 1406] and so on till amplitudes at = [155, 311, 467, 623, 779, 935,1091,1247, 1403]. After that, we again get the same sets over and over again.
3. All the amplitude values in each such pattern set are then divided by the maximum value of sample amplitude in the set. Therefore, now the maximum amplitude value in the pattern set becomes 1, and the rest of the amplitude values in that set become $\leq$1. This step is named 'pattern-normalization' in this document.
4. Now we center the sample points so that the highest sample is in the middle of the row. In our case, we will take a 15-point row. So, the highest amplitude value (which is 1 after pattern normalization) is kept at index 7 (indexing starts at 0). All other sample values are mapped accordingly. Where there is no value to map, it is padded with zero. This step is named 'pattern centering' in this document (Table 1).

5. Then, the centroid of this normalized and centered set is computed by the formula:

Table 1 Pattern-normalized and pattern-centered pixel amplitude of FBG sensor spectrum in Fig. 2

Pixel no. relative to the peak	Normalized and centered pixel amplitude
0	0
1	0
2	0
3	0
4	0
5	0
6	0.366
7	1
8	0.878
9	0.223
10	0
11	0
12	0
13	0
14	0

$$centroid = \frac{\sum_{n} \{n * max(0, (A_n - A_{mp}))\}}{\sum_{n} max(0, (A_n - A_{mp}))}, n = 0, 1, \ldots 14$$

where n is the index number, A_n is the Amplitude of the normalized and centered pattern set, and A_{mp} is the minimum positive amplitude in the pattern set

In our example :

$$\begin{aligned} centroid = & \{0 \times 0 + 1 \times 0 + 2 \times 0 + 3 \times 0 + 4 \times 0 + 5 \times 0 \\ & + 6 \times (0.366 - 0.223) + 7 \times (1 - 0.223) \\ & + 8 \times (0.878 - 0.223) + 9 \times (0.223 - 0.223) \\ & + 10 \times 0 + 11 \times 0 + 12 \times 0 \\ & + 13 \times 0 + 14 \times 0 / \{0 + 0 + 0 + 0 + 0 + 0 + (0.366 - 0.223) \\ & + (1 - 0.223) + (0.878 - 0.223) + (0.223 - 0.223) \\ & + 0 + 0 + 0 + 0 + 0\} \\ & = 11.537/1.575 = 7.325 \end{aligned}$$

Now, finally, '1' at index seven is replaced by centroid –6 = 7.325–6 = 1.325. Therefore, now the final pattern set is given in Table 2.

6. The target value for a particular sample set is derived by subtracting the corresponding wavelength for the pixel with the highest amplitude from the central

Table 2 Normalized and centered pixel amplitude with centroid

Pixel no. relative to the peak	Normalized and centered pixel amplitude
0	0
1	0
2	0
3	0
4	0
5	0
6	0.366
7	1.325
8	0.878
9	0.223
10	0
11	0
12	0
13	0
14	0

wavelength of the FBG spectrum. For the case illustrated in Fig. 2, the sample (pixel) with the highest amplitude has its corresponding wavelength at 691 pm, and the FBG spectrum has a central wavelength at 750 pm. The target for this set of samples is 750–691 = +59. This means that the central wavelength of the FBG sensor spectrum is 59 pm more than the wavelength corresponding to the 7th index of the table.

In the following figure (Fig. 3), another training sample is illustrated. The sample (pixel) with the highest amplitude has its corresponding wavelength at 811 pm, and the FBG spectrum has a central wavelength of 750 pm. For this set of samples, the target is $750 - 811 = -61$. This means that the central wavelength of the FBG sensor spectrum is 61 pm less than the wavelength corresponding to the seventh index of the table.

While generating the network's training data, the lower amplitudes were systematically dropped in some pattern sets so that the model is more sensitive to higher amplitudes. Lower amplitudes are often prone to noise; hence, they are discarded unless they are above a certain threshold. However, while dropping the lower amplitudes, it is ensured that at least two of the highest amplitudes are present in the pattern set.

Proposed an Artificial Neural Network. The ANN has two layers for this purpose. The input layer comprises 15 nodes corresponding to the 15 elements of normalized, centered, and padded rows, with the central element (index 7) replaced with a centroid value. The input layer is densely connected to the nodes of the output layer. The output layer has 156 nodes, each corresponding to a 1 pm shift left or right of the pixel with maximum amplitude. Adam optimizer was used with softmax activation function to train this neural network, and the learning rate was 0.01 (Fig. 4).

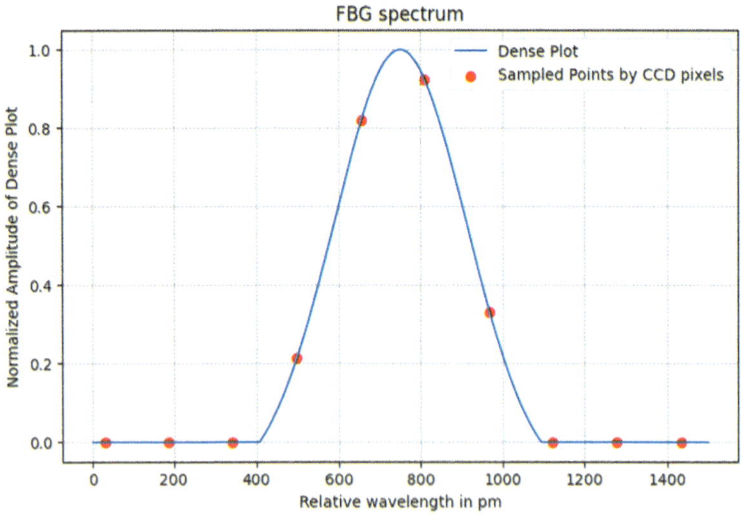

Fig. 3 FBG sensor spectrum (blue line) sampled by CCD pixels at discrete points (red dots)

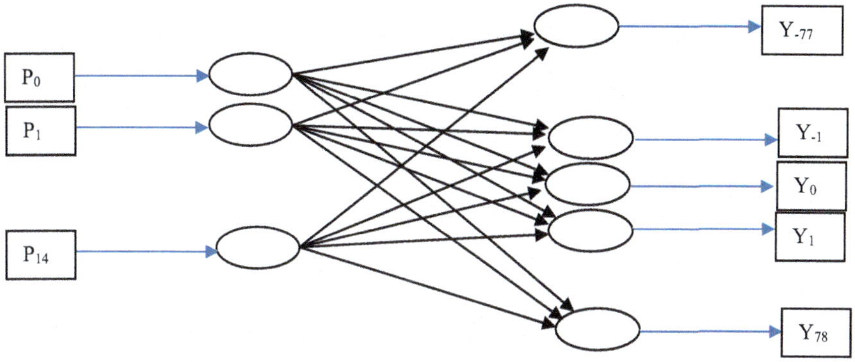

Fig. 4 Architecture of ANN formed

3.2 Bragg Wavelength Prediction Using the Proposed Trained ANN for CCD Array-Based FBG Interrogator

After training the ANN model, it can be employed to predict the central wavelength of an FBG sensor spectrum in the following manner. When an optical signal from an FBG is incident on a CCD array, a specific set of pixels is illuminated, resulting in voltage buildup on this set. The voltage values of all pixels in this set are normalized by dividing them by the maximum voltage in the set, making the maximum normalized pixel value 1, which is assigned to the pixel with the highest registered voltage.

The wavelength corresponding to this particular pixel is denoted as W_{max} in picometres. The normalized amplitude values are then centered. In this process, the sample with a value of 1 (i.e., the max value) is mapped to index 7 of another array capable of holding 15 elements (index 0 to index 14). Other sample values are mapped accordingly in this array to keep their relative positions unchanged. Empty positions in this array, where no corresponding element from the FBG spectrum, are filled with zero values.

Next, the centroid of this set is computed, and six is subtracted from this centroid value. Then, this value is placed in index 7 of the array, replacing the previous value of 1. These elements of this new array are input into the trained model. The trained model activates one of the 156 output nodes. Depending on which output node is activated, a specific wavelength (denoted as X in pico-metres) is added or subtracted from W_{max} to determine the central wavelength of the FBG spectrum.

Algorithm for determining the central wavelength from the CCD pixel values: The algorithm for predicting the central wavelength of an FBG spectrum is illustrated below with an example.

Say we obtain the following voltages in the CCD pixels. The pixel numbers and respective voltages are shown in the diagram and the Table 3 (Fig. 5).

These pixel amplitudes are first stored in a 15-element row as follows. Appropriate zero-padding is done for pixels whose amplitude is either zero or less than a threshold (Table 4).

Table 3 Voltage values on CCD pixels

Pixel no. of CCD array	Voltage recorded on pixel
204	1.697
205	4.635
206	4.068
207	1.036

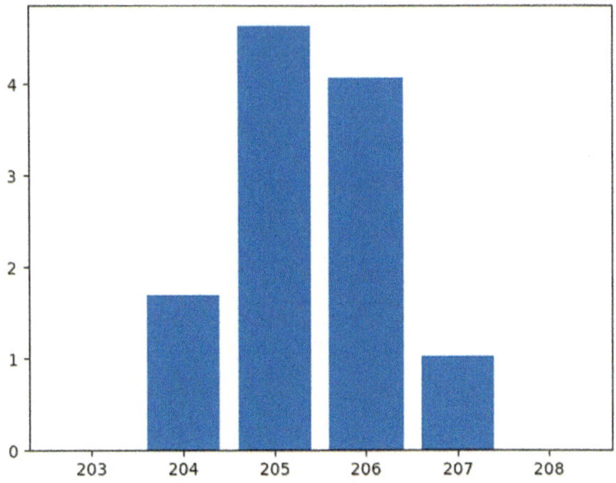

Fig. 5 Voltages on CCD pixel array due to optical signal from FBG

Table 4 Pixel amplitude value

Index no	Pixel amplitude value
0	1.697
1	4.635
2	4.068
3	1.036
4	0
5	0
6	0
7	0
8	0
9	0
10	0
11	0
12	0
13	0
14	0

Then, the values are normalized by dividing all the elements by the maximum element value (Table 5).

Then, the values are centered, so the max value 1 is at index seven (Table 6).

Table 5 Normalized value

Index no	Normalized value
0	0.366
1	1
2	0.878
3	0.223
4	0
5	0
6	0
7	0
8	0
9	0
10	0
11	0
12	0
13	0
14	0

Table 6 Centered normalized value

Index No	Centered normalized value
0	0
1	0
2	0
3	0
4	0
5	0
6	0.366
7	1
8	0.878
9	0.223
10	0
11	0
12	0
13	0
14	0

Then, the centroid of this normalized and centered set is computed by the centroid formula:

$$centroid = \frac{\sum_n \{n * max(0, (A_n - A_{mp}))\}}{\sum_n max(0, (A_n - A_{mp}))}, n = 0, 1, \ldots 14$$

In our example:

$$\begin{aligned} centroid = \{&0 \times 0 + 1 \times 0 + 2 \times 0 + 3 \times 0 + 4 \times 0 + 5 \times 0 + 6 \times (0.366 - 0.223) \\ &+ 7 \times (1 - 0.223) + 8 \times (0.878 - 0.223) \\ &+ 9 \times (0.223 - 0.223) + 10 \times 0 + 11 \times 0 \\ &+ 12 \times 0 + 13 \times 0 + 14 \times 0\}/\{0 + 0 + 0 + 0 + 0 + 0 \\ &+ (0.366 - 0.223) + (1 - 0.223) + (0.878 - 0.223) + (0.223 - 0.223) \\ &+ 0 + 0 + 0 + 0 + 0\} = 11.537/1.575 = 7.325 \end{aligned}$$

Now, finally, the one at index seven is replaced by centroid –6 = 7.325–6 = 1.325.

So, now the final training set is as follows (Table 7).

Now, these values are fed into our trained ANN model, and we get the output as +59. From the calibration data of the CCD array sensor, we can see that the 205th pixel (i.e., the pixel with the highest voltage) corresponds to wavelength 1542.122 nm. So, the central wavelength of the CCD sensor array is 1542.122 nm + 59 pm =

Table 7 Final training set

Index no	Training value
0	0
1	0
2	0
3	0
4	0
5	0
6	0.366
7	1.325
8	0.878
9	0.223
10	0
11	0
12	0
13	0
14	0

1542.181 nm. We have used the classification instead of regression in the proposed model. Therefore, instead of a single output, our model has 156 output nodes each in the 1 pm step. Only one of the output nodes is activated, and the wavelength corresponding to that activated node is considered the output of ANN. This output wavelength is used to compute the central wavelength of the FBG sensor.

3.3 Implementation and Performance of the Model

After the training algorithm generates the training samples, the ANN model is built using Tensorflow and the Keras package. It is a two-layer, densely connected model. The computation was carried out on the Google Colab platform. In the ANN model, the Softmax activation function is used. For training the model, the Adam optimizer is used along with the 'Sparse Categorical Crossentropy' loss function.

While training the model, the low-valued samples were systematically truncated so that the output of the ANN is primarily governed by higher-valued samples. This will allow us to set a threshold level to truncate low-valued samples, which are more affected by noise. The data is trained for 1,30,000 epochs.

The training samples are divided into training and validation sets. The training set is used to train the ANN model. Then, using the trained ANN model, the Bragg wavelength of the FBGs is predicted for validation set samples. The Error in the predicted value of FBG wavelength is computed, and the following observations are made.

- The error data showed that the correct wavelength is predicted in 56.56% of the cases.
- The Error is within ±1 pm in 95.84% of predictions.
- Error is limited within ±2 pm in 99.51% of predictions
- Error is limited within ±3 pm in 99.83% of predictions
- The maximum range of predicted Error is ±4 pm, which happens in less than 0.02% of predictions
- The mean absolute Error observed is less than 1 pm.

4 Comparison with Other FBG Central Wavelength Detection Algorithms

As discussed earlier, the two most common traditional algorithms for central wavelength detection of an FBG sensor are curve fitting algorithm-based detection and centroid finding algorithm-based detection. Both have advantages and limitations, which are discussed below with respect to our proposed algorithm. The issues are illustrated with reference to a distorted FBG spectrum.

4.1 Curve Fitting Algorithm Based Detection

There are several types of curve fitting algorithms, and polynomial curve fitting and Gaussian curve fitting are more popular than others. Curve fitting algorithms assume that the shape of the FBG sensor spectrum profile matches the shape of some family of mathematical functions (like Gaussian or Polynomial). Then, with the available sample points, the algorithm tries to fit a particular function of that family to the sampled points iteratively. Curve-fitting algorithms are accurate when the FBG profile shape matches the assumed family of functions. However, quite often, the spectral shape of the FBG sensor, when captured on an Optical Spectrum Analyzer, is separate from a form of a mathematical function. In cases where sparse sampling has been done, and the actual spectral shape of the FBG sensor is mathematically different from the one assumed by the fitting algorithm, the predicted curve is not accurate. In such cases, when the FBG sensor spectrum shifts by some value, the central peak of the predicted curve shifts by a different value.

In the illustration below, the central value of the FBG spectrum is shifted by 80 pm. However, the Gaussian fit of the sampled plot shifts by 90 pm (Figs. 6 and 7).

However, this curve fitting-based FBG Bragg wavelength detection technique has been reported to be more accurate than centroid algorithm-based Bragg wavelength detection [4].

However, the main limitation of this curve-fitting approach stems from the fact that it is an iterative process [1] and, therefore, is unsuitable for real-time applications with stringent needs for efficient time performance.

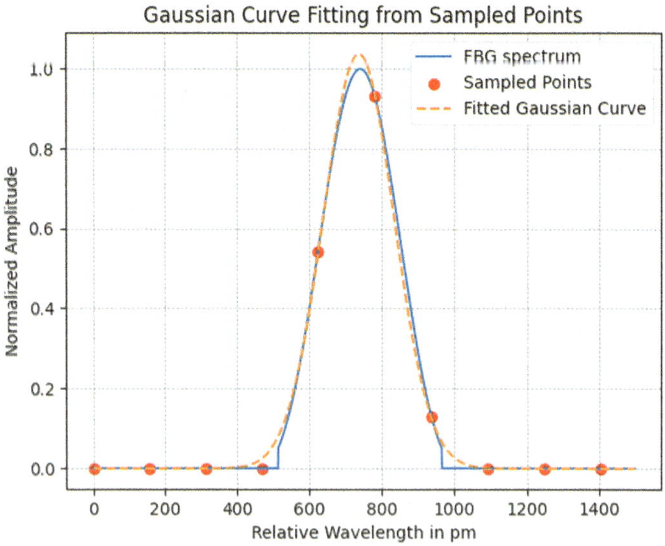

Fig. 6 A simulated FBG peak with central wavelength 740 is sampled, and the Gaussian curve is fitted. The fitted curve's central wavelength is 735

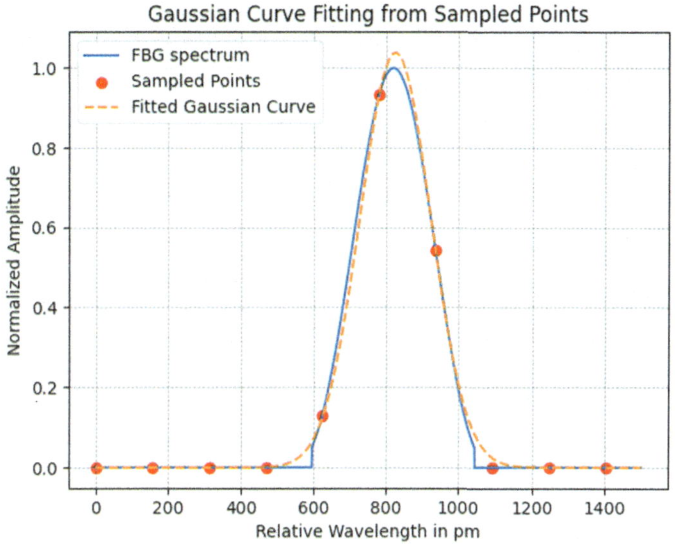

Fig. 7 A simulated FBG peak with central wavelength 820 is sampled, and the Gaussian curve is fitted. The fitted curve's central wavelength is 825

In contrast, the proposed ANN-based algorithm for central wavelength detection is iterative only during the model's training. Once the model parameters are determined by a supervised learning technique, this model can be deployed in the field for prediction by an algorithm that is not iterative. Hence, the proposed ANN-based algorithm overcomes the limitation regarding the real-time processing of the FBG sensor data.

4.2 *Centroid Finding Algorithm Based Detection*

The centroid finding algorithm has the advantage that it is not an iterative algorithm. It can be implemented on an FPGA for real-time applications requiring efficient time performance. However, the centroid finding algorithm-based FBG sensor spectrum central wavelength detection suffers from accuracy-related issues when sparse sampling of a distorted spectral peak is done. When we find the centroid of a spectral peak, it is the centroid of the polygon formed by joining the sampled points by straight lines. When the sampling is dense, the shape of the polygon is very similar to that of the spectrum. However, when the sampling is sparse, the shape of the polygon is quite different from the spectrum's shape.

So, in the case of sparse sampling, while doing CCD array-based interrogation, when the FBG spectrum shifts in wavelength, the shape of the polygon made by sampling points on the FBG sensor spectrum changes. As a result, the centroid of

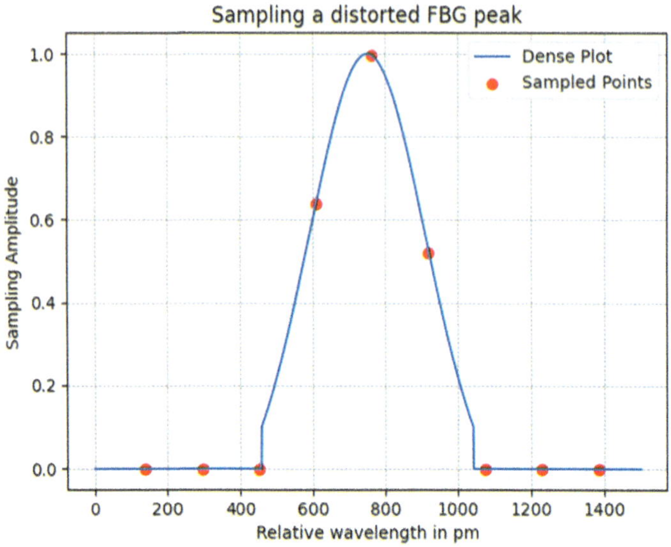

Fig. 8 A simulated FBG peak with central wavelength 750 is sampled at intervals of 156 pm

the polygon may shift by a different value compared to the shift of the FBG sensor spectrum.

In the following section, we have illustrated that when an FBG peak is sampled 156 pm apart, and central wavelength detection is done using a centroid finding algorithm, the Error can reach up ± 8 pm with a mean absolute error of 2.92 pm. The FBG peak used for this computation has a distorted profile because amplitude samples less than 0.1 are ignored (Figs. 8 and 9).

Our proposed ANN algorithm takes the centroid value as one of the input features. However, to address this Error in central wavelength determination for a given FBG sensor spectrum using the centroid method, the values of all the samples capturing the spectrum are also fed in as inputs in the training data. As a result, the prediction accuracy improves in our ANN-based algorithm, and it is limited to ± 2 pm with a mean absolute error of 0.513 pm for the same distorted profile FBG peak (Fig. 10).

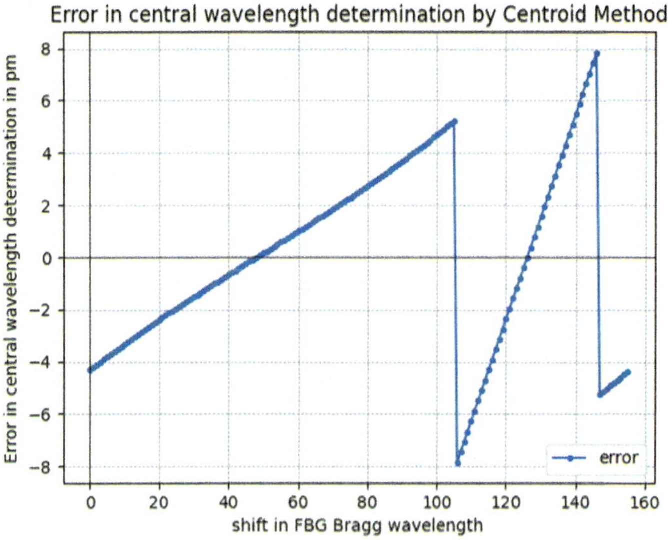

Fig. 9 Error in central wavelength computed using centroid algorithm when the peak shifts left in wavelength and sampling is done at intervals of 156 pm. The leftward shift in wavelength in pm is the x-axis

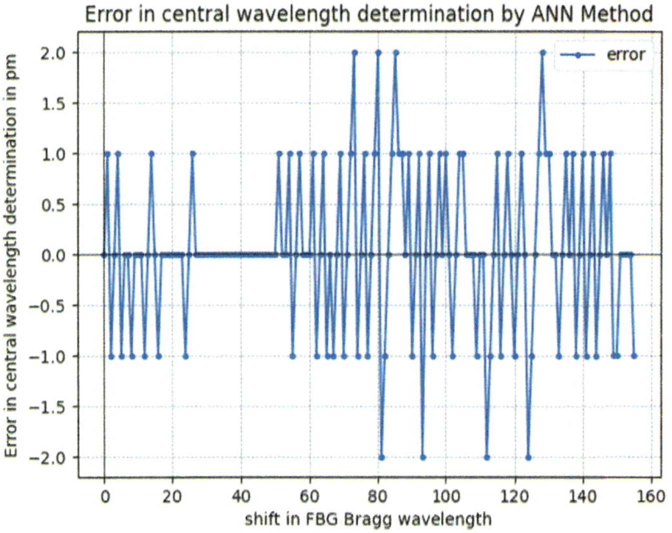

Fig. 10 The error in central wavelength is computed using the proposed ANN algorithm when the peak shifts left in wavelength. Sampling is done at intervals of 156 pm. The leftward shift in wavelength in pm is the x-axis

5 Conclusion

Our proposed ANN-based algorithm combines the advantages of both techniques (curve fitting and centroid finding). This is non-iterative and FPGA implementable, like the centroid finding algorithm, and it is also fairly accurate, like curve fitting algorithms. In cases where the spectral peak shape differs considerably from the fitting function, in those cases, the proposed ANN-based algorithm outperforms the curve fitting algorithms because the training is done on the actual shape of the spectrum obtained by dense sampling rather than any mathematical approximation of the spectrum shape. Moreover, the training data is generated based on the spectrum shape in our proposed ANN algorithm. Hence, this method remains valid even if the FBG sensor spectral peak undergoes a translation in wavelength, keeping the shape the same. Such translation/shift in FBG Bragg wavelength may be caused by pre-strain variation and changes in ambient nominal conditions and the proposed ANN-based algorithm adequately takes care of this.

Acknowledgements This study was carried out jointly by the Computer Science & Engineering Department of Jadavpur University Kolkata and the Fiber Optics & Photonics Division of CSIR-Central Glass & Ceramic Research Institute Kolkata. The authors acknowledge and thank the heads of both departments and the heads of the institutes, the Director CSIR-CGCRI, and the VC of Jadavpur University for their valuable support and encouragement.

Disclosure of Interests The authors have no competing interests to declare relevant to this article's content.

References

1. Alves J, Santos JL, Carvalho A, Lage A (2003) Fiber Bragg, sensor interrogation system, based on a CCD spectrometer. Sensors (IEEE Xplore)
2. Jiang Y (2008) High-resolution interrogation technique for fiber optic extrinsic Fabry–Perot interferometric sensors by the peak-to-peak method. Opt Soc Am (2008)
3. Muthuramalingam A, Himavathi S, Srinivasan E (2008) Neural network implementation using FPGA: issues and application. World Academy of Science, Engineering and Technology
4. Bodendorfer T, Muller MS, Hirth F, Koch AW (2009) Comparison of different peak detection algorithm with regards to spectrometric fiber Bragg grating interrogation systems (IEEE Xplore)
5. Negri L, Nied A, Kalinowski H, Paterno A (2011) Benchmark for peak detection algorithms in fiber Bragg grating interrogation and a new neural network for its performance improvement MDPI sensors
6. Tosi D (2017) Review and analysis of peak tracking techniques for fiber bragg grating sensors. MDPI Sensors
7. Chen Z, Zhang C (2022) Analysis of tunable F-P filter and peak-detection algorithm in FBG demodulation system. In: International conference on networking systems of AI
8. Agliullin T et al (2023) Application of neural network algorithms for central wavelength determination of fiber optic sensors. MDPI Appl Sci

Performance Assessment of Sequential Models for Solar Radiation Forecasting Over Varying Forecast Horizon

Asif Iqbal Middya, Sarbani Roy, and Agnish Paul

Abstract A precise projection of solar system output is critical for efficient power grid operation and effective energy management. Now, forecasting solar radiance is a crucial part of this projection process. It may be done in a number of ways, such as by using AI-based techniques and remote sensing data with physical models. The recent literature explored the potential various sequential models such as LSTM and its variants for solar radiation forecasting. However, there is not much research done on how well the sequential models perform when forecasting solar radiation across different prediction horizons. In this context, this study performs the performance assessment of some state-of-the-art sequential models such as LSTM, BiLSTM, Hybrid LSTM (proposed model), Simple RNN and Stacked LSTM for solar radiation forecasting over varying forecast horizon (h = 5 h to h = 45 h). The experimental results highlight that although the stacked LSTM, Hybrid LSTM and BiLSTM generally demonstrated better performance for most horizons and performance metrics, the proposed hybrid model has been dominated most of the time.

Keywords Solar radiation forecasting · Sequential models · Deep learning

1 Introduction

Accurate forecasting of solar radiation is crucial for the efficient operation and planning of solar power plants. However, forecasting model building has various challenges since solar radiation is a complicated spatio-temporal phenomenon. In the literature, there are several approaches for predicting solar radiation [1]. These techniques fall into two main categories: machine learning models and cloud imaging

A. I. Middya (✉) · S. Roy
Department of Computer Science and Engineering, Jadavpur University, Kolkata, India
e-mail: asifim.rs@jadavpuruniversity.in

S. Roy
e-mail: sarbani.roy@jadavpuruniversity.in

A. Paul
Department of Information Technology, Jadavpur University, Kolkata, India

© The Author(s), under exclusive license to Springer Nature Singapore Pte Ltd. 2025

N. Chaki et al. (eds.), *Applied Computing for Software and Smart Systems*, Lecture Notes in Networks and Systems 980, https://doi.org/10.1007/978-981-97-9762-2_4

coupled with physical models. Traditional forecasting methods frequently fail to reflect the complex temporal dynamics and non-stationary nature of solar radiation data. Studying and modelling these complexities is critical for constructing reliable forecasting models that can respond to variations in solar radiation patterns. In recent years various deep learning based sequential models are also employed in the tie series forecasting of solar radiation. However, little research has been conducted on how well sequential models function when forecasting solar radiation over different prediction horizons. In order to address this problem we have explored a total of five state-of-the-art deep learning based sequential models. Specifically, this study evaluates sequential models for solar radiation forecasting, such as LSTM (Long Short-Term Memory) [2], BiLSTM [3], Hybrid LSTM (proposed model), Simple RNN (Recurrent Neural Network) [4] and Stacked LSTM [5], over varying prediction horizons. Evaluating the performance of these methods proves challenging due to varying forecasting horizons and performance indicators. The following points summarize the paper's significant contributions.

- The research paper attempted to find out the efficient sequential model(s) for solar radiation forecasting over multiple forecast horizons (h = 5 h to h = 45 h). A Hybrid LSTM model is proposed and compared with other sequential models (e.g. LSTM, BiLSTM, Simple RNN and Stacked LSTM) for forecasting of solar radiation.
- Rigorous experiments have been conducted to assess using a range of performance measures, including MAE, RMSE, R2, correlation and MAPE. A real-world solar radiation dataset with 9504 samples has been utilized in this analysis.

The remaining article is structured as follows. Section 2 provides a brief description of the existing works. The problem statement is presented in Sect. 3. The details of the methodology are provided in Sect. 4. The evaluation and the results are presented in Sect. 5. The conclusion and some future directions are discussed in Sect. 6.

2 Related Works

The accurate forecasting of solar radiation is crucial for optimizing the integration of solar energy into the electrical grid and maximizing the efficiency of solar photovoltaic (PV) systems [6]. Due to the inherent stochastic nature of solar radiation, driven by diverse meteorological factors, finding robust forecasting models has been a significant research challenge. In recent years, Long Short-Term Memory (LSTM) networks, a type of recurrent neural network (RNN) adept at learning long-term dependencies, have emerged as a promising tool for solar radiation forecasting due to their ability to capture the complex temporal dynamics of the data.

Several studies have demonstrated the superiority of LSTMs over traditional statistical and machine learning approaches like persistence models, ARIMA and regression techniques [7–16]. The ability of LSTMs to capture long-term dependencies in sequential data makes them well-suited for tasks like time series forecasting, especially when accurate predictions are required for longer forecast horizons. Research has explored various LSTM architectures and hyperparameter tuning strategies to optimize forecasting performance. Studies by Ghimire et al. [5], Tong et al. [17] and Sharda et al. [18] investigated the effectiveness of stacked LSTMs, encoder-decoder structures and attention mechanisms in improving prediction accuracy. Forecasting accuracy generally decreases with increasing forecast horizon as the influence of unpredictable weather events grows. However, LSTMs consistently outperform other methods even for longer horizons [19].

While recent research has investigated the application of various sequential models, including LSTM and its variants, for solar radiation forecasting, a critical knowledge gap remains regarding their performance across diverse prediction horizons. This significantly hinders practical implementation, as accurate forecasts are required for both short-term (e.g. grid operation) and long-term (e.g. energy planning) needs. This study specifically tackles this gap by conducting a comprehensive performance assessment of five state-of-the-art sequential models: LSTM [2], BiLSTM [3], Hybrid LSTM, Simple RNN [4] and Stacked LSTM [5]. We systematically evaluate their forecasting accuracy across a broad range of prediction horizons, spanning from 5 to 45 h. This allows for a nuanced understanding of how each model's strengths and weaknesses manifest at different timescales, enabling practitioners to make informed choices based on their specific forecasting needs.

3 Problem Statement

While various sequential models like LSTM and its variants have shown promise in solar radiation forecasting, there is a lack of research on their performance across different prediction horizons. This poses a challenge for efficient power grid operation and energy management, as accurate forecasts are crucial for both short-term (e.g. 5 h) and long-term (e.g. 45 h) planning.

4 Methodology

Figure 1 describes a framework of the study with various major components. The raw solar radiation data is first smoothed with an exponential smoothing function. The smoothed data is then divided into training and test sets. Since we are dealing with the time series data it is not shuffled before train-test split. Approximately first 90% data are considered for model training (Number of training samples = 8500) and rest of the samples are used for testing. Following that, we apply a standard scaler exclusively to the training data and subsequently utilize the same scaler to standardize both the training and test datasets. Subsequently, the time-series sequence is transformed into a supervised form via a process known as batching. This involves splitting the data sequence into input-output vectors, with a set of n successive values designated as an input vector and the next t successive values designated as an output

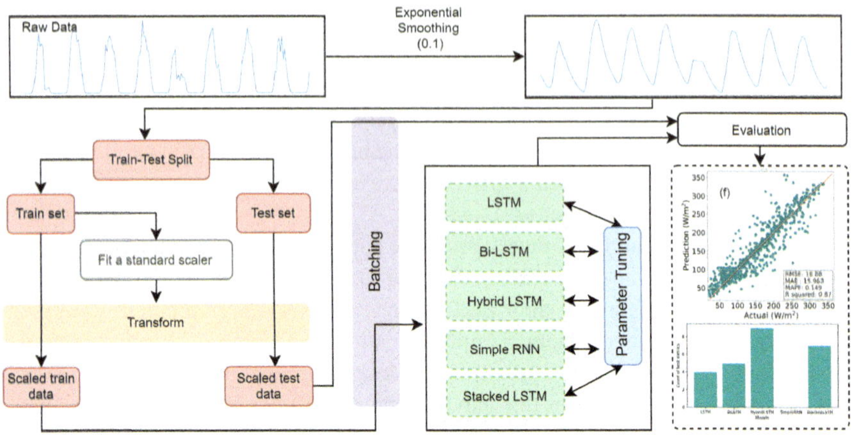

Fig. 1 Framework for solar radiation forecasting

Table 1 Dataset description

Statistics	Value
Count	9504
Mean	141.178139
Std	188.083719
Min	8.03
25th percentile	8.57
50th percentile	15.175
75th percentile	257.4525
Max	749.42

vector. After batching various LSTM-based models namely vanilla LSTM, BiLSTM, Hybrid LSTM, Simple RNN, Stacked LSTM will be explored for multi-horizon solar radiation forecasting. Finally the performance assessment of the above-mentioned sequential models will be assessed for varying forecast horizon in terms of various performance metrics.

A real-world dataset of hourly solar radiation data is considered for model building and analysis. The solar radiation data used in this study is collected from a static monitoring station located at Sri Aurobindo Marg, Delhi, India. The data is acquired from CPCB's website (https://cpcb.nic.in). The time series solar radiation data is recorded in W/mt^2. The dataset consists of a total of 9504 time series hourly samples of solar radiation data. The collected data spans from January 2019 to February 2020. Various descriptive statistics of the dataset (e.g. mean, standard deviation (std), minimum (i.e. min), etc.) are provided in Table 1.

This section explains the architectures of various LSTM-based models for modelling multi-horizon forecasting of solar radiation. Figure 2a shows the architecture of vanilla LSTM. The vanilla LSTM model having a single LSTM layer having number of units $U_n = h$, where h is the forecast horizon. Note that we have separate model for every $h \in \{5, 15, 25, 35, 45\}$. We have used activation function as `tanh`. The LSTM layer is followed by a Dense layer with the same number of units as the LSTM layer (i.e. h) and a Reshape layer. Figure 2b shows the architecture of StackedLSTM. This model has two consecutive LSTM layers followed by a dense and a Reshape layer. For all the layers, we have set the number of units $U_n = h \in \{5, 15, 25, 35, 45\}$. Figure 2c shows the architecture of BiLSTM. The model architecture consists of a BiLSTM layer and a dense layer with $U_n = h \in \{5, 15, 25, 35, 45\}$ followed by a Reshape layer. Fig. 2d shows the architecture of SimpleRNN. The model consists of a RNN layer, a dense and a Reshape layer.

Finally we have develop a HybridLSTM model as shown in Fig. 2e. Six different sub-models (m1, m2, m3, m4, m5 and m6) are created and their outputs are concatenated. m1 and m4 have LSTM layers with 32 and 50 units respectively. Each of the LSTM layers is followed by a dense layer with $U_n = h$ units. m2 and m5 have BiLSTM layers with 32 and 50 units respectively followed by a dense layer $U_n = h$ units. Similarly, the sub-models m3 and m6 have SimpleRNN layers with 32 and 50 units followed by a dense layer of $U_n = h$. For the all the sub-models $h \in \{5, 15, 25, 35, 45\}$. The concatenated layer is followed by a dense layer and a reshape layer.

The dataset is divided into training and test sets using a train-test split, with around 90% of the data allocated to the training set and approximately 10% to the test set. The hyperparameter optimization in this study is conducted using the grid-search method. For different models, various combinations of optimizer, number of units and total epochs are explored through grid search. For all the models the optimal performances achieve around 70 epochs with Adam as the optimizer.

5 Evaluation and Results

This section presents the evaluation and results of the performance assessments of sequential models for solar radiation forecasting. The models were crafted utilizing the Tensorflow Keras API and fine-tuned with grid-search approach. Google Colaboratory was employed for model building with Python as the programming language. Throughout the model building process, crucial Python libraries like Pandas, Numpy, Scikit-learn and others were utilized.

Diverse performance indicators such as coefficient of determination (R^2), root mean squared error (RMSE), mean absolute error (MAE), mean absolute percentage error (MAPE) and correlation coefficient (corr) are employed to evaluate the performance of Sequential Models in forecasting solar radiation across varying forecast horizons.

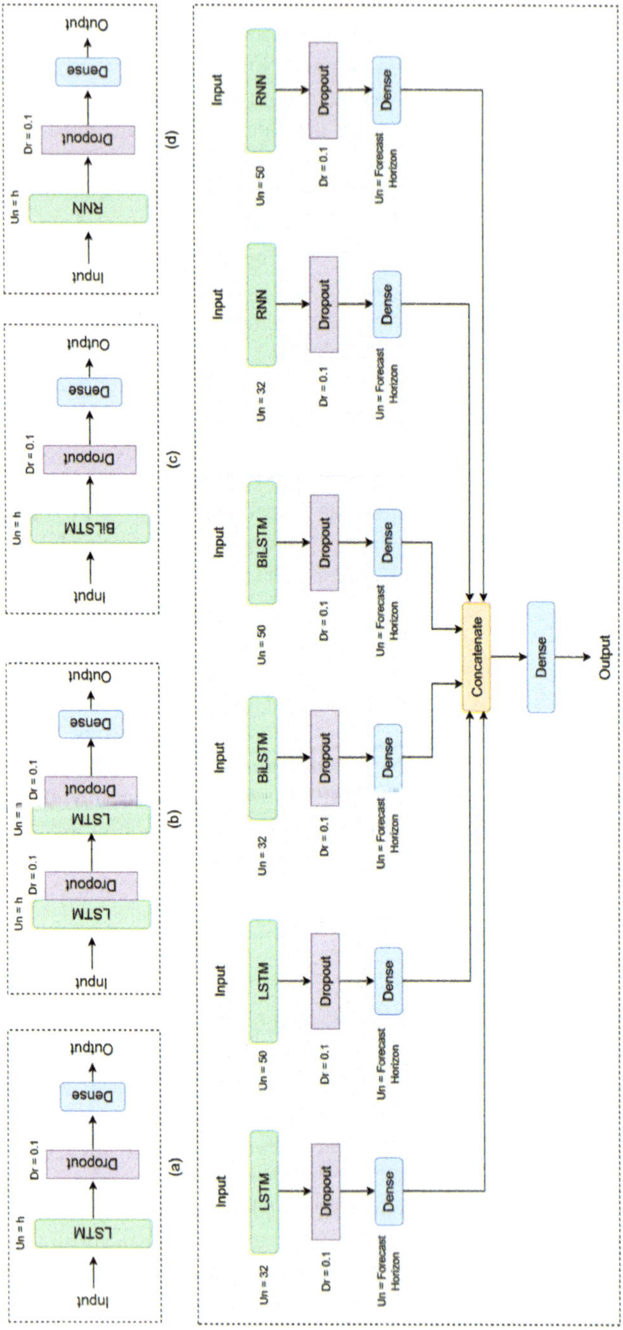

Fig. 2 Model architectures of various sequential models **a** Vanilla LSTM **b** Stacked LSTM **c** BiLSTM **d** SimpleRNN **e** HybridLSTM

Table 2 Accuracy of the models for various forecast horizons

Horizon	Metric	SimpleRNN	LSTM	BiLSTM	StackedLSTM	HybridLSTM
5	MAE	20.487	15.963	17.293	15.350	19.053
	RMSE	23.810	18.880	20.032	18.173	21.981
	MAPE	0.220	0.149	0.161	0.141	0.205
	R2	0.835	0.870	0.870	0.876	0.844
	Corr	0.917	0.935	0.933	0.938	0.924
15	MAE	22.900	23.211	21.195	22.425	22.811
	RMSE	27.332	27.536	25.443	26.742	26.906
	MAPE	0.212	0.203	0.182	0.192	0.201
	R2	0.760	0.762	0.781	0.769	0.761
	Corr	0.881	0.881	0.890	0.884	0.885
25	MAE	26.146	26.715	26.854	28.254	23.496
	RMSE	31.440	32.532	32.678	34.225	29.056
	MAPE	0.249	0.251	0.252	0.264	0.219
	R2	0.698	0.675	0.658	0.630	0.724
	Corr	0.856	0.848	0.839	0.830	0.870
35	MAE	30.012	29.896	29.496	28.575	28.345
	RMSE	36.236	35.741	36.088	35.133	34.909
	MAPE	0.263	0.260	0.253	0.241	0.236
	R2	0.661	0.671	0.646	0.665	0.671
	Corr	0.832	0.836	0.826	0.831	0.841
45	MAE	33.316	30.916	31.116	30.873	33.987
	RMSE	41.190	38.861	39.499	39.856	43.259
	MAPE	0.326	0.289	0.303	0.284	0.296
	R2	0.585	0.636	0.613	0.606	0.531
	Corr	0.790	0.817	0.806	0.805	0.784

Table 2 provides the accuracies of the models for various forecast horizons (h). It is interesting to observe that no single model dominates for all the forecast horizons. However the models like Stacked LSTM, Hybrid LSTM and BiLSTM provide relatively better performance for most of the forecast horizons and the performance indications.

The observation versus prediction plots for the models are provided in Fig. 3. For a shorter forecast horizon, most of the data points (the blue dots) are following the diagonal line (in red) and eventually producing better results. However for a longer forecast horizon, scattered away from the diagonal line. This is to be expected, as it is generally more difficult to make accurate predictions further into the future. The Simple RNN model performs poorly for all forecast horizons, with a large number of points scattered far away from the diagonal line. This suggests that the model is not able to capture the underlying relationships in the data well. The Hybrid LSTM model performs similarly to the Stacked LSTM model, with most of the points clustered

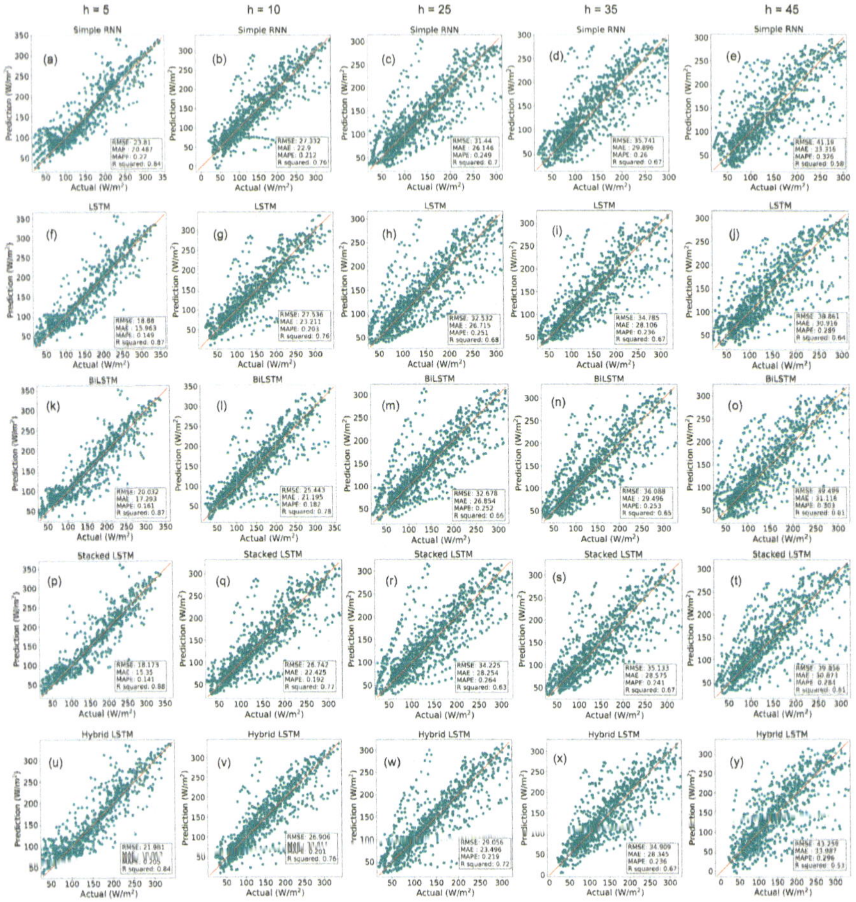

Fig. 3 Observation versus prediction of various models for varying forecast horizons (h ∈ { 5, 15, 25, 35, 45 })

close to the diagonal line. This suggests that the model is able to learn the most complex relationships in the data.

The models are also compared based on the number of times they outperformed for different forecast horizon and performance metrics. Fig. 4 shows the proposed hybrid model has been dominated most of the time followed by the stacked LSTM and the BiLSTM models. The HybridLSTM model achieved the best metric nearly twice as many times as the StackedLSTM model. The Simple RNN model has been dominated none of the times. It is important to note that the number of times a model achieved the best metric is an important measure of its performance and to some extent captures the genericness of the model.

The computational efficiency of the models is also assessed over the 100 runs. In Fig. 5, the execution times of the models are presented with the help of box plots.

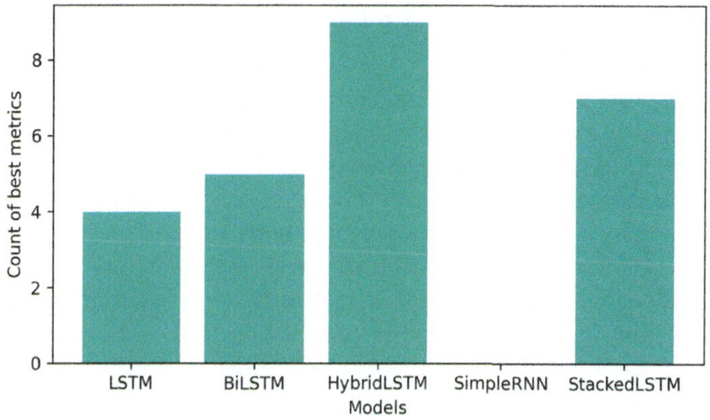

Fig. 4 Count of best metrics

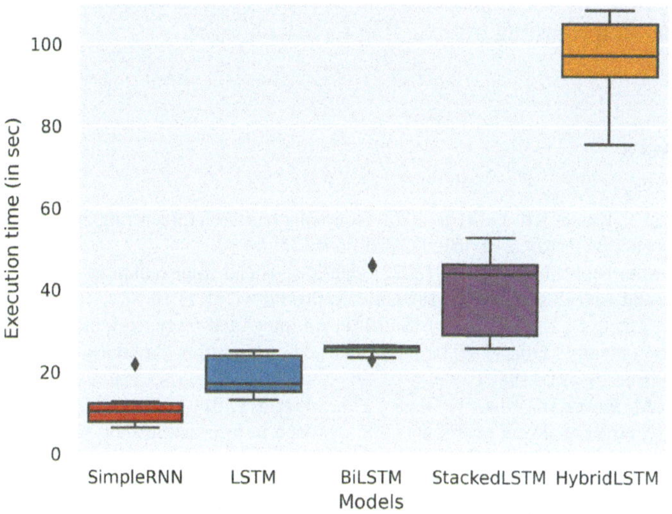

Fig. 5 Execution time of the models

The box plot shows the distribution of the execution time for five different machine learning models: SimpleRNN, LSTM, BiLSTM, StackedLSTM and HybridLSTM. The x-axis of the plot shows the models, and the y-axis shows the execution time in seconds. The box in the box plot represents the interquartile range (IQR), which is the middle 50% of the data. It is observed that the execution time is much higher for the hybrid model compared to the other models. The box plot also shows that the execution time of the models is variable with IQR is smallest for the BiLSTM model and largest for the StackedLSTM model.

6 Conclusion and Future Works

In this paper, we focus on the assessment of the performance of sequential models for solar radiation forecasting over varying forecast horizons. Our research found that no single sequential model consistently outperforms others across all forecast horizons for solar radiation prediction. However, Stacked LSTM, Hybrid LSTM and BiLSTM generally demonstrated better performance for most horizons and metrics. As anticipated, forecast accuracy decreased with increasing horizon length, with shorter horizons (5h) yielding higher accuracy than longer horizons (45h). While the Hybrid LSTM model achieved superior accuracy in most cases, its execution time was significantly higher. Stacked LSTM and BiLSTM, on the other hand, presented a favourable balance between accuracy and computational efficiency.

As our future works, we would like to investigate forecasting capabilities of sequential models at various timescales, such as daily or weekly, in addition to hourly predictions. We would also like to come up with an end-to-end system for solar radiation forecasting over varying forecast horizon.

References

1. Krishnan N, Kumar KR, Inda CS (2023) How solar radiation forecasting impacts the utilization of solar energy: a critical review. J Clean Prod:135860
2. Yildirim A, Bilgili M, Ozbek A (2023) One-hour-ahead solar radiation forecasting by MLP, LSTM, and ANFIS approaches. Meteorol Atmos Phys 135(1):10
3. Peng T, Zhang C, Zhou J, Nazir MS (2021) An integrated framework of Bi-directional long-short term memory (BiLSTM) based on sine cosine algorithm for hourly solar radiation forecasting. Energy 221:119887
4. Jaihuni M, Basak JK, Khan F, Okyere FG, Sihalath T, Bhujel A, Park J, Lee DH, Kim HT (2022) A novel recurrent neural network approach in forecasting short term solar irradiance. ISA Trans 121:63–74
5. Ghimire S, Deo RC, Wang H, Al-Musaylh MS, Casillas-Pérez D, Salcedo-Sanz S (2022) Stacked LSTM sequence-to-sequence autoencoder with feature selection for daily solar radiation prediction: a review and new modeling results. Energies 15(3):1061
6. Díaz-Bedoya D, González-Rodríguez M, Clairand J-M, Serrano-Guerrero X, Escrivá-Escrivá G (2023) Forecasting univariate solar irradiance using machine learning models: a case study of two Andean cities. Energy Convers Manage 296:117618
7. Nath P, Middya AI, Roy S (2023) Empirical assessment of transformer-based neural network architecture in forecasting pollution trends. Int J Data Sci Anal:1–17
8. Choudhury A, Middya AI, Roy S (2022) A comparative study of machine learning and deep learning techniques in forecasting air pollution levels. In: Proceedings of international conference on data science and applications: ICDSA 2021, vol 1. Springer, pp 607–619
9. Middya AI, Roy S (2022) Spatio-temporal variation of covid-19 health outcomes in India using deep learning based models. Technol Forecast Soc Change 183:121911. https://doi.org/10.1016/j.techfore.2022.121911
10. Das R, Middya AI, Roy S (2021) High granular and short term time series forecasting of PM2.5 air pollutant—a comparative review. Artif Intell Rev 55(2):1253–1287. https://doi.org/10.1007/s10462-021-09991-1

11. Nath P, Saha P, Middya AI, Roy S (2021) Long-term time-series pollution forecast using statistical and deep learning methods. Neural Comput Appl 33(19):12 551–12 570. https://doi.org/10.1007/s00521-021-05901-2
12. Nandi A, De A, Mallick A, Middya AI, Roy S (2022) Attention based long-term air temperature forecasting network: ALTF net. Knowl Based Syst 252:109442. https://doi.org/10.1016/j.knosys.2022.109442
13. Nath P, Roy B, Saha P, Middya AI, Roy S (2022) Hybrid learning model for spatio-temporal forecasting of PM2.5 using aerosol optical depth. Neural Comput Appl 34(23):21 367–21 386. https://doi.org/10.1007/s00521-022-07616-4
14. Choudhury A, Middya AI, Roy S (2022) Attention enhanced hybrid model for spatiotemporal short-term forecasting of particulate matter concentrations. Sustain Cities Soc 86:104112. https://doi.org/10.1016/j.scs.2022.104112
15. De A, Nandi A, Mallick A, Middya AI, Roy S (2023) Forecasting chaotic weather variables with echo state networks and a novel swing training approach. Knowl Based Syst 269:110506. https://doi.org/10.1016/j.knosys.2023.110506
16. Middya AI, Roy S (2022) Pollutant specific optimal deep learning and statistical model building for air quality forecasting. Environ Pollut 301:118972. https://doi.org/10.1016/j.envpol.2022.118972
17. Tong J, Xie L, Fang S, Yang W, Zhang K (2022) Hourly solar irradiance forecasting based on encoder-decoder model using series decomposition and dynamic error compensation. Energy Convers Manage 270:116049
18. Sharda S, Singh M, Sharma K (2020) RSAM: robust self-attention based multi-horizon model for solar irradiance forecasting. IEEE Trans Sustain Energy 12(2):1394–1405
19. Zheng H, Yuan J, Chen L (2017) Short-term load forecasting using EMD-LSTM neural networks with a Xgboost algorithm for feature importance evaluation. Energies 10(8):1168

Distributed Systems

Advancing Smartphone Sensor-Based Keystroke Dynamics for Implicit and Active Authentication: Addressing Challenges and Enhancing Usability Control

Soumen Roy, Utpal Roy, Devadatta Sinha, and Rajat Kumar Pal

Abstract Smartphone sensors play a pivotal role in delineating user behaviour, particularly in the context of keystroke dynamics-based implicit and active/continuous authentication, as well as liveness detection. The acquisition of multifaceted sensory data during typing serves as a fundamental source of behavioural attributes, yet the field faces several intricate challenges requiring careful consideration to recalibrate the delicate balance between security and usability. Notable challenges encompass the capricious nature of anomaly detectors when confronted with unavoidable influencing factors, the substantial battery drainage associated with prolonged sensor operation, the computational inefficiency intrinsic to smartphones, and the imperative need for usability control in frequently switching applications and balancing security in both device and application levels, especially in scenarios with varying security requirements. This study endeavours to mitigate these challenges through the implementation of cutting-edge anomaly detection techniques, incorporating four distinct schemes—namely, *column medians template formation*, *synthetic samples*, *soft biometrics* and *score-level fusion*. The study further leverages advanced sensory patterns, including gyroscope, accelerometer and rotational information, captured during abbreviated typing intervals at a low sampling rate to mitigate battery consumption. In the pursuit of enhanced usability control, a *Watchman score* is assigned to each running application, dictating security adjustments based on genuineness evaluations of micro-behaviours within predefined temporal windows. A novel rule-based mechanism is introduced to govern device and application security within a continuous authentication mode. This empirical investigation, supported by robust statistical evidence, aspires to yield more precise, energy-efficient, and user-friendly

S. Roy (✉)
Department of Computer Science, Bagnan College, Khalore, Bagnan, Howrah 711303, India
e-mail: soumen.roy_2007@yahaoo.co.in

U. Roy
Department of Computer and System Sciences, VisvaBharati, Santiniketan 731235, India
e-mail: roy.utpal@gmail.com; utpal.roy@visva-bharati.ac.in

D. Sinha · R. K. Pal
Department of Computer Science and Engineering, University of Calcutta, Acharya Prafulla Chandra Roy Siksha Prangan, JD-2, Sector-III, Saltlake City, Kolkata 700106, India

© The Author(s), under exclusive license to Springer Nature Singapore Pte Ltd. 2025

N. Chaki et al. (eds.), *Applied Computing for Software and Smart Systems*, Lecture Notes in Networks and Systems 980, https://doi.org/10.1007/978-981-97-9762-2_5

implicit and active user identity verification mechanisms. The findings present an avenue for future exploration and formulation of smartphone user identity verification issues based on the established methodologies and research facets elucidated in this study.

Keywords Active authentication · Anomaly detector · Keystroke dynamics · Smartphone security · Smartphone sensors

1 Introduction

Smartphones have become the most powerful computing devices, and they contain a wealth of sensitive information. Therefore, safeguarding them against unauthorized access has become imperative [53]. Conventional security measures exhibit inherent limitations [50, 57], with session hijacking emerging as a notable concern [49]. Behavioural biometric methodologies, exemplified by Keystroke Dynamics (KD), are garnering attention for their potential in perpetuating continuous user authentication, leveraging the distinctiveness inherent in user-specific activities such as holding, typing and swiping [27, 48, 57]. Noteworthy strides have been made through the employment of Machine Learning (ML) models harnessing sensor data generated during typing [63], albeit there remains a requisite for supplementary mechanisms to fortify both stability and security [37].

In recent years, smartphones, which are increasingly equipped with embedded sensors, have become a primary source of biometric systems for different applications [10, 47, 57]. Behavioural biometric attributes, encompassing diverse interactions with the device such as receiving calls, typing, swapping, holding and scrolling, engender unique patterns that can function as a clandestine key for user authentication, either supplanting or augmenting conventional knowledge-based authentication methodologies like PINs or passwords [26]. As an exemplary approach to implicitly, covertly and continuously verifying users, KD relies on monitoring, measuring and analysing typing behaviours without requiring direct involvement on the device [24].

1.1 Keystroke Dynamics as Active/Continuous User Authentication

KD emerges as a versatile and economically viable behavioural biometric attribute, amenable to facile capture across an array of devices, including smartphones, computer keyboards and touch-pads. Its efficacy lies in facilitating active authentication seamlessly throughout a user session, ensuring continuous implicit verification without interrupting the session and mitigating unproductive time associated with password-related interactions. Active authentication, synonymous with continuous, implicit or transparent authentication, can be implemented either on-device or off-device, embodying characteristics of continuity and transparency. The integral phases of active authentication encompass enrolment and identity verification, involving sequential steps such as data acquisition, feature extraction, data preprocessing, template formation, classification, decision-making and re-authentication [22]. Notably, the re-authentication mechanism should adhere to attributes such as implicitness, knowledge independence, resistance to observation, fine-grained protection, continuity, usability and cost-effectiveness, as underscored in diverse studies [17, 32, 41]. Prudent considerations during model design are imperative to ensure the efficacy of the re-authentication technique while concurrently upholding on-device/off-device security, application-level/device-level security, and the nuanced security requirements associated with low-secure and high-secure applications.

1.2 Research Gaps and Motivation

Compared with alternative biometric methods, touch interaction offers covert data capture and cost-effectiveness as a promising avenue for continuous authentication [47, 55]. However, challenges, including high intra-class variation across sessions, multifactorial influences, battery consumption for operating sensors continuously throughout a session, frequently switching applications with varying security needs, and usability control in both device and application levels, pose significant impediments to the efficacy of touch-based authentication [12, 28, 38]. Leveraging the built-in sensors of contemporary smartphones to capture orientation and forces acting on the device holds potential for enhancing usability and reliability, albeit concerns regarding battery consumption necessitate careful consideration for sustained sensor operation. This study seeks to address identified research gaps in this domain, with a multifaceted approach encompassing the creation of standard and realistic datasets, the exploration of multi-class classification strategies, the judicious selection of anomaly detectors and a comprehensive examination of the impact of battery power overhead on continuous user authentication using KD [22].

The lack of standard and realistic datasets specifically collected by touch typing on smartphone touch screens is an integral aspect of this study. Existing studies predominantly rely on datasets derived from conventional keyboards, often characterized by limited diversity and the inclusion of a narrow demographic range, leading to potential model biases. To address this limitation, the study advocates for the creation of a comprehensive dataset incorporating advanced multi-sensory features, such as gyroscope, rotation and acceleration, during typing interactions on smartphones, ensuring inclusivity across diverse user profiles.

Furthermore, the study delves into the intricacies of anomaly detection in the continuous authentication paradigm, emphasizing the need for a meticulous evaluation of various detectors in diverse settings. The selection of a suitable anomaly detector is deemed pivotal, given the demonstrated variability in performance across different datasets. The study advocates for a comprehensive assessment of multiple detectors using a standard dataset, considering factors such as technology, scenario and operational variations. This multifaceted evaluation approach seeks to elucidate the optimal anomaly detector for continuous user authentication using smartphone-derived sensory data, recognizing the nuanced challenges posed by diverse evaluation scenarios.

The overarching goals and contributions of this study are underscored by the imperative to design an implicit and active smartphone user authentication system employing sensory data. With a focus on evaluating and comparing the performance of various anomaly detectors, the study encompasses technology, scenario and operational evaluations. Addressing critical research gaps in dataset creation, classification strategies, detector selection, battery power considerations and usability concerns, the study aspires to advance the understanding and implementation of effective continuous user authentication systems grounded in KD.

1.3 Objectives and Contributions

This study articulates a comprehensive set of research objectives aimed at advancing the field of active user authentication through KD on smartphones. The study focuses on evaluating and demonstrating the reliability and applicability of twelve anomaly detectors in four distinct schemes. This evaluation is conducted using typing samples collected at six different interaction durations on smartphones, aligning with objectives for time efficiency, battery conservation and enhanced accuracy in active user authentication. The framework proposed seeks to establish usability control by harmonizing device and application-level security considerations without compromising accuracy or battery power. Further objectives involve the creation of a dataset featuring diverse short typing durations and associated personal information for validation and exploration of influencing factors, a comparative analysis of state-of-the-art detectors in static and dynamic modes, and the proposition and evaluation of a template formation method for distance-based anomaly detection. Additionally, the study investigates data augmentation's impact on reducing acquisition time and enhancing anomaly detection through synthetic data. Further objectives encompass demonstrating the influence of incorporating soft biometric attributes into continuous typing pattern analysis, evaluating the efficacy of score-level fusion methods, employing Analysis of Variance (ANOVA) for pairwise performance comparisons, applying the Linear Mixed-effect Model (LMM) to scrutinize factors affecting detector performance, and comparing different schemes to identify optimal performance. This array of research objectives collectively contributes to the development and empirical validation of enhanced active user authentication mechanisms on smartphones.

2 Related Works and Analysis

The landscape of user authentication has witnessed an abundance of studies focused on entry-point security measures that lack continuity throughout a session [8, 50]. Despite this, a considerable portion of users continues to rely on traditional one-time authentication attributes such as passwords, PINs, sketches and fingerprints [43]. The ubiquity of smartphones in handling highly secure applications, including corporate email, social networking, online examinations and e-banking, underscores the need for continuous and robust security measures, given the devices' centrality in storing private and sensitive data. Previous research has explored diverse attributes, including mouse click, mouse movement, KD, face, touch interaction, speech, device usage and body movement, for active user authentication in both desktop and smartphone environments. While KD-based active authentication for IBM PCs was introduced in 1995 [56], recent attention has gravitated towards smartphones, driven by their widespread application scenarios.

Studies leveraging smartphone sensors for active authentication have explored various features such as image, touch sensors, location, accelerometer, gyroscope, WiFi, app usage, and texture and shape features, yielding diverse performance outcomes [1, 2, 16, 36, 65]. Multisensory data generated by sensors attached to smartphones provides a rich source of features for continuous authentication, with datasets like *HMOG* and *Touchalytics* capturing patterns related to holding, tapping, swiping and scrolling activities [6, 21, 44]. Despite challenges related to intra-class variation and data quality, KD's covert data capture during normal user activities renders it a promising attribute for active authentication. To address accuracy limitations, studies have proposed multi-modal approaches, incorporating contextual factors, and leveraging various pattern recognition methods, including traditional statistics, deep learning and binary classifiers [19, 23, 34, 62].

Recent endeavours have explored binary classifiers for mobile terminal identity authentication, applying classification methods in various studies [7, 25, 34, 51, 52, 59]. An emerging trend involves utilizing anomaly detectors as one-class classifiers for implicit and active authentication of smartphone users, with studies highlighting the need for a sound comparison framework considering diverse datasets and characteristics [1, 5, 13, 16, 18, 33, 35, 54, 66]. The intricacies of these methods, coupled with the variances in datasets, underscore the evolving landscape of active user authentication on smartphones, calling for continued research to refine and optimize these approaches.

The following are the classification approaches (detectors) for user authentication that have been reported and found suitable in this study. Table 1 presents the list of twelve detectors and the results in unique environments and settings. Table summarizes the state-of-the-art anomaly detectors across various feature arrangements and inputs, along with their performance metrics. The detectors are evaluated based on their effectiveness in different settings, including desktop and smartphone environments, using features such as timing, sensory data and statistical analysis. Table also

Table 1 State-of-the-art anomaly detectors on several feature arrangements for a variety of inputs and results have been identified in the literature

Anomaly detectors	Study	Inputs	Devices	Features	EER (%)
Autoencoder (AE)	[40]	Free text	Desktop	Timing	7.82
	[42]	Complex	Desktop	Timing	6.51
Canberra distance (CD)	[64]	Free text	Desktop	Timing	5.11
Cosine distance (COD)	[39]	Free text	Desktop	Timing	4.4
	[14]	Simple	Smartphone	Timing	Acc. 91.84
Dynamic time wrapping (DTW)	[61]	Complex	Desktop	Timing	2.8 to 5.1
Euclidean distance (ED)	[4]	Free text	Smartphone	Timing and sensory	0.086
	[14]	Simple	Smartphone	Timing	Acc. 90.82
k-means (KM)	[15]	Free text	Desktop	Timing	13.2
Lorentzian distance (LD)	[45]	Simple	Desktop	Timing	4.4
Mahalanobis distance (MD)	[9]	Free text	Desktop	Timing	22.3
Minkowski distance (MKD)	[45]	Simple	Desktop	Timing	21.9
	[58]	Free text	Desktop	Timing	2.94
One-class support vector machine (OCSVM)	[3]	Free text	Desktop	Time to barcode	9.88
	[33]	Free text	Smartphone	Sensory and statistical	4.66
Outlier-count (OC)	[25]	Numeric	Smartphone	Piezoelectric force	2.4
Scaled Manhattan (SMD)	[9]	Free text	Desktop	Timing	17.6

provides insights into the detection performance, represented by Equal Error Rates (EER), for each detector and scenario.

3 Proposed Approaches

This section explains the main ideas of the four approaches to enhance the performance of the state-of-the-art detectors and the schemes for better usability control of active authentication design.

3.1 Proposed Design Settings for Improved Performance

To enhance the performance of the active authentication system, the following four design settings have been proposed which have not been yet tested in the topic area earlier.

Approach 1: Median template formation. Data acquisition module produces a series of sensory feature vales $< g_x, g_y, g_z >$ for gyroscope, $< a_x, a_y, a_z >$ for accelerometer, etc. for each event e of a time frame. For a small duration (lets 3.5 sec.) of a series of typing activities $\{e_1, e_2, \ldots\}$, there is a stream of feature values (lets $g_x = \{g_{x1}, g_{x2}, \ldots\}$). Due to the sufficiently large feature values, it is difficult to analyse. Therefore, a windowing technique is desirable for each type of feature. We utilized a non-overlapping sliding window technique to partition the continuous typing stream into multiple fixed-size samples. This technique creates a feature table where a number of the columns represent the data points collected in a small period and the number of rows represents the number of samples. Here, each row is replaced by a sorted sequence of each respective row. Several feature tables are merged to create an input table that is ready to train the model/template. Instead of using the mean, it has been suggested to employ the median of each column for each row of the input table to mitigate uncertain changes in the data points, a common occurrence in KD. This approach is only for similarity measure detectors. Figure 1 shows the steps of this proposed scheme.

Approach 2: Augmentation (adversarial Noise samples) scheme. A large number of samples is effective while developing ML models. However, collecting huge samples requires more time and effort. Additionally, it consumes a huge battery for operating sensors for a longer time. Therefore, synthetic data samples have been suggested. To develop the synthetic data, adversarial noise samples are used. It expands

Raw samples for template formation

Sub.	β1	β2	β3	γ1	γ2	γ3	α1	α2	α3
S001	257	255	255	18	20	18	-7	-2	-7
S001	248	248	247	27	26	27	2	1	2
S001	244	245	245	31	31	30	1	1	1
S001	245	244	245	31	30	31	0	-2	-2

Sorted raw samples for template formation

Sub.	β1	β2	β3	γ1	γ2	γ3	α1	α2	α3
S001	255	255	257	18	18	20	-7	-7	-2
S001	247	248	248	26	27	27	1	2	2
S001	244	245	245	30	31	31	1	1	1
S001	244	245	245	30	31	31	-2	-2	0

Column median template (X)

Sub.	β1	β2	β3	γ1	γ2	γ3	α1	α2	α3
S001	245.5	246.5	246.5	28	29	29	-0.5	-0.5	0.5

Sorted claim sample (Y)

β1	β2	β3	γ1	γ2	γ3	α1	α2	α3
243	243	244	30	30	31	-1	-1	1

Claim sample

β1	β2	β3	γ1	γ2	γ3	α1	α2	α3
244	243	243	31	30	30	1	-1	-1

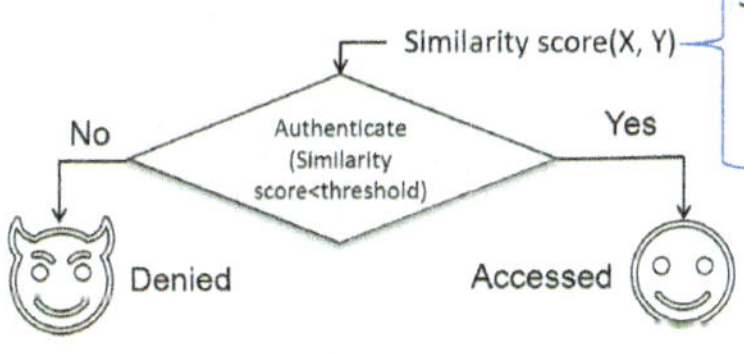

Fig. 1 **Steps of approach 1**. Gyroscope information in three directions (β, γ and α) are sorted and then used to develop user's template. Then similarity score is calculated with the template and claim for the decision

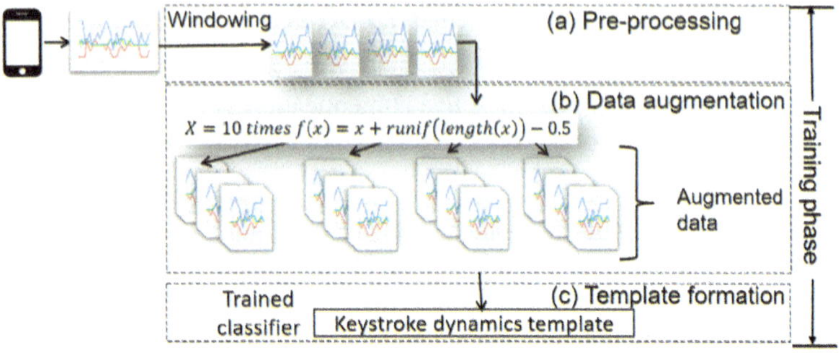

Fig. 2 **Steps of approach 2**. Synthetic sample development involves the introduction of adversarial noise to aid in template formation or model building. The function *runif()* generates random numbers between 0 and 1 with the size of the input table (x)

the size of the training set by adding random noise to the input table. This is the simple scheme of data augmentation that is introduced. It not only reduces the time of data acquisition, but it also reduces battery consumption and over-fitting problems. In addition, it has generalisation capability. Figure 2 presents the method of the proposed scheme. Here, the pre-processing step transforms the continuous patterns into ML ready form. Then, data augmentation develops synthetic samples from the raw patterns with a fixed size by adding noise. This process continued for 10 times for 10 synthetic patterns. All these patterns are used to build a model.

Approach 3: Soft biometric scheme. It is a feature-level fusion scheme that incorporates users' traits, including age group, gender, handedness, hand (s) used, etc. with the sensory data as a behavioural pattern [46]. This scheme uses several predictive models to recognise the user's traits that consume extra time. However, it enhances accuracy and reduces the impact of some of the influencing factors. This technique has been applied in fixed-text KD authentication earlier, but soft biometrics in active authentication is proposed as a more accurate system design. Figure 3 presents how the predicted labels can be found and how the predicted labels can be incorporated for a concatenated pattern. For the primary investigation for this study, we manually set the binary values (i.e. female = 0, male = 1). In this way, we have incorporated four common users' traits—age group (below 18 and 18+), gender (male and female), handedness (left and right) and hands-used (one and both). Therefore, four unique predictive models are required to get the predicted label or score.

Approach 4: Score-level fusion scheme. Due to poor data quality, detector performance changes, sometimes significantly, when changing datasets. Therefore, the confidence level of the authentication score is less. To enhance the confidence level and enhance the performance of the authentication model, more than one detector score has been suggested. It fuses the scores of multiple detectors generated in multiple ways. It is less time-efficient, but it is a more reliable approach. Figure 4 presents how the similarity scores are fused for the final decision. In this study, only two

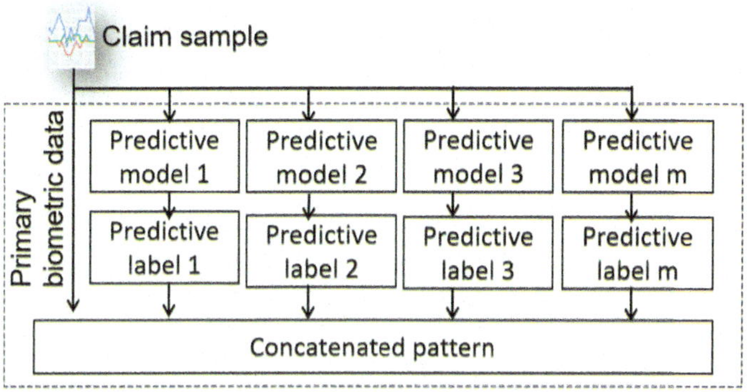

Fig. 3 **Steps of approach 3**. Recognising extra information for concatenated sample using multiple predictive models

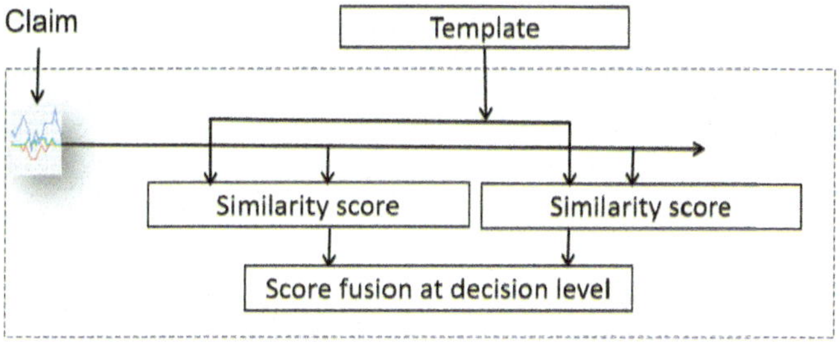

Fig. 4 **Steps of approach 4**. Score-level fusion for more confident score

similarity-based detectors' scores have been considered for the effectiveness of this approach in the domain area.

3.2 Proposed Framework for Usability Control

A user may interact with any device, not limited to smartphones, personal digital assistants or tablets. Each device has its own context information that includes the name of the operating system, version, date of installation, Global Positioning System (GPS) information, memory information, number of applications and browser information. At the start of a session, this context information is recorded for future authentication as well as re-authentication. Present and past captured contextual information are matched. Depending on the matching score (we have given the name *Watchman score*), t is assigned. If the matching score is 100% accurate, the value of t is t_{max}; otherwise, it is 0. We used this integer value for the usability of the model in the continuous scenario presented in Fig. 5. With each successfully recognised valid user, the t is increased. The incremented value can be the maximum of some limit

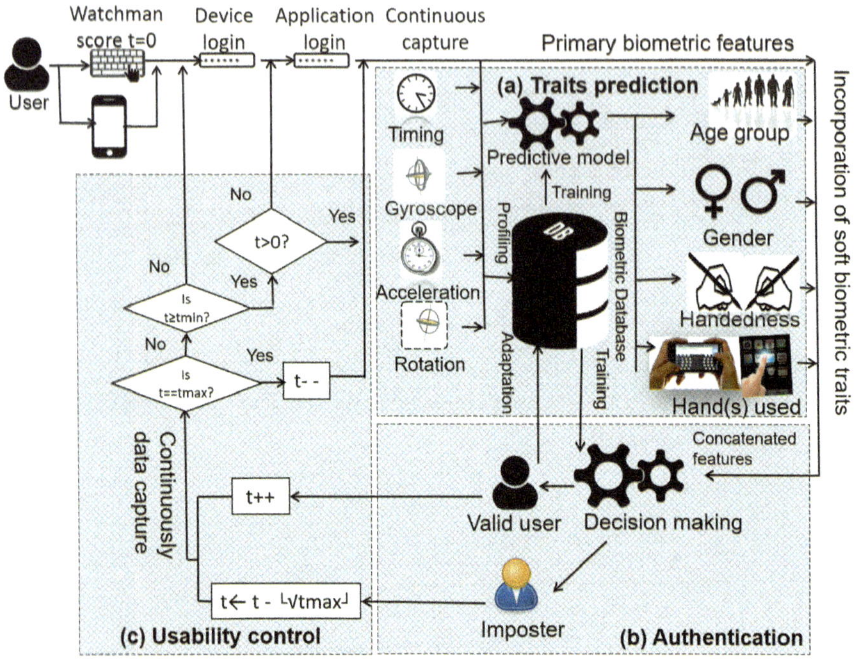

Fig. 5 **Overview of the proposed authentication framework and its usability control.** It shows how the user's traits can be predicted and incorporated and updates the *Watchman score (t)* for smooth usability without compromising security. At starting of any session, t is assigned by 0 and it will be incremented by 1 after recognizing a valid user and t will be decremented while imposter. The value of t will be within the range t_{min} and t_{max}. It indicates the starting point of the re-authentication phase

t_{max}. The t_{max} is designed based on the security level of any application. For example, if the application is sensitive, like an e-commerce or banking application, the t_{max} value will be a tiny positive integer. Larger t_{max} indicates the security needs of the particular application are less. To put it another way, the high-security application has a lower t_{max}.On the other hand, at each unsuccessful attempt, the t value will be decreased. We will adjust the value of t as per the genuineness. The range of t is in between t_{min} and t_{max}. Absolute value of t_{min} is same as t_{max} but it is a negative integer that has been presented in Fig. 6.

The selection of system usability should be predicated upon the speed with which the system can verify user identity. Consequently, achieving a balance between security and usability is paramount in the design of an authentication system.

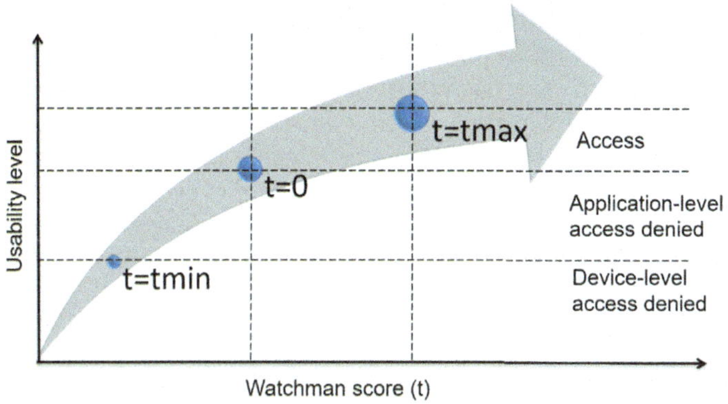

Fig. 6 Usability levels of the system at *Watchman score*. Genuineness will be continued for each positive value of t

4 Dataset, Methodology and Techniques

This section explains the details of the dataset used in this study and the implementation of the proposed designs.

4.1 Dataset: Data Acquisition Tools and Protocols

To ensure the robustness and efficacy of our study, the acquisition of a versatile and well-balanced dataset, accommodating diverse variations, is imperative. The absence of standardized protocols for data acquisition and benchmarking has hindered the concrete comparison of various evaluation processes, as highlighted in prior research [29]. To address this, we developed a web-based application (accessible at https://keystrokeanalysis.shinyapps.io/project/->Data Acquisition tab) for collecting a KD dataset via a touchscreen phone. The data, captured at a sampling rate of 2 Hz, encompasses activities such as typing short texts, drop box selection, swiping and scrolling. The low sampling rate was chosen to mitigate energy consumption and CPU usage, aligning with established considerations in the literature [11, 60]. The application, implemented using Hypertext Markup Language (HTML) and JavaScript, covertly collected gyroscope, accelerometer and rotation information for each subject. The dataset features typing patterns from users across diverse categories, encompassing gender, age group, handedness, hand(s) used and qualifications, catering to a broader population. Soft biometric traits were collected to implement a soft biometric technique and explore their impact as influencing factors. Additionally, predefined text ("Kolkata") was used in our usability design.

The dataset was exclusively obtained using a single smartphone to maintain consistency in typing patterns, recognizing potential variations in timing resolution, weight and screen size across different devices. All data were collected in portrait mode with a single device, facilitating the generation of potential imposter samples, albeit unsuitable for cross-device validation. Subjects, comprising 30 children (aged below 18) and 57 adults, participated in two sessions with four repetitions each, engaging with a *Moto G Plus 4th Gen.* smartphone with a 5.5-inch screen. Prior to data collection, subjects received instructions and a brief overview of the experimental process. Activities such as typing, zooming, and browsing were permitted during data capture, ensuring a realistic and user-friendly environment for dataset creation.

4.2 Feature Extraction

The feature extraction process is a pivotal step aimed at identifying universal and distinctive characteristics present in all users' typing patterns. This approach facilitates accelerated training, reduces model complexity and enhances interpretability. The selection of an appropriate feature subset improves the overall evaluation performance of the model. In our experimentation, gyroscope information in the beta, gamma and alpha directions, acceleration in the x, y and z directions, acceleration including gravity in the x, y and z directions, and rotation information in the beta, gamma and alpha directions were employed, concurrently captured at a sampling rate of 2 Hz. The deliberate choice of this low sampling rate is motivated by the imperative to conserve battery life, as affirmed by the findings in [60], which underscore the substantial impact of sampling rates on power usage. Thus, maintaining a low sampling rate is prudent to optimize energy efficiency.

The feature vector considered for model building spanned a duration of 14 s (four times the 3.5-second duration for each of the four repeated samples) for every user. In the evaluation phase, data were collected for six distinct short time frames (1.0 sec., 1.5 sec., 2.0 sec., 2.5 sec., 3.0 sec., 3.5 sec.) during typing activities, enabling a comprehensive assessment of the model's performance across varying temporal durations.

4.3 Pre-processing

The processing and normalization of the dataset in our study involve two distinct subparts. The first subpart focuses on aligning the feature vector, while the second subpart is dedicated to normalizing the data. Following the completion of feature extraction, the feature values for each feature type are systematically rearranged into a sorted sequence. This arrangement facilitates a one-to-one comparison of minimum values and maximum values, ensuring a comprehensive assessment of feature values. Given the consideration of small time frames in our study, statistical features are excluded from the normalization process. This decision is based on the recognition that statistical features are less effective for small patterns characterized by unpredictable activities that may occur frequently but not over an extended duration.

Subsequently, the data is normalized within the range [0, 1] using min-max normalization, as expressed by Equation (1), to expedite the computing process. In this equation, x_i represents the value undergoing normalization.

$$x_i' = \frac{x_i - \min(x)}{\max(x) - \min(x)} \tag{1}$$

4.4 Anomaly Detector Implementation

Upon completing the preprocessing stage, various anomaly detection methods were employed in our study. Initially, four instances with the same window length, collected during the first session, were utilized to generate the template/model. In the case of the similarity detector, a mean sample, following the approach presented in the study by Killourhy et al. [29], was employed for template formation. Conversely, to assess the efficacy of our methods, the median feature vector was also applied. The user profile, serving as the mathematical representation of the training samples, encapsulates either the mean or median sample of each feature vector in distance or similarity measure detectors. Subsequently, the second set of four instances, with an identical window length from other sessions, was employed for model testing. A myriad of anomaly detectors were systematically applied to evaluate their performance in the continuous domain.

During the testing phase, claim samples, obtained from both legitimate users and imposters, were tested against a previously stored reference template to generate a detection score/anomalous score/similarity score, wherein a lower score indicates greater similarity. In our experiment, samples collected during the second session were deemed genuine, and templates were formed based on a user. Conversely, samples collected during the first session by other users were regarded as impostor samples. The detectors were tested by comparing a new vector against the user's profile, yielding a detection score that signifies the degree of similarity or dissimilarity. This evaluation process was conducted across all subjects, and the average error rates were recorded to gauge the detectors' performance.

In KD, the system interprets the current behaviour as either normal or anomalous. The continuous data acquisition process, running in the background during detection, continually gathers data on KD activities, allowing for a real-time comparison with the legitimate user's template generated during training. This ongoing data acquisition ensures that decisions regarding the current user's identity are based on a sequence of activities, transforming the performance of anomaly detectors with changing datasets. Identifying a suitable detector poses a particular challenge in the KD domain due to various factors influencing detector performance. We considered several common and top-performing anomaly detectors, predominantly based on similarity/distance measures for efficient mathematical calculations. Notable methods included OCSVM for low computational cost, deep learning Autoencoders for their scalability and various cluster analysis or statistical methods. While a comprehensive description of each detector is beyond the scope of this text, a concise overview is provided, acknowledging potential differences in reimplementation from the original to adapt them effectively to common datasets.

4.5 Classification and Evaluation

The present investigation incorporates a practical perspective, considering the constraints typical in real-world scenarios. At the onset, only samples from a single user are available for constructing the model. Consequently, the evaluation process is tailored to accommodate a one-class classification, aligning with the initial scenario of limited user data. To comprehensively assess model performance, a total of eighty-seven evaluations are conducted, corresponding to the number of subjects under consideration. In each iteration, one user is designated as the genuine user, while the remaining users serve as imposters. Half of the samples from the genuine user are utilized for template development (in the case of distance-based detectors) or model construction (for Autoencoder and OCSVM), with the remaining half, along with imposter samples, allocated for training purposes. The owner's samples are specifically employed for calculating False Rejection Rate (FRR), among other metrics, while anomaly scores for each training instance are recorded. This iterative process is repeated for every subject, and the resulting average Equal Error Rates (EERs) are meticulously documented. The outcomes of the training and testing phases for all detectors are presented in the subsequent results section.

In leveraging the study of KD as a metric for user authentication, we have employed widely recognized ML tools and techniques. The KD data, captured through smart devices, is analysed utilizing R (version 3.6.3) statistical tools, facilitated by the RStudio cloud platform (https://rstudio.cloud/). This approach ensures a robust examination of KD biometrics, employing contemporary ML methodologies to assess authentication mechanisms and model performance.

4.6 Samples Separation

Training dataset. The samples collected in the first session are used to train the model for each subject. In the case of the availability of multiple sessions' data, we used half of the session for training the user's specific model, and the rest of the session was used for a false acceptance rate. The growing window technique is used for the user's template.

Testing dataset. The remaining half of the session has been used for the testing set for false rejection rate. However, samples developed in the first session other than the user's samples have been used for a false acceptance rate. Here, we considered only the first six consecutive samples for testing. This separation has been done for each user, so average results have been considered.

4.7 Template Formation

To develop the template, we used the "miscTools" R package for the function *colMeans()* to train the model and the "philentropy" package for distance() to generate anomaly scores in case of similarity measure detectors. However, different ML models use different strategies. The package "kernlab" was used for the function *ksvm(trainset, type = 'one-svc', nu = .05, kpar = list(sigma = 0.05))* to train the OCSVM model, while the package "h2o" is used for *h2o.deeplearning(x = trainset, activation = "Tanh", autoencoder = T, hidden = c(3, 2, 3), epochs = 11)* to train the deep learning model, and package "DTW" is used for dynamic time wrapping distance. To train k-means model, we used *kmeans(trainset, centers=3, iter.max=20, nstart=10)* from the package "stats". All the implementation was uploaded with the help of the Shiny framework and uploaded to https://www.shinyapps.io/ with the link https://keystrokeanalysis.shinyapps.io/project/ -> **Anomaly Detection** tab for future experiments with a unique strategy.

4.8 Similarity Score Generation

For similarity score, we used the "philentropy" R package. It calculates the similarity between the template and test sample and generates a similarity score/anomalous score. A unique similarity score has been created for each test sample. Since the user's template and user's test sample are quite similar, the similarity score for the user's samples is lower than the similarity score for the sample other than the genuine user.

4.9 Decision Making

Decision has been made depending on the threshold value (Th) and the similarity score. Claim sample will be treated as genuine while *Anomalous-score* is less than Th, otherwise, the claim will be treated as imposter as per the Eq. 2.

$$user = \begin{cases} Genuine & \text{if Anomalous-score } < Th \\ Imposter & \text{Otherwise} \end{cases} \tag{2}$$

4.10 Anomaly Detection and Evaluation

All detectors were evaluated in the same evaluation setting for sound comparison. For each subject, the true positive is calculated with the subject's patterns and the rest of the samples rather than the subject are used for true negative points. This evaluation has been conducted for all subjects, and average EERs were recorded. This evaluation process has been maintained for all datasets considering different typing durations.

4.11 Statistical Analysis and Effects

An ANOVA was conducted to find the significant difference, and Tukey's Honest Significant Differences (TukeyHSD) was used for pairwise comparison. We have used LMM to understand the effect of changing detectors. It depicts the impact of the baseline performance clearly.

5 Experimental Results

In this section, the results of the state-of-the-art anomaly detectors and the same detectors applying four proposed schemes have been presented for different interaction durations to deeply understand the effects of the proposed schemes.

5.1 Performances of the State-of-the-Art Detectors

Performances of twelve state-of-the-art detectors on collected samples in six different interaction durations have been recorded under the same evaluation procedure to compare the detectors' performances in different short interaction durations soundly. Average EERs and standard deviations (Sds) of each detector in different window lengths using a typical scheme (in the case of similarity measure detectors, the column mean template formation) have been presented in Table 2. These results indicate at different times what would be the performance of the detectors that were not tested earlier. Along with this, it indicates how much a result can change from a result through statistical methods. These results can play a major role in assessing the performance of the detectors and selecting detectors for future research. Results indicate that the error rates are decreasing with the increase of the interaction duration. In all cases, the EER is lower in SMD. It is also observed that performance is significantly changed starting from 2.5 sec. in most of the detectors.

Table 2 Average EERs and standard deviations (Sds) within brackets from the evaluation results of state-of-the-art detectors in different short typing activities. The symbol, '*' indicates a significantly ($p < 0.05$) different performance compared to AE, where 'v' indicates significantly ($p < 0.05$) better performance compared to the duration of 1.0 sec. The top performance for each column is indicated in bold-face and the top performance for each row is indicated in Italic-face

Detector	Average EER (Sd)						
	Fixed-text	1.0 sec.	1.5 sec.	2.0 sec.	2.5 sec.	3.0 sec.	3.5 sec.
AE (baseline)	0.0088 (0.0370)	0.0959 (0.1080)	0.0716 (0.0862)	0.0577 (0.0914)	0.0435 (0.0829)v	*0.0393* *(0.0796)*v	0.0439 (0.0867)v
CD	0.0326 (0.0702)	0.1600 (0.1610)	0.1370 (0.1570)	0.1060 (0.1250)	0.1060 (0.1260)	0.0954 (0.1130)v	*0.0859* *(0.0988)*v
COD	0.0417 (0.0655)	0.1530 (0.1650)	0.1380 (0.1610)	0.1200 (0.1530)	0.1110 (0.1540)*	*0.1000* *(0.1450)*	0.1010 (0.1400)
DTW	0.0447 (0.0944)	0.1420 (0.1330)	0.1140 (0.1360)	0.0859 (0.1210)v	0.0765 (0.1120)v	*0.0682* *(0.1020)*v	0.0750 (0.1090)v
ED	0.0397 (0.0650)	0.1480 (0.1610)	0.1320 (0.1560)	0.1140 (0.1500)	0.1070 (0.1510)	*0.0994* *(0.1420)*	0.1010 (0.1390)
KM	0.4520 (0.0966)	*0.2920* *(0.1640)**	0.3120 (0.1760)*	0.3320 (0.1790)*	0.3390 (0.1770)*	0.3490 (0.1710)*	0.3900 (0.1550)v*
LD	0.0182 (0.0475)	0.1260 (0.1290)	0.1070 (0.1290)	0.0826 (0.1160)	0.0779 (0.1140)	0.0705 (0.1070)v	*0.0675* *(0.0993)*v
MD	0.3380 (0.1840)	0.2920 (0.1780)*	0.3100 (0.1770)*	0.3150 (0.1770)*	0.3140 (0.1860)*	0.3300 (0.1960)*	0.3300 (0.1970)*
MKD	0.1100 (0.1090)	0.1710 (0.1830)*	0.1650 (0.1810)*	0.1570 (0.1840)*	0.1470 (0.1810)*	*0.1310* *(0.1690)**	0.1350 (0.1650)*
OCSVM	0.0453 (0.0740)	0.1820 (0.1540)*	0.1650 (0.1550)*	0.1390 (0.1370)*	0.1190 (0.1310)v*	0.1080 (0.1270)v*	*0.1040* *(0.1240)*v*
OC	0.0276 (0.0701)	0.1380 (0.1080)	0.1060 (0.1110)	0.0800 (0.0886)v	0.0684 (0.0863)v	0.0644 (0.0850)v	*0.0621* *(0.0798)*v
SMD	**0.0067** **(0.0365)**	**0.0827** **(0.0909)**	**0.0634** **(0.0820)**	**0.0507** **(0.0764)**	**0.0418** **(0.0724)**v	***0.0370*** ***(0.0674)***v	***0.0370*** ***(0.0665)***v

We arrange the names of the detectors in alphabetic order throughout this paper. In the same table, we compare the performances of the detectors with the performance of AE (baseline). We observe the performances of some detectors (KM, MD, MKD, OCSVM) are significantly different (indicated by '*') than the baseline detector. The performance of SMD is all-time high (indicated by bold-face). However, these differences are not significant with AE. Similarly, the performances of CD, COD, DTW, ED, LD and OC are not significantly different from AE. However, the state-of-the-art scheme is found to be worse for KM, MD, MKD and OCSVM.

As per the results, if we consider the window length of 1.0 sec., then we can observe 9.59% of EER for AE. With the increment of window length, EER is decreased. We observe 4.39% of EER while considering the window length of 3.5 sec. The performance of AE is significantly changed (indicated by 'v') when the window length is set to 2.5 sec. The performance of the SMD is 8.27% while considering the window length of 1.0 sec. and it is 3.7% while considering 3.5 sec. This is the lowest EER compared to other detectors. It is also identified that if we increase the duration, then the EER will be reduced gradually. This difference is significant when the duration is 2.5 sec. or 3.0 sec. in most cases. Interaction duration in authentication should be less to minimise the battery power consumption and enable effective continuous authentication. But considering less interaction increases the error rate. Therefore, the operational evaluation of the detectors needs to be understood. On the other hand, variations (Sd) are huge due to the various influencing factors that need to be identified.

5.2 Performances of the Detectors with Approach 1

The training phase of the authentication model creates a reference template that represents the statistical measurement of the repeated samples. These samples, in the form of feature vectors, are translated into mean vectors for future authentication. The distance between the claim and this mean vector creates a score for decision-making. In the case of the mean vector, the median vector has been proposed for distance-based detectors. In the cases of AE, KM and OCSVM, the template is built in different ways. Therefore, the performance of these detectors is not changed with median template formation. The performances of the similarity measure detectors (except AE, KM and OCSVM indicated by symbol '#') while considering median template formation have been presented in Table 3. We observe SMD is a suitable detector in this approach.

The overall performances of the distance-based detectors are slightly better than the mean template formation (results in Table 2). In some cases, like CD and OC, these differences are significant. CD reduces the performance of 16.0%, 13.7%, 10.6%, 10.6%, 9.54% and 8.59% of EER for different durations to 12.9%, 10.6%, 7.63%, 7.71%, 7.33% and 6.61%. Similarly, OC also reduces the EER with this approach. With median template formation, we observe OC outperforms AE while the duration is 1.5 sec. to 3.5 sec.

5.3 Performances of the Detectors with Approach 2

To reduce the data acquisition time and improve the reliability of the detectors, synthetic samples generated from the original samples have been proposed. With this approach, the detector performances have been recorded in Table 4. This approach is very effective for AE, KM, LD, MD and OCSVM. However, this approach is not always suitable for some of the state-of-the-art detectors (CD, COD, ED and MKD).

Table 3 Average EERs and Sds within brackets from the evaluation results of state-of-the-art detectors with **Approach 1 (median template formation)** in different short duration of typing activities. The symbol, '*' indicates a significantly ($p < 0.05$) different performance compared to AE, where 'v' indicates significantly ($p < 0.05$) better performance compared to the duration of 1.0 sec. The top performance for each column is indicated in bold-face and the top performance for each row is indicated in Italic-face

Detector	Average EER (Sd)						
	Fixed-text	1.0 sec.	1.5 sec.	2.0 sec.	2.5 sec.	3.0 sec.	3.5 sec.
AE (baseline)	0.0085 (0.0369)	0.0946 (0.1020)	0.0694 (0.0885)	0.0532 (0.0888)v	0.0435 (0.0813)v	*0.0407* *(0.0813)*v	0.0448 (0.0880)v
CD	0.0223 (0.0570)	0.1290 (0.1510)	0.1060 (0.1320)	0.0763 (0.1060)v	0.0771 (0.1020)v	0.0733 (0.0946)v	*0.0661* *(0.0858)*v
COD	0.0408 (0.0676)	0.1400 (0.1690)	0.1290 (0.1650)	0.1100 (0.1580)	0.1030 (0.1580)	0.0946 (0.1500)	*0.0942* *(0.1450)*
DTW	0.0351 (0.0830)	0.1140 (0.1310)	0.0918 (0.1250)	0.0687 (0.1160)	0.0606 (0.1010)v	*0.0564* *(0.0959)*v	0.0613 (0.0943)v
ED	0.0393 (0.0676)	0.1380 (0.1670)	0.1240 (0.1620)	0.1060 (0.1570)	0.1000 (0.1560)	*0.0932* *(0.1480)*	0.0954 (0.1460)
$KM^{\#}$	0.452 (0.0966)	*0.2920* *(0.1640)**	0.3120 (0.1760)*	0.3320 (0.1790)*	0.3390 (0.1770)*	0.3490 (0.1710)*	0.3900 (0.1550)V*
LD	0.0152 (0.0448)	0.1080 (0.1340)	0.0921 (0.1290)	0.0684 (0.1140)	0.0655 (0.1080)	0.0620 (0.1050)	*0.0573* *(0.0946)*v
MD	0.3200 (0.1760)	*0.2750* *(0.1820)**	0.2920 (0.1790)*	0.2970 (0.1740)*	0.2980 (0.1760)*	0.3160 (0.1890)*	0.3300 (0.1980)*
MKD	0.1140 (0.1130)	0.1660 (0.1950)	0.1620 (0.1910)*	0.1590 (0.1850)*	0.1440 (0.1850)*	*0.1330* *(0.1780)**	0.1340 (0.1740)*
$OCSVM^{\#}$	0.045 (0.0740)	0.1820 (0.1540)*	0.1650 (0.1550)*	0.1390 (0.1370)*	0.1190 (0.1310)v*	0.1080 (0.1270)v*	*0.1040* *(0.1240)*v*
OC	0.0150 (0.0507)	0.0968 (0.0984)	0.0680 (0.0900)v	0.0476 (0.0710)v	0.0413 (0.0705)v	0.0375 (0.0644)v	*0.0351* *(0.0624)*v
SMD	**0.0065** **(0.0366)**	**0.0738** **(0.0907)**	**0.0573** **(0.0800)**	**0.0431** **(0.0705)**	**0.0369** **(0.0681)**v	**0.0341** **(0.0666)**v	***0.0340*** ***(0.0639)***v

5.4 Performances of the Detectors with Approach 3

Incorporating predicting user traits with the primary biometric samples is used to enhance the performance of the biometric models in accuracy and time efficiency. The detector performances with this approach have been recorded in Table 5. In some cases (like COD and ED), this approach is effective. However, the performances of KM and MD are not effective. Soft biometric technique with the ED is all time suitable approach. In some cases (1.0 sec.), it is significantly better than AE.

As per the detector performances in different schemes, it is clear that in the case of mean or median template formation SMD outperforms others. While consider-

Table 4 Average EERs and Sds within brackets from the evaluation results of state-of-the-art detectors with **Approach 2 (augmented samples)** in different short bursts of typing activities. The symbol, '*' indicates a significantly ($p < 0.05$) worst performance compared to AE, where 'v' indicates a significantly ($p < 0.05$) better performance compared to the duration of 1.0 sec. The top performance for each column is indicated in bold-face and the top performance for each row is indicated in Italic-face

Detector	Average EER (Sd)						
	Fixed-text	1.0 sec.	1.5 sec.	2.0 sec.	2.5 sec.	3.0 sec.	3.5 sec.
AE (baseline)	0.0094 (0.0382)	**0.0812** **(0.0948)**	**0.0626** **(0.0815)**	**0.0410** **(0.0774)**v	**0.0331** **(0.0768)**v	***0.0314*** ***(0.0780)***v	**0.0355** **(0.0816)**v
CD	0.4080 (0.1720)	0.4220 (0.1440)*	0.4030 (0.1350)*	0.4420 (0.1410)*	0.4280 (0.1540)*	*0.3850* *(0.1450)**	0.4010 (0.1380)*
COD	0.2450 (0.1820)	0.2880 (0.2750)*	0.2460 (0.2770)*	0.2200 (0.2670)*	0.2050 (0.2560)*	0.1910 (0.2420)*	*0.1860* *(0.2320)**
DTW	0.0859 (0.1070)	0.1140 (0.1590)	0.1030 (0.1600)	0.0932 (0.1600)	0.0866 (0.1540)	0.0768 (0.1490)	*0.0727* *(0.1500)*
ED	0.1570 (0.1420)	0.1480 (0.1970)	0.1340 (0.1980)*	0.1270 (0.2010)*	0.1220 (0.1980)*	*0.1160* *(0.1870)**	0.1170 (0.1820)*
KM	0.0078 (0.3690)	0.0854 (0.0944)	0.0674 (0.0859)	0.0492 (0.0831)	0.0393 (0.0877)v	*0.0351* *(0.0777)*v	0.0388 (0.0807)v
LD	0.101 (0.0443)	0.1020 (0.1070)	0.0788 (0.1000)	0.0614 (0.0901)v	0.0534 (0.0824)v	0.0493 (0.0800)v	*0.0469* *(0.0793)*v
MD	0.1590 (0.1660)	0.0841 (0.1070)	0.0589 (0.0990)	0.0425 (0.1050)	0.0428 (0.1200)	*0.0435* *(0.1100)*	0.0483 (0.0999)
MKD	0.2040 (0.1580)	0.1660 (0.2030)*	0.1510 (0.2040)*	0.1460 (0.2100)*	0.1430 (0.2030)*	*0.1350* *(0.1870)**	0.1360 (0.1850)*
OCSVM	0.1300 (0.1630)	0.0827 (0.0908)	0.0660 (0.0850)	0.0468 (0.0769)	*0.0382* *(0.0731)*v	0.0452 (0.0942)	0.0601 (0.1110)
OC	0.0279 (0.0674)	0.1390 (0.1120)	0.1060 (0.1080)	0.0852 (0.0944)v	0.0697 (0.0860)v	0.0630 (0.0815)v	*0.0620* *(0.0807)*v
SMD	0.0064 (0.0363)	0.0827 (0.0874)	0.0656 (0.0823)	0.0501 (0.0742)v	0.0407 (0.0695)v	0.0374 (0.0662)v	*0.0358* *(0.0635)*v

ing the augmented approach, Autoencoder is suitable. However, Euclidean distance outperforms others using a soft biometric scheme.

5.5 Performances of the Detectors with Approach 4

The ensemble approach is common to enhance the performance of the model and increase the confidence level of the model. Here, more than one detector's performance has been considered. The performance of the ensemble approach of Scale Manhattan and Outlier-count has been presented in Table 6. As per the results, the ensemble approach with the soft biometric technique is a suitable model because

Table 5 Average EERs and Sds within brackets from the evaluation results of state-of-the-art detectors with **Approach 3 (soft biometric technique)** in different short bursts of typing activities. (Notes—'*' indicates a significantly ($p < 0.05$) worst performance compared to AE. 'v' indicates significantly ($p < 0.05$) better performance compared to the duration of 1.0 sec. The top performance for each column is indicated in bold-face and the top performance for each row is indicated in Italic-face)

Detector	Average EER (Sd)						
	Fixed-text	1.0 sec.	1.5 sec.	2.0 sec.	2.5 sec.	3.0 sec.	3.5 sec.
AE (baseline)	0.0083 (0.0368)	0.0743 (0.1010)	0.0633 (0.0869)	0.0472 (0.0854)	0.0397 (0.0808)	*0.0370* *(0.0787)*	0.0420 (0.0857)
CD	0.0189 (0.0545)	0.0414 (0.0613)	0.0436 (0.0638)	0.0404 (0.0659)	0.0444 (0.0685)	*0.0465* *(0.0707)*	0.0472 (0.0721)
COD	0.0172 (0.0440)	**0.0269** **(0.0485)***	0.0263 (0.0512)	0.0252 (0.0591)	0.0248 (0.0591)	*0.0271* *(0.0649)*	0.0308 (0.0662)
DTW	0.0252 (0.0675)	0.0488 (0.0773)	0.0429 (0.0814)	0.0364 (0.0819)	0.0339 (0.0729)	*0.0324* *(0.0675)*	0.0385 (0.0732)
ED	0.0168 (0.0442)	**0.0262** **(0.0489)***	**0.0252** **(0.0505)**	**0.0242** **(0.0585)**	**0.0247** **(0.0587)**	***0.0271*** ***(0.0626)***	**0.0315** **(0.0648)**
KM	0.449 (0.0988)	*0.2580* *(0.1710)**	0.2900 (0.1800)*	0.3160 (0.1790)	0.3260 (0.1800)*	0.3390 (0.1770)*	0.3820 (0.1610)v*
LD	0.0102 (0.0359)	0.0368 (0.0599)	0.0357 (0.0639)	0.0312 (0.0630)	0.0329 (0.0640)	*0.0351* *(0.0673)*	0.0380 (0.0712)
MD	0.3200 (0.1760)	*0.2920* *(0.1780)**	0.3100 (0.1770)*	0.3150 (0.1770)	0.3140 (0.1860)*	0.3300 (0.1960)*	0.3300 (0.1970)*
MKD	0.0292 (0.0487)	*0.0293* *(0.0486)*	0.0298 (0.0536)	0.0325 (0.0645)	0.0322 (0.0654)	0.0307 (0.0653)	0.0367 (0.0694)
OCSVM	0.0184 (0.0442)	0.0300 (0.0425)	0.0274 (0.0397)	0.0253 (0.0467)	*0.0241* *(0.0460)*	0.0258 (0.0518)	0.0298 (0.0536)
OC	0.0127 (0.0453)	0.0883 (0.0924)	0.0701 (0.0908)	0.0555 (0.0746)	0.0518 (0.0801)v	*0.0495* *(0.0760)*v	0.0496 (0.0720)v
SMD	0.0064 (0.0364)	0.0707 (0.0851)	0.0557 (0.0778)	0.0451 (0.0729)	0.0379 (0.0682)v	0.0348 (0.0654)v	*0.0345* *(0.0645)*v

of the low error rates. In this study, only a single pair of the score has been fused and observed the performance in other settings. Due to the space limit, we have not considered other pairs of detectors. The following are the responses to the research gaps for score-level fusion.

- Score-level fusion with column mean and median template formations have been tested and achieved 3.58% and 2.98% of EER respectively.
- Score-level fusion with synthetic samples has been tested and produced 3.54% of EER.
- Score-level fusion with soft biometric technique has been tested and obtained most impressive performance, which achieved 2.76% of EER.

Table 6 Average EERs and Sds within brackets from the evaluation results of state-of-the-art detectors with **Approach 4 (ensemble approach of Scale Manhattan and Outlier-count)** in different short bursts of typing activities. The symbol, '*' indicates a significantly ($p < 0.05$) worst performance compared to the Mean template, where 'v' indicates a significantly ($p < 0.05$) better performance compared to the duration of 1.0 sec. The top performance for each column is indicated in bold-face and the top performance for each row is indicated in Italic-face

Approach	Average EER (Sd)						
	Fixed-text	1.0 sec.	1.5 sec.	2.0 sec.	2.5 sec.	3.0 sec.	3.5 sec.
Mean template	0.0069 (0.0371)	0.0819 (0.0896)	0.0642 (0.0825)	0.0506 (0.0761)v	0.0411 (0.0712)v	0.0378 (0.0680)v	*0.0358* *(0.0648)*v
Median template	0.0066 (0.0367)	0.0696 (0.0850)	0.0536 (0.0766)	0.0400 (0.0685)v	0.0343 (0.0659)v	0.0312 (0.0617)v	*0.0298* *(0.0586)*v
Augmented samples	0.0067 (0.0365)	0.0843 (0.0878)	0.0655 (0.0818)	0.0499 (0.0736)v	0.0410 (0.0699)v	0.0382 (0.0663)v	*0.0354* *(0.0627)*v
Soft biometric	**0.0063** **(0.0365)**	**0.0543** **(0.0760)**	**0.0434** **(0.0686)**	**0.0343** **(0.0661)**v	**0.0301** **(0.0619)**v	**0.0277** **(0.0579)**v	***0.0276*** ***(0.0566)***v

6 Discussions

6.1 Comparative Analysis of Detectors in Active Authentication

6.1.1 Performance Comparison

An ANOVA was conducted to analyse the variation in the estimated EERs. The ANOVA summary shows the significant differences between the average EERs. However, it cannot identify which detector differs from the others. TukeyTSD, which is based on the Studentized Range Statistics, has been used to calculate the significance of the differences (95% family-wise confidence level) between the estimated EERs of different anomaly detectors. Table 7 shows pair-wise statistical significance scores and differences. The lower diagonal of the matrix represents p-values for the null hypothesis (average EERs are the same), if this is small, it means there is a difference in the performance. Besides, if this score is below 0.05, it means there is a significant difference between the distributions. The upper diagonal of the matrix represents the difference between the estimated means. The table shows that the Autoencoder's performance has not changed significantly with Outlier-count (only difference is 2.78% of EER) and Scaled Manhattan (only 0.7% of EER). Scale Manhattan's subsequent performances are always noticeably different from others.

Table 7 TukeyHSD results. Pair-wise significance differences (upper diagonal) and significance level (lower diagonal) of estimated EERs. Texts in bold are significantly different

Detector	AE	CD	COD	DTW	ED	KM	LD	MD	MKD	OCSVM	OC	SMD
AE	–	**0.0563**	**0.0619**	**0.0350**	**0.0583**	**0.2770**	**0.0299**	**0.2565**	**0.0924**	**0.0776**	0.0278	−0.007
CD	0.0000	–	0.0056	−0.021	0.0021	**0.2208**	−0.026	**0.2002**	**0.0361**	0.0214	**−0.028**	**−0.063**
COD	0.0000	1.0000	–	−0.026	−0.004	**0.2152**	**−0.032**	**0.1946**	**0.0305**	0.0158	**−0.034**	**−0.068**
DTW	0.0028	0.3586	0.0774	–	0.0233	**0.2420**	−0.005	**0.2215**	**0.0574**	**0.0426**	−0.007	**−0.042**
ED	0.0000	1.0000	1.0000	0.2211	–	**0.2187**	**−0.028**	**0.1982**	**0.0341**	0.0193	**−0.030**	**−0.065**
KM	0.0000	0.0000	0.0000	0.0000	0.0000	–	**−0.247**	**−0.021**	**−0.185**	**−0.199**	**−0.249**	**−0.284**
LD	0.0253	0.0924	0.0113	1.0000	0.0456	0.0000	–	**0.2266**	**0.0624**	**0.0477**	−0.002	**−0.037**
MD	0.0000	0.0000	0.0000	0.0000	0.0000	0.4143	0.0000	–	**−0.164**	**−0.179**	**−0.229**	**−0.263**
MKD	0.0000	0.0016	0.0201	0.0000	0.0044	0.0000	0.0000	0.0000	–	−0.015	**−0.064**	**−0.099**
OCSVM	0.0000	0.3478	0.7980	0.0000	0.5144	0.0000	0.0000	0.0000	0.8634	–	**−0.050**	**−0.084**
OC	0.0555	0.0447	0.0044	0.9996	0.0203	0.0000	1.0000	0.0000	0.0000	0.0000	–	**−0.034**
SMD	0.9998	0.0000	0.0000	0.0001	0.0000	0.0000	0.0013	0.0000	0.0000	0.0000	0.0037	–

Table 8 Average training and verification time (in sec.) was taken by the detectors. Texts in bold are the fastest detectors in verification

Detectors	Training time	Testing time	Detectors	Training time	Testing time
AE	1.2984	0.5470	LD	0.0012	0.5395
CD	0.0011	0.7338	MD	0.0020	**0.0012**
COD	0.0012	0.7338	MKD	0.0012	0.5583
DTW	0.0012	**0.0040**	OCSVM	0.0025	**0.0025**
ED	0.0011	0.3876	OC	0.0027	**0.0016**
KM	0.0023	**0.0045**	SMD	0.0014	**0.0020**

6.1.2 Comparison of Time Complexity

To enhance the battery life of active authentication and re-authentication, a short interaction duration (minimum of 2.5 s.) has been recommended in the previous section. However, the testing time taken by the detector is also an important parameter. On the other hand, in the adaptive method, training and testing times are both important. The average time in milliseconds (ms) taken by the detectors to train the model and to verify the claim samples has been recorded. Table 8 shows the evaluation time taken by each detector. We depicted the arrangement of the detectors in the table depending on the average verification time, which is much shorter than the average training time. Since building the model is a rare activity (except for adaptive training), however, verification time in continuous mode is the most frequent activity. The Outlier-count, Scaled Manhattan, OCSVM, Mahalanobis, K-means and DTW are the best in time efficiency that is suitable specifically for the smartphone since the computational capability and battery life of smartphones are typically lower than desktops. These detectors reduce the re-authentication delay, which is important and needs to be considered in designing a user-active authentication model.

6.2 *Factors Analysis*

To understand the experimental results and the influencing factors, a popular statistical technique—LMM has been used. It enables us to understand the sense of detector behaviour and the source of uncertainty. Due to the statistical nature of this technique, it does not always give perfect results. There is some residual uncertainty that needs to be understood.

LMM results for fixed-effects and random-effects have been presented in Tables 9 and 10 for fixed text. The EER obtained for AE for fixed text is attractive (only 0.8%) and this performance is not much varied in different schemes for fixed text. On the other hand, SMD outperforms AE using all schemes and found an EER of 0.2% less than the intercept. However, SMD and OC are suitable alterna-

Table 9 LMM results for fixed text in different settings. Estimated/adjusted fixed effects for the detectors indicate how the error rate can be changed in changing the detectors. The symbol, '+', indicates the worst and '−' indicates better than AE (Intercept), where 'v' indicates the significantly ($p < 0.05$) different

Detector	Estimated/adjusted EER			
	State-of-the-art	Median template	Augmented samples	Soft biometric
AE (Intercept)	0.0088	0.0088	0.0094	0.0083
CD	+0.0238v	+0.0138	+0.3983v	+0.0105
COD	+0.0329v	+0.0322v	+0.2357v	+0.0089
DTW	+0.0359v	+0.0265v	+0.0766v	+0.0169
ED	+0.0309v	+0.0308v	+0.1473v	+0.0085
KM	+0.4432v	+0.4435v	−0.0016	+0.4404v
LD	+0.0094	+0.0067	+0.0007	+0.0019
MD	+0.3290v	+0.3109v	+0.1494v	+0.3111v
MKD	+0.1009v	+0.1053v	+0.1949v	+0.0209v
OCSVM	+0.0365v	+0.0368v	+0.1211v	+0.0101
OC	+0.0188	+0.0064	+0.0185	+0.0044
SMD	**−0.0021**	**−0.0020**	**−0.0030**	**−0.0019**

Table 10 Random effects for the different groups for fixed text. It shows the effects for each group

Group	# Groups	State-of-the-art		Median template		Augmented samples		Soft biometric	
		Sd	ICC	Sd	ICC	Sd	ICC	Sd	ICC
Subjects	87	0.0474	0.2940	0.0435	0.2694	0.0609	0.2450	0.0322	0.2013
Qualification	6	0.0081	0.0086	0.0050	0.0036	0.0000	0.0000	0.0042	0.0035
Age group	3	0.0000	0.0000	0.0000	0.0000	0.0000	0.0000	0.0000	0.0000
Hand(s) used	2	0.0001	0.0000	0.0000	0.0000	0.0000	0.0000	0.0000	0.0000
Handedness	2	0.0008	0.0001	0.0000	0.0000	0.0000	0.0000	0.0001	0.0000
Gender	2	0.0072	0.0068	0.0032	0.0015	0.0000	0.0000	0.0000	0.0000
Residual		0.0726		0.0714		0.1070		0.0640	

tives since the differences are not significant. Random effects present how different groups influence the detector performance. Detectors' behaviour with the proposed approaches has been presented to understand the factors in different approaches. Infusing factors for each approach and the uncertainty they introduce have also been presented. Due to this multitude of factors, the performance of a detector on different datasets is varied and has been presented with an Intraclass Correlation Coefficient (ICC). It is found the 4.74% of EER can be varied for subject selection in the case of state-of-the-art approach, which is changed for proposed schemes. The results indicated that subject-based variability is less for the soft biometric method. It is also found that the age group, on the other hand, does not affect the detector performance.

For free text, LMM results for fixed-effects and random-effects have been presented in Tables 11 and 12. It is discovered that AE achieves 7.54% EER for the

Table 11 LMM results for free text in different settings. The estimated or adjusted fixed effects for the detectors. The symbol, '+' indicates worst and '−' indicates better than AE (Intercept), where '*' indicates the significantly ($p < 0.05$) worst and 'v' indicates significantly ($p < 0.05$) better performance compared to AE (Intercept)

Detector	Estimated/adjusted EER			
	State-of-the-art	Median template	Augmented samples	Soft biometrics
AE (Intercept)	0.0754	0.0705	**0.0723**	0.0503
CD	+0.0563*	+0.0301*	+0.3661*	−0.0067
COD	+0.0619*	+0.0543*	+0.1751*	−0.0238v
DTW	+0.0350*	+0.0177*	+0.0436*	−0.0118v
ED	+0.0583*	+0.0519*	+0.0799*	−0.0241v
KM	+0.2770*	+0.2780*	+0.0051	+0.2679*
LD	+0.0299*	+0.0179*	+0.0178*	−0.0156v
MD	+0.2565*	+0.2437*	+0.0059	+0.2645*
MKD	+0.0924*	+0.0921*	+0.0986*	−0.0187v
OCSVM	+0.0776*	+0.0786*	+0.0090	−0.0235v
OC	+0.0278*	−0.0033	+0.0400*	+0.0102*
SMD	**−0.0066**	**−0.0112**	+0.0046	**−0.0042**

state-of-the-art technique. Where SMD is chosen as the best AE alternative. In this case, SMD obtains 0.66% less EER than AE. In the case of an augmented approach, AE is most suited, achieving 7.23% EER. However, COD, DTW, ED, LD, MKD and OCSVM surpass AE substantially when employing soft biometric techniques. It demonstrates that detector behaviour varies depending on the scheme. As a result, for more accurate findings, the right scheme for each detector must be established; otherwise, promising results will not be obtained.

We discovered the explanation for the detector performance variations while changing datasets with random effects. The subject-specific variance is 8.9%, which may change depending on the scheme. Similarly, in this study, a collection of groups has been found for the same reason. However, the degree of this variance is different across groups.

6.3 Comparison of Approaches

A comparison of the proposed approaches with the state-of-the-art detectors has been presented in Table 13. It is observed that median template information is effective when using a similarity measure detector as a classifier. However, data augmentation is not appropriate for all similarity detectors. But the soft biometric technique is always suited to all detectors. It gives the idea of which approach is suitable for

Table 12 Random effects for the different groups for free text. It shows the effects for each group

Group	# Groups	State-of-the-art		Median template		Augmented samples		Soft biometrics	
		Sd	ICC	Sd	ICC	Sd	ICC	Sd	ICC
Subject	87	0.0891	0.3986	0.0907	0.4288	0.0821	0.3099	0.0509	0.2764
Gender	2	0.0001	0.0000	0.0000	0.0000	0.0000	0.0000	0.0000	0.0000
Age group	3	0.0208	0.0218	0.0129	0.0086	0.0182	0.0153	0.0029	0.0009
Hand(s) used	2	0.0000	0.0000	0.0000	0.0000	0.0000	0.0000	0.0000	0.0000
Handedness	2	0.0347	0.0604	0.0258	0.0347	0.0451	0.0935	0.0000	0.0000
Qualification	6	0.0000	0.0000	0.0000	0.0000	0.0177	0.0144	0.0000	0.0000
Window length	6	0.0157	0.0124	0.0131	0.0090	0.0181	0.0151	0.0017	0.0003
Residual		0.1005		0.0997		0.1095		0.0823	

Table 13 Performance of the detectors in different approaches to enhance the performance of the state-of-the-art models. Texts in bold-face indicate the best performance of the detector. 'v' indicates the significant difference with state-of-the-art performances

Detector	Average EER (Sd)			
	State-of-the-art	Median template	Augmented samples	Soft biometrics
AE	0.0586 (0.0914)	0.0577 (0.0902)	**0.0475 (0.0836)**	0.0506 (0.0874)
CD	0.1150 (0.1340)	0.0878 (0.1160)v	0.4140 (0.1440)v	**0.0439 (0.0669)**v
COD	0.1200 (0.1540)	0.1120 (0.1580)	0.2220 (0.2600)v	**0.0268 (0.0583)**v
DTW	0.0936 (0.1220)	0.0754 (0.1130)	0.0911 (0.1550)	**0.0388 (0.0757)**v
ED	0.1170 (0.1500)	0.1100 (0.1560)	0.1270 (0.1930)	**0.0265 (0.0574)**v
KM	0.3360 (0.1720)	0.3360 (0.1720)	**0.0525 (0.0866)**v	0.3180 (0.1780)
LD	0.0886 (0.1170)	0.0756 (0.1160)	0.0653 (0.0921)v	**0.0349 (0.0647)**v
MD	0.3150 (0.1850)	0.3010 (0.1830)	**0.0534 (0.1070)**v	0.3150 (0.1850)
MKD	0.1510 (0.1770)	0.1500 (0.1840)	0.1460 (0.1980)	**0.0319 (0.0614)**v
OCSVM	0.1360 (0.1410)	0.1360 (0.1410)	0.0565 (0.0902)v	**0.0271 (0.0468)**v
OC	0.0865 (0.0973)	**0.0544 (0.0800)**v	0.0874 (0.0982)	0.0608 (0.0822)v
SMD	0.0521 (0.0778)	0.0465 (0.0750)	0.0520 (0.0759)	**0.0464 (0.0735)**

which detector. The augmented approach is impressive for AE, KM and MD. While median template formation, OC is quite impressive, which is a time-efficient model as well. The following are the responses against the unanswered questions.

In our experiments, we found that using column median for template formation significantly improved the performance of all similarity-based detectors, reducing EER for CD and OC. SMD was identified as one of the best detectors with median template formation. Synthetic samples generated using AE, KM and OCSVM also reduced EER for all three detectors. However, the augmented approach had variable impact on different detectors, and careful assessment is needed. Soft biometric technique showed impressive performance in terms of accuracy, and all detectors performed well with this feature-level score fusion method. ED was the top performer with an EER of 2.65%.

6.4 Comparison of Our Proposed Selected Models with Existing Literature

Comparison of the detectors with four unique approaches has been compared in Sect. 5. In Table 14, we tried to compare the selected approaches with other existing literature in the topic area. Since, the feature arrangements, pattern lengths, and metrics are unique, it is not possible for sound comparison. Therefore, our approaches with small periods are effective in various issues like time complexity, battery consumption and usability control.

7 Conclusions

This study conducts a thorough investigation into the performance of twelve anomaly detectors, evaluating patterns derived from smartphone data within short time frames. The findings contribute insights into the enhancement of these detectors, paving the way for a robust continuous authentication model based on KD to bolster smartphone security and usability. Analysing the training and testing times under scenarios involving frequent application switches, the study strives to establish a secure KD-based model for active authentication throughout user sessions. To evaluate the proposed approach without compromising security, a novel usability metric termed the *Watchman score* is introduced.

Identified influencing factors affecting detector behaviour are addressed in the formulation of a secure KD-based model. The study offers innovative procedures for sensor-based active authentication, achieving an impressive EER ranging from 2.7% to 3.7%. Proposals encompass reducing data acquisition and re-authentication times for lowered energy consumption, coupled with algorithmic enhancements for device and application-level security. Results of anomaly detection on the developed dataset shed light on detector performance variations with changing window lengths, singling out Scaled Manhattan, Autoencoder and Outlier-count as more suitable for continuous sensory data determination.

Four distinct approaches, focusing on improving detector performance and battery power efficiency, are presented and subjected to comprehensive analysis. Statistical evidence is provided to elucidate the influencing factors affecting detector performance variations across datasets. The study recommends potential features for future investigations, including pause duration, touchpoint coordinates preference, swipe movement, pressure and magnetometer information, to enhance practicality and align with European user identity verification standards. The data collection at a low sam-

Table 14 Comparison of our selected approaches and previous free text KD-based models in smartphone environment. Several methods on sensory features have been proposed. But due to the uneven feature arrangements and evaluation parameters models were not compared

Study	Year	Method	Features	EER (%)
[31]	2017	OCSVM, KNN, K-Means	T	0.44
[16]	2017	KNN, OCSVM and similarity	G, A, R	2.40 and 5.50
[54]	2017	HMM	G, A	FRR 5.03 & FAR 3.98
[67]	2017	OCSVM	G, A, R, TP, S	1.20
[66]	2018	OCSVM and iForest	T, TP	Acc. 95.85
[13]	2018	One-class MLP	A, G, S	Acc. 85.77
[35]	2018	OCSVM	T, P, G, A, R, TP, S	5.50
[20]	2018	iForest	T, TP	FAR 7.50 & FRR 6.40
[33]	2018	OCSVM	G, A, S	4.66
[18]	2019	Distance measure	T, P, TP	Acc. 100
[1]	2019	OCSVM	T, G, A, R, TP	Acc. 82.20 to 97.10
[30]	2020	Kolmogorov-Smirnov (KS) statistic	T, G, A, R, TP, S	1.00
[25]	2020	Outlier-count	Piezoelectric force	2.40
[14]	2021	Cosine distance	T	Acc. 91.84
[4]	2021	Euclidean distance	T & Sensory	0.08
[14]	2021	Euclidean distance	T	Acc. 90.82
Proposed	Recent	Scale Manhattan (Conventional); 3.5 sec.	G, A, R	3.70 (6.65)
Proposed	Recent	Autoencoder (Approach 2); 3.0 sec.	G, A, R	3.14 (7.80)
Proposed	Recent	Euclidean (Approach 1 and 3); 2.0 sec.	G, A, R	2.42 (5.85)
Proposed	Recent	Score-fusion (Approach 4); 3.5 sec.	G, A, R	3.58 (6.48)

Table 14 (continued)

Proposed	Recent	Score-fusion (Approach 1 and 4); 3.5 sec.	G, A, R	2.98 (5.86)
Proposed	Recent	Score-fusion (Approach 1 and 3); 3.5 sec.	G, A, R	2.76 (5.66)
Proposed	Recent	Score-fusion (Approach 2 and 4); 3.5 sec.	G, A, R	3.54 (6.27)
Proposed	Recent	Scale Manhattan (Approach 1); 3.5 sec.	G, A, R	3.40 (6.39)

HMM->Hidden Markov Model, MLP->Multi Layer Perception, T->Timing, G->Gyroscope, A->Accelerometer, R->Rotation, TP->Touchpoint, S->Statistical features, P->Pressure

pling rate of 2 *Hz*, deemed suitable for low-time resolution devices and standard browsers, is underscored for its role in minimizing battery consumption.

Furthermore, the study underscores the need for adaptation mechanisms to counter biometric template ageing and emphasizes the importance of anti-spoofing measures to thwart adversarial attacks deploying generative models. Acknowledging potential privacy implications, the study advocates careful consideration of the suggested methodology in real-world applications for practical deployment.

Declarations

Ethical approval All procedures performed in studies involving human participants were in accordance with the ethical standards of the institutional and/or national research committee and with the 1964 Helsinki Declaration and its later amendments or comparable ethical standards.

Conflict of interest Informed consent was obtained from all individual participants included in the study.

Data availability The datasets generated during and/or analysed during the current study are available from the corresponding author on reasonable request.

Informed consent The authors declare that they have no conflict of interest.

References

1. Acien A, Morales A, Vera-Rodriguez R, Fierrez J, Tolosana R (2019) MultiLock: mobile active authentication based on multiple biometric and behavioural patterns. In: Proceedings of the 1st international workshop on multimodal understanding and learning for embodied applications (MULEA 2019), pp 53–59. https://doi.org/10.1145/3347450.3357663
2. Ahmad J, Sajjad M, Jan Z, Mehmood I, Rho S, Baik SW (2017) Analysis of interaction trace maps for active authentication on smart devices. Multimed Tools Appl 76(2017):4069–4087. https://doi.org/10.1007/s11042-016-3450-y
3. Alpar O (2021) Biometric keystroke barcoding: a next-gen authentication framework. Expert Syst Appl 177:1–12. https://doi.org/10.1016/j.eswa.2021.114980
4. Alsuhibany SA, Almuqbil AS (2021) Analyzing the effectiveness of touch keystroke dynamic authentication for the Arabic language. Wirel Commun Mob Comput 2021(15):1–15. https://doi.org/10.1155/2021/9963129
5. Amraoui N, Besrour A, Ksantini R, Zouari B (2020) Implicit and continuous authentication of smart home users. In: Proceedings of the 33rd international conference on advanced information networking and applications (AINA 2019), pp 1228–1239. https://doi.org/10.1007/978-3-030-15032-7_103
6. Antal M, Bokor Z, ZsoltSzabó L (2015) Information revealed from scrolling interactions on mobile devices. Pattern Recogn Lett 56(2015):7–13. https://doi.org/10.1016/j.patrec.2015.01.011
7. Anusas-Amornkul T (2019) Strengthening password authentication using keystroke dynamics and smartphone sensors. In: Proceedings of the 9th international conference on information communication and management (ICICM 2019). ACM International Conference Proceeding Series, Association for Computing Machinery, pp 70–74. https://doi.org/10.1145/3357419.3357425
8. Ayeswarya S, Norman J (2019) A survey on different continuous authentication systems. Int J Biom 11(1):67–99. https://doi.org/10.1504/IJBM.2019.096574
9. Ayotte B, Banavar M, Hou D, Schuckers S (2020) Fast free-text authentication via instance-based keystroke dynamics. IEEE Trans Biom Behav Identity Sci 2(4):377–387
10. Barra S, De Marsico M, Nappi M, Narducci F, Riccio D (2019) A hand-based biometric system in visible light for mobile environments. Inf Sci 479:472–485. https://doi.org/10.1016/J.INS.2018.01.010
11. Basar OE, Alptekin G, Volaka HC, Isbilen M, Incel OD (2019) Resource usage analysis of a mobile banking application using sensor-and-touchscreen-based continuous authentication. In: Proceedings of the 16th international conference on mobile systems and pervasive computing (MobiSPC 2019). Procedia Computer Science, vol 155. ScienceDirect, pp 185–192. https://doi.org/10.1016/j.procs.2019.08.028
12. Bours P, Ellingsen J (2018) Cross keyboard keystroke dynamics. In: Proceedings of the 1st international conference on computer applications and information security (ICCAIS 2018), pp 1–6. https://doi.org/10.1109/CAIS.2018.8441945
13. Buriro A, Gupta S, Crispo B, Frari FD (2018) DIALERAUTH: a motion-assisted touch-based smartphone user authentication scheme. In: Proceedings of the 8th ACM conference on data and application security and privacy (CODASPY 2018). Association for Computing Machinery, Tempe, AZ, USA, pp 267–276. https://doi.org/10.1145/3176258.3176318
14. Casanova A, Cascone L, Castiglione A, Meng W, Pero C (2021) User recognition based on periocular biometrics and touch dynamics. Pattern Recogn Lett 148(2021):114–120. https://doi.org/10.1016/j.patrec.2021.05.006
15. Chang TY, Tsai CJ, Yeh JY, Peng CC, Chen PH (2020) New soft biometrics for limited resource in keystroke dynamics authentication. Multimed Tools Appl 79(2020):23295–23324. https://doi.org/10.1007/s11042-020-09042-x
16. Chen Y, Shen C, Wang Z, Yu T (2017) Modeling interactive sensor-behavior with smartphones for implicit and active user authentication. In: Proceedings of the IEEE international conference

on identity, security and behavior analysis (ISBA 2017), pp 1–6. https://doi.org/10.1109/ISBA.2017.7947694
17. Crawford H, Renaud K, Storer T (2013) A framework for continuous, transparent mobile device authentication. Comput Secur 39(2013):127–136. https://doi.org/10.1016/j.cose.2013.05.005
18. Dee T, Richardson I, Tyagi A (2019) Continuous transparent mobile device touchscreen soft keyboard biometric authentication. In: Procedings of the 32nd international conference on VLSI design and 18th international conference on embedded systems (VLSID 2019). IEEE, pp 539–540. https://doi.org/10.1109/VLSID.2019.00125
19. Feng T, Zhao X, Desalvo N, Liu TH, Gao Z, Wang X, Shi W (2015) An investigation on touch biometrics: behavioral factors on screen size, physical context and application context. In: Proceedings of the IEEE international symposium on technologies for homeland security (HST 2015). https://doi.org/10.1109/THS.2015.7225318
20. Filippov AI, Iuzbashev AV, Kurnev AS (2018) User authentication via touch pattern recognition based on isolation forest. In: Proceedings of the IEEE conference of Russian young researchers in electrical and electronic engineering (ElConRus 2018). IEEE, pp 1485–1489. https://doi.org/10.1109/EIConRus.2018.8317378
21. Frank M, Biedert R, Ma E, Martinovic I, Song D (2013) Touchalytics: on the applicability of touchscreen input as a behavioral biometric for continuous authentication. IEEE Trans Inf Forensics Secur 8(1):136–148. https://doi.org/10.1109/TIFS.2012.2225048
22. Gupta S, Buriro A, Crispo B (2018) Demystifying authentication concepts in smartphones: ways and types to secure access. Mob Inf Syst. https://doi.org/10.1155/2018/2649598
23. Hayashi E, Das S, Amini S, Hong J, Oakley I (2013) CASA: context-aware scalable authentication. In: Proceedings of the 9th symposium on usable privacy and security (SOUPS 2013), pp 1–10. https://doi.org/10.1145/2501604.2501607
24. He L, Li Z, Shen C (2017) Performance evaluation of anomaly-detection algorithm for keystroke-typing based insider detection. In: ACM international conference proceeding series. https://doi.org/10.1145/3063955.3063987
25. Huang A, Gao S, Chen J, Xu L, Nathan A (2020) High security user authentication enabled by piezoelectric keystroke dynamics and machine learning. IEEE Sens J 20(21):13037–13046. https://doi.org/10.1109/JSEN.2020.3001382
26. Iapa AC, Cretu VI (2021) Shared data set for free-text keystroke dynamics authentication algorithms. Preprints at May, pp 1–28. https://doi.org/10.20944/preprints202105.0255.v1
27. Khan H, Hengartner U, Vogel D (2020) Mimicry attacks on smartphone keystroke authentication. ACM Trans Priv Secur 23(1):1–34. https://doi.org/10.1145/3372420
28. Killourhy KS (2012) A scientific understanding of keystroke dynamics. Ph.D. thesis, Carnegie Mellon University
29. Killourhy KS, Maxion RA (2009) Comparing anomaly-detection algorithms for keystroke dynamics. In: Proceedings of the international conference on dependable systems and networks, pp 125–134. https://doi.org/10.1109/DSN.2009.5270346
30. Kim J, Kang P (2020) Freely typed keystroke dynamics-based user authentication for mobile devices based on heterogeneous features. Pattern Recogn 108(2020):107556. https://doi.org/10.1016/j.patcog.2020.107556
31. Kim J, Kim H, Kang P (2018) Keystroke dynamics-based user authentication using freely typed text based on user-adaptive feature extraction and novelty detection. Appl Soft Comput 62:1077–1087. https://doi.org/10.1016/j.asoc.2017.09.045
32. Li L, Zhao X, Xue G (2013) Unobservable re-authentication for smartphones. In: Proceedings of the 20th annual network and distributed system security symposium. The Internet Society, pp 1–16
33. Li Y, Hu H, Zhou G (2019) Using data augmentation in continuous authentication on smartphones. IEEE Internet Things J 6(1):628–640. https://doi.org/10.1109/JIOT.2018.2851185
34. Liang X, Zou F, Li L, Yi P (2020) Mobile terminal identity authentication system based on behavioral characteristics. Int J Distrib Sens Netw 16(1):1–12. https://doi.org/10.1177/1550147719899371

35. Liu X, Shen C, Chen Y (2018) Multi-source interactive behavior analysis for continuous user authentication on smartphones. In: Zhou J, Wang Y, Sun Z, Jia Z, Feng J, Shan S, Ubul K, Guo Z (eds.) Proceedings of the Chinese conference on biometric recognition (CCBR 2018). Lecture Notes in Computer Science, vol 10996. Springer, Cham, pp 669–677. https://doi.org/10.1007/978-3-319-97909-0_71
36. Mahbub U, Sarkar S, Patel VM, Chellappa R 2016) Active user authentication for smartphones: a challenge data set and benchmark results. In: Proceedings of the IEEE 8th international conference on biometrics theory, applications and systems (BTAS 2016), pp 1–8. https://doi.org/10.1109/BTAS.2016.7791155
37. Meng W, Li W, Wong DS (2018) Enhancing touch behavioral authentication via cost-based intelligent mechanism on smartphones. Multimedia Tools and Applications 77(2018):30167–30185. https://doi.org/10.1007/s11042-018-6094-2
38. Mhenni A, Cherrier E, Rosenberger C, Essoukri Ben Amara N (2019) Analysis of Doddington zoo classification for user dependent template update: application to keystroke dynamics recognition. Future Gener Comput Syst 97:210–218. https://doi.org/10.1016/j.future.2019.02.039
39. Mokoena T, Sabatta D (2020) User classification by keystroke dynamics using text retrieval methods. In: Proceedings of the international SAUPEC/RobMech/PRASA conference, SAUPEC/RobMech/PRASA 2020, pp 1–6. https://doi.org/10.1109/SAUPEC/RobMech/PRASA48453.2020.9040956
40. Monaco JV (2016) Robust keystroke biometric anomaly detection, pp 1–32. Preprint at arXiv:1606.09075
41. Mostafa H, Elkorany AM, El-Ramly M, Shaban H (2019) Behavio2Auth: sensor-based behavior biometric authentication for smartphones. In: Proceedings of the ArabWIC 6th annual international conference research track (ArabWIC 2019). Association for Computing Machinery, pp 1–6. https://doi.org/10.1145/3333165.3333176
42. Patel Y, Ouazzane K, Vassilev VT, Faruqi I, Walker GL (2019) Keystroke dynamics using auto encoders. In: Proceedings of the international conference on cyber security and protection of digital services, cyber security, pp 1–8. https://doi.org/10.1109/CyberSecPODS.2019.8885203
43. Rasnayaka S, Sim T (2018) Who wants continuous authentication on mobile devices? In: Proceedings of the IEEE 9th international conference on biometrics theory, applications and systems (BTAS 2018). IEEE, pp 1–9. https://doi.org/10.1109/BTAS.2018.8698599
44. Roy A, Halevi T, Memon N (2014) An HMM-based behavior modeling approach for continuous mobile authentication. In: Proceedings of the IEEE international conference on acoustics, speech and signal processing (ICASSP 2014), pp 1–5. https://doi.org/10.1109/ICASSP.2014.6854310
45. Roy S, Roy U, Sinha D (2015) Distance based models of keystroke dynamics User authentication. Int J Adv Eng Res Sci 2(9):89–94
46. Roy S, Roy U, Sinha D (2017) User authentication: Keystroke dynamics with soft biometric features. In: Internet of Things (IOT) technologies, applications, challenges and solutions, pp 105–124
47. Roy S, Roy U, Sinha D (2018) The probability of predicting personality traits by the way user types on touch screen. Innov Syst Softw Eng 15(1):27–34. https://doi.org/10.1007/s11334-018-0317-6
48. Roy S, Roy U, Sinha D, Pal RK (2022) AI for Stress Diagnosis at Home Environment. In: Next generation healthcare informatics. Studies in Computational Intelligence, vol 1039, pp 173–195. https://doi.org/10.1007/978-981-19-2416-3_10
49. Roy S, Roy U, Sinha D, Pal RK (2023) Imbalanced ensemble learning in determining Parkinson's disease using Keystroke dynamics. Expert Syst Appl 217(11):119522. https://doi.org/10.1016/j.eswa.2023.119522
50. Roy S, Sinha D, Pal R, Roy U (2024) A unique approach towards keystroke dynamics based entry-point user access control. Int J Biom 16(1). https://doi.org/10.1504/IJBM.2024.10056899

51. Saini BS, Singh P, Nayyar A, Kaur N, Bhatia KS, El-Sappagh S, Hu JW (2020) A three-step authentication model for mobile phone user using keystroke dynamics. IEEE Access 8(2020):125909–125922. https://doi.org/10.1109/ACCESS.2020.3008019
52. Salem A, Sharieh A, Sleit A, Jabri R (2019) Enhanced authentication system performance based on keystroke dynamics using classification algorithms. KSII Trans Internet Inf Syst 13(8). https://doi.org/10.3837/tiis.2019.08.014
53. Sciarretta G, Carbone R, Ranise S, Viganò L (2020) Formal analysis of mobile multi-factor authentication with single sign-on login. ACM Trans Priv Secur 23(3):13–37. https://doi.org/10.1145/3386685
54. Shen C, Li Y, Chen Y, Guan X, Maxion RA (2017) Performance analysis of multi-motion sensor behavior for active smartphone authentication. IEEE Trans Inf Forensics Secur 133(1):48–62. https://doi.org/10.1109/TIFS.2017.2737969
55. Shen C, Zhang Y, Guan X, Maxion RA (2015) Performance analysis of touch-interaction behavior for active smartphone authentication. IEEE Trans Inf Forensics Secur 11(3):1–16. https://doi.org/10.1109/TIFS.2015.2503258
56. Shepherd S (1995) Continuous authentication by analysis of keyboard typing characteristics. In: European convention on security and detection, vol 1995. IEEE, pp 111–114. https://doi.org/10.1049/cp:19950480
57. Stragapede G, Vera-Rodriguez R, Tolosana R, Morales A, Acien A, El G, Lan L, Le Lan G (2022) Mobile behavioral biometrics for passive authentication. Pattern Recogn Lett 157:35–41. https://doi.org/10.1016/J.PATREC.2022.03.014
58. Sun Y, Ceker H, Upadhyaya S (2017) Anatomy of secondary features in keystroke dynamics-achieving more with less. In: IEEE international conference on identity, security and behavior analysis (ISBA 2017). https://doi.org/10.1109/ISBA.2017.7947691
59. Sun Y, Ceker H, Upadhyaya S (2016) Shared keystroke dataset for continuous authentication. In: Proceedings of the 8th IEEE international workshop on information forensics and security (WIFS 2016), pp 1–6. https://doi.org/10.1109/WIFS.2016.7823894
60. Tobola A, Streit FJ, Espig C, Korpok O, Sauter C, Lang N, Schmitz B, Hofmann C, Struck M, Weigand C, Leutheuser H, Eskofier BM, Fischer G (2015) Sampling rate impact on energy consumption of biomedical signal processing systems. In: IEEE 12th international conference on wearable and implantable body sensor networks, BSN 2015, pp 1–6. https://doi.org/10.1109/BSN.2015.7299392
61. Toosi R, Akhaee MA (2021) Time-frequency analysis of keystroke dynamics for user authentication. Future Gener Comput Syst 115:438–447. https://doi.org/10.1016/j.future.2020.09.027
62. Van hamme T, Preuveneers D, Joosen W (2018) Managing distributed trust relationships for multi-modal authentication. J Inf Secur Appl 40:258–270. https://doi.org/10.1016/j.jisa.2018.01.003
63. Wang H, He H, Song C, Tang H, Sun Y, Qiao Y, Zhang W (2022) Who is using the phone? Representation-learning-based continuous authentication on smartphones. Secur Commun Netw 2022:1–13
64. Whiskerd N, Körtge N, Jürgens K, Lamshöft K, Ezennaya-Gomez S, Vielhauer C, Dittmann J, Hildebrandt M (2020) Keystroke biometrics in the encrypted domain: a first study on search suggestion functions of web search engines. EURASIP J Inf Secur 2020(1):1–16. https://doi.org/10.1186/s13635-020-0100-8
65. Wu G, Wang J, Zhang Y, Jiang S (2018) A continuous identity authentication scheme based on physiological and behavioral characteristics. Sensors 18(1):179. https://doi.org/10.3390/s18010179
66. Yang Y, Guo B, Wang Z, Li M, Yu Z, Zhou X (2019) BehaveSense: continuous authentication for security-sensitive mobile apps using behavioral biometrics. Ad Hoc Netw 84(2019):9–18. https://doi.org/10.1016/j.adhoc.2018.09.015
67. Zhu H, Hu J, Chang S, Lu L (2017) ShakeIn: secure user authentication of smartphones with single-handed shakes. IEEE Trans Mob Comput 16(10):2901–2912. https://doi.org/10.1109/TMC.2017.2651820

A Prototype for Peer-to-Peer Transient Microservices with Wi-Fi Direct

Deepak Gohil, Dharmik Jethva, Krish Patel, Ujjwal Singh, Ved Vaghela, and Novarun Deb

Abstract This research explores the implementation of a peer-to-peer transient service architecture, focusing on decentralized microservices served by the nearest peers using Wi-Fi Direct. Utilizing Wi-Fi Direct technology, the research facilitates direct communication between devices without relying on traditional networks. This approach allows users to seamlessly exchange data and services in transient contexts. Two distinct services were integrated into the architecture: a sensor-based interaction service and a resource-based computation service. Results demonstrate the feasibility of the proposed architecture, showcasing the potential for decentralized, proximity-driven services in various application domains. By showcasing the practicality of peer-to-peer transient services, this research opens avenues for decentralized communication models, contributing to advancements in mobile technology.

Keywords Wi-Fi Direct technology · Peer-to-peer transient service architecture · Decentralized microservices · Proximity-driven services

D. Gohil · D. Jethva · K. Patel · U. Singh · V. Vaghela (✉) · N. Deb
Indian Institute of Information Technology Vadodara, Gujarat, India
e-mail: 202152343@iiitvadodara.ac.in

D. Gohil
e-mail: 202152311@iiitvadodara.ac.in
URL: https://iiitvadodara.ac.in/

D. Jethva
e-mail: 202152309@iiitvadodara.ac.in

K. Patel
e-mail: 202152318@iiitvadodara.ac.in

U. Singh
e-mail: 202151158@iiitvadodara.ac.in

N. Deb
e-mail: novarun_deb@iiitvadodara.ac.in

© The Author(s), under exclusive license to Springer Nature Singapore Pte Ltd. 2025

N. Chaki et al. (eds.), *Applied Computing for Software and Smart Systems*, Lecture Notes in Networks and Systems 980, https://doi.org/10.1007/978-981-97-9762-2_6

1 Introduction

In modern mobile computing, the demand for efficient and robust peer-to-peer (P2P) communication systems has become increasingly pronounced. This report explores P2P transient services, focusing on a decentralized architecture where each peer contributes services, eliminating the need for a central server.

The transient service embraces a dynamic and adaptive network, distributing tasks among connected peers to ensure continuous service delivery, adapting to changing resource availability. This innovative approach enhances fault tolerance by preventing system breakdowns if any peer becomes unavailable, dispersing services across the network.

Scalability is a key advantage, with the network seamlessly accommodating additional peers without the constraints imposed by a centralized server. This scalability is complemented by efficient resource utilization, intelligently distributing tasks and services across the network to optimize computing power and bandwidth usage.

This research addresses these challenges through the exploration of a peer-to-peer transient service architecture, aiming to redefine the paradigm of microservice delivery. The overarching problem lies in the limitations of centralized servers to efficiently handle dynamic and transient interactions between devices within close proximity.

The chosen focus on Wi-Fi Direct as the underlying peer-to-peer technology stems from the imperative to harness the power of direct connections between devices in a local environment. The aim of this research is to design, implement, and evaluate a decentralized architecture where microservices are served by the nearest peers using Wi-Fi Direct, facilitating a more responsive and resource-efficient model. This shift toward proximity-based interactions aligns with the contemporary demand for real-time and context-aware services.

The scope of this work encompasses the comprehensive exploration of Wi-Fi Direct as a peer-to-peer solution, its integration into an Android application, and the development of two distinct services: a geolocation service for sensor-based interactions and a smart reply service for computation-based interactions. By addressing the challenges associated with decentralization and proximity-based services, this research makes significant contributions to the evolving landscape of peer-to-peer architectures. The subsequent sections of this paper delve into the methodology, implementation details, and outcomes, shedding light on the practical implications and benefits of adopting Wi-Fi Direct for peer-to-peer transient service architectures.

2 Literature Survey

This section encompasses key research contributions relevant to peer-to-peer networks and transient services, providing valuable insights into various technologies and implementations that have shaped the landscape of decentralized interactions.

2.1 WebRTC Technology Overview and Signaling Solution Design and Implementation

Sredojev, Samardzija, and Posarac present a comprehensive study on WebRTC technology, addressing the challenges associated with real-time communication. This work is particularly relevant to our investigation as it focuses on the design and implementation of a novel signaling mechanism, a critical aspect in peer-to-peer networks. The exploration of WebSocket servers for signaling aligns with our interest in optimizing communication flow between peers and servers [1].

2.2 Research of NAT Traversal Based on RTP Relay Server Under Mobile Internet Environment

Wang et al. delve into the complexities of SIP/NAT traversal, proposing a scheme that leverages SIP and RTP relay servers. This paper is pertinent to our research as it addresses obstacles in communication under specific network conditions. The emphasis on the cooperation of servers for efficient traversal aligns with the challenges of decentralized microservices in peer-to-peer networks [2].

2.3 Wi-Fi-Aware as a Connectivity Solution for IoT Pairing: Enabling New Proximity-Based Services

Saloni and Hegde explore the role of Wi-Fi Aware (NAN) in the context of the Internet of Things (IoT). This work is relevant to our investigation as it highlights the importance of connectivity solutions for proximity-based services. The discussion on pairing Wi-Fi-enabled devices with IoT resonates with our focus on utilizing Wi-Fi Direct for proximity-aware microservice delivery in peer-to-peer networks [3].

2.4 Device-To-Device Communications with Wi-Fi Direct: Overview and Experimentation

Camps-Mur, Garcia-Saavedra, and Serrano provide a thorough overview of Wi-Fi Direct, emphasizing its potential impact on direct device-to-device communications. This article is crucial to our research as it offers insights into the practical implications of Wi-Fi Direct, including delays during device discovery and connection establishment. The experimental evaluation informs our understanding of the performance of Wi-Fi Direct in real-world scenarios, aligning with our goal of optimizing communication between nearby peers [4].

2.5 Egalitarian Transient Service Composition in Crowdsourced IoT Environment

This paper explores a groundbreaking framework designed for the unique dynamics of transient services in the context of peer-to-peer architecture. The focus lies in addressing the inherent dynamism of both service consumers and providers in space and time. By leveraging a Dynamic Bayesian Network, the paper models the dynamic behavior of service providers, offering a novel perspective on the composition of transient services. The framework transforms this composition into a multi-objective temporal optimization problem, prioritizing the continuous provision of services to maximize consumer coverage while minimizing the cost of service usage over an extended period. This paper contributes to the understanding of peer-to-peer architecture's role in optimizing transient service provision, offering valuable insights for future developments in this domain [5].

3 Present Investigation

In this section, we detail the methodology, procedures, and techniques adopted for implementing the peer-to-peer transient service architecture. The technology selection played a crucial role, considering WebRTC, relay servers, Wi-Fi Aware, and Wi-Fi Direct. The goal was to establish a self-sustained network without reliance on external servers, leading to the elimination of WebRTC and relay servers. Wi-Fi Aware was excluded due to its early stage and limited device support, leaving Wi-Fi Direct as the optimal choice for creating small LAN networks independently.

To demonstrate the peer-to-peer transient service architecture, we developed a Flutter application utilizing the "flutter_p2p_connection" library. The deployment involved executing the Flutter application on Wi-Fi Direct-supported devices, allowing them to autonomously establish connections as peers without centralized servers.

1. Technology Selection: Wi-Fi Direct was chosen for its server-independent, self-sustained network capability.
2. Flutter Application Development: Utilizing the "flutter_p2p_connection" library, the Flutter application provided Wi-Fi Direct controls, device discovery, and message-based communication through Web sockets.
3. Custom Protocol Design: The following points outline the key motivations behind our decision to develop a custom JSON protocol for our peer-to-peer transient service architecture:

    (a) Standardized Communication: The JSON protocol defines a common language for all devices, ensuring everyone understands the format of messages exchanged. This eliminates confusion and simplifies how devices interact.

(b) Focus on Services: Developers can concentrate on building the core functionalities of their services (like "smart reply") because the protocol handles how data is packaged and delivered. It's like having a pre-defined way to wrap and send payload, allowing developers to focus on what's inside.
(c) Simplified Service Discovery: The protocol includes a field identifying the requested service (e.g., "smart reply"). This makes it easy for receiving devices to understand what kind of action is needed, similar to having a label on a box indicating its contents.
(d) Lightweight Data Exchange: Our custom JSON protocol facilitates a lightweight approach to data exchange. JSON's simple structure and efficient parsing align well with the real-time, bidirectional communication offered by WebSockets. This contrasts with protocols like REST and GraphQL, which are typically optimized for client–server architectures and may require additional overhead for use with WebSockets.

Our current implementation utilizes a basic JSON protocol for communication between devices. This protocol defines the message structure as follows:

```
JSON:
     {
          ``action''  :  ``<request/response>'',
          ``service'':  ``<service_name>'',
          ``data''  :  {<data>}
     }
```

(a) action (string): Denotes the communication type ("request" or "response").
(b) service (string): Identifies the targeted service (e.g., "smart reply").
(c) data (object): Carries any relevant service data (often empty for requests).

This structure provides a foundation for efficient communication, but it represents a primitive approach. Here's how the protocol could be extended for broader data type support:

- Data Type Identification: A new field within the data object could specify the data type (e.g., "string", "number", "boolean"). This allows the receiving device to interpret the data correctly.
- Base64 Encoding for Files: For files, the data object could house a Base64 encoded string containing the file content. Base64 encoding represents binary data in a text-based format suitable for JSON transmission.
- Metadata for Complex Data: For complex data structures, the data object could hold metadata describing the data format (e.g., JSON schema). This would enable the receiving device to properly parse and utilize the data.

While these are potential extensions, the current implementation focuses on a simple structure for basic communication. Future development will explore incorporating these enhancements to establish a more versatile protocol capable of handling diverse data types within the peer-to-peer transient service architecture.

4 Implementation

Two services, geolocation and smart reply, were implemented using the custom protocol. Geolocation services exchanged geographical coordinates, while smart reply services used a machine learning model for text responses.

4.1 Peer-To-Peer Transient Service App Workflow

The operation of our peer-to-peer transient service app involves a series of steps to seamlessly connect providers and consumers within a Wi-Fi Direct network.

Provider Setup:

- The user, acting as a provider, initiates the process by enabling the Wi-Fi functionality on their device.
- Upon obtaining the user's permission, the app creates a Wi-Fi Direct group on the provider's device.
- Simultaneously, the app initiates the listening process on a specific port to await connection requests from potential consumers.

Consumer Joining Process:

- Users intending to consume services also enable their Wi-Fi functionality.
- The app on the consumer's device scans the vicinity for available Wi-Fi Direct groups.
- A list of discovered Wi-Fi Direct groups is presented to the consumer, who selects the desired group based on proximity or service preference.

Connection Establishment:

- Upon selecting a Wi-Fi Direct group, the consumer's device initiates a connection request to the provider's device.
- The consumer's device connects to the specified port on the provider's device, establishing a direct and secure connection.

Communication Over Established Connection:

- With the Wi-Fi Direct connection successfully established, the provider and consumer devices can now communicate directly.
- The app utilizes a custom JSON-based protocol for streamlined service identification and data transfer.

- Consumers send service requests in the prescribed JSON format, specifying the type of service they seek (e.g., "smart reply") and relevant content.

Service Processing:

- The provider's app, upon receiving a service request, processes the request based on the specified service type and content.
- In the case of "smart reply", for instance, the provider's machine learning model computes a response to the provided content.

Response Transmission

- The provider's app sends back the computed response to the consumer's device over the established Wi-Fi Direct connection.
- The response adheres to the same JSON format, ensuring a standardized and efficient communication flow (Fig. 1).

Service Consumption

- The consumer's app receives and interprets the response, completing the service interaction.
- The consumer can now utilize the obtained information or continue to engage in additional service requests as needed.

By following this workflow, our app enables a dynamic and decentralized peer-to-peer transient service architecture, allowing users to seamlessly connect, request services, and exchange information within a Wi-Fi Direct network. The emphasis on simplicity and autonomy empowers users to create self-sustained networks for various transient services.

5 Results and Discussions

The outcomes of the investigation into the peer-to-peer transient service architecture using Wi-Fi Direct are presented in this section, accompanied by an in-depth discussion and analysis. The discussion logically unfolds, drawing inferences, outlining contributions, and providing insights into potential future work.

5.1 Wi-Fi Direct Connection Establishment

The initial phase of the communication process establishes a direct connection between provider and consumer devices using Wi-Fi Direct technology. As depicted in Figs. 2, 3, and 4, the provider device takes the lead by creating a Wi-Fi Direct group. This essentially creates a private network specifically for the peer-to-peer interaction. The provider then starts "listening" on a designated port within this network, similar to how servers listen for incoming connections.

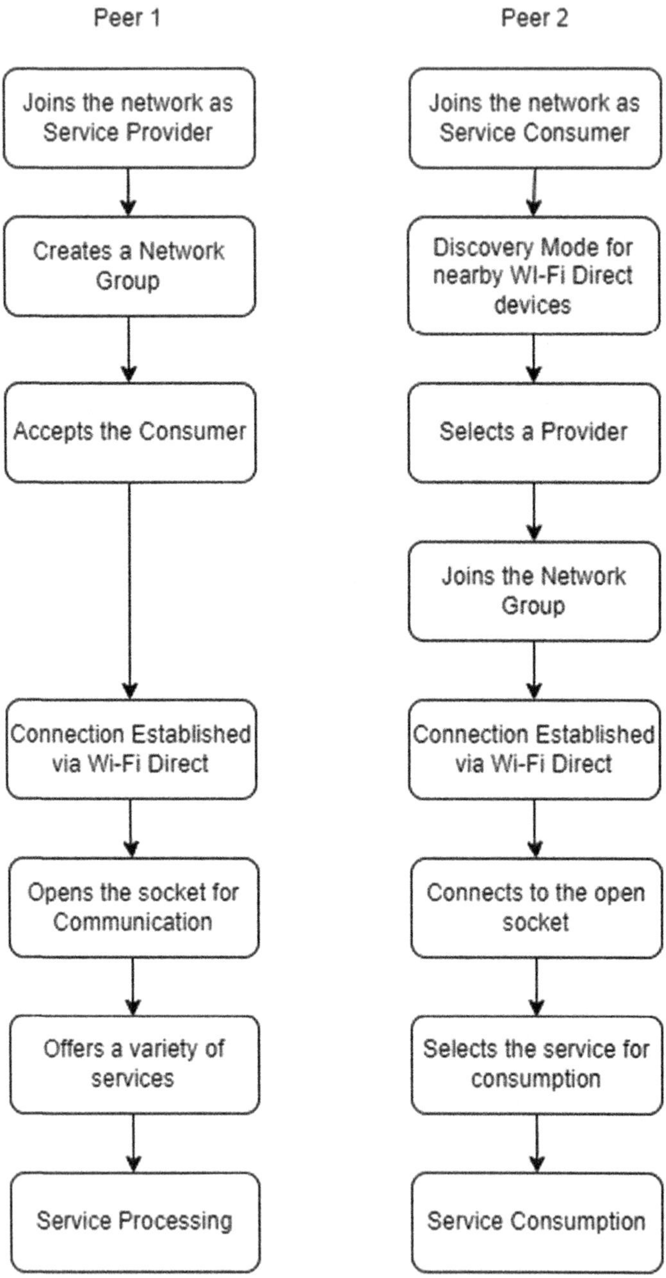

Fig. 1 Workflow

Fig. 2 Peer joins the network

Fig. 3 Provider opens socket after joining the network group

Fig. 4 Consumer connects to the socket

Meanwhile, the consumer device detects the available Wi-Fi Direct group and joins it. Once connected, the consumer initiates communication by sending an HTTP request to the specific port where the provider is listening. This initial request serves a purpose beyond simply sending data. It acts as a handshake, indicating the consumer's desire to avail a service and thus upgrading the communication to a more efficient method—WebSockets.

Following the successful HTTP handshake, both devices transition to using WebSockets for real-time, bidirectional communication. This is where the custom JSON protocol comes into play. Data exchange between devices occurs using messages formatted according to this protocol, ensuring a standardized and efficient flow of information. As the communication progresses, the provider and consumer exchange data payloads containing relevant service requests and responses, all structured within the custom JSON format.

5.2 Service Interaction—Smart Reply

Figures 5, 6, and 7 capture a service interaction, specifically a "smart reply" request and response. The provider received the consumer's request, processed it using a machine learning model, and sent back the computed smart reply. This demonstrates the successful implementation of computation-based services over the Wi-Fi Direct connection.

This successful exchange demonstrates the architecture's ability to not only facilitate direct communication between devices via Wi-Fi Direct but also offload computational tasks. In this case, the provider device leverages a machine learning model to process the request and generate the smart reply, highlighting the potential for computation-intensive services within the peer-to-peer network.

Fig. 5 Consumer selects the smart reply service for consumption

Fig. 6 Consumer sends a message to the provider for a smart reply

Fig. 7 Consumer receives the smart reply from the provider

5.3 Service Interaction—Geolocation

In addition to computation-based services, Figs. 8, 9, and 10 illustrate the successful implementation of sensor-based geolocation services. Peers exchanged geographical coordinates, demonstrating the versatility of the peer-to-peer transient service architecture.

These figures depict the successful exchange of geographical coordinates between peers, showcasing the architecture's ability to leverage device sensors. This expands the potential applications beyond services requiring significant processing power

Fig. 8 Consumer selects the geolocation service for consumption

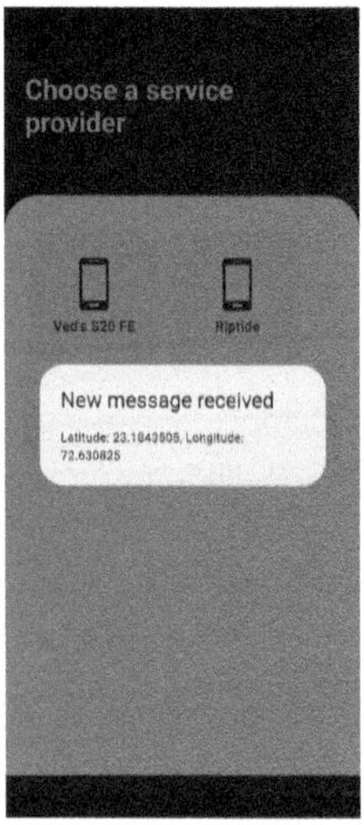

Fig. 9 Consumer receives the geolocation of the provider

and opens the door for functionalities that utilize a device's built-in hardware capabilities. The successful implementation of sensor-based geolocation services further highlights the adaptability and diverse range of services supported by this innovative architecture.

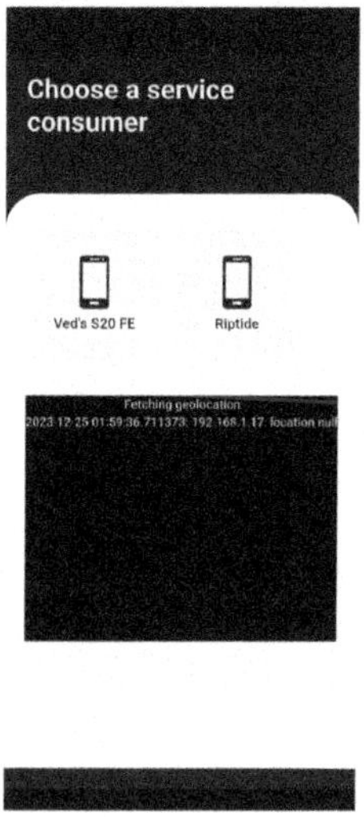

Fig. 10 Provider with the logs of the consumer requests

5.4 *Discussion*

Connection Stability and Latency:

- The establishment of Wi-Fi Direct connections proved to be stable and efficient, facilitating direct communication between peers.
- Minimal latency was observed during the connection initiation and data transfer phases, ensuring a responsive user experience.

Protocol Effectiveness:

- The custom JSON-based protocol demonstrated effectiveness in standardizing service interactions.
- Its structured format facilitates seamless communication, contributing to the overall efficiency of the peer-to-peer transient service architecture.

Service Interaction Performance:

- Service interactions showcased reliable performance for both computation-based and sensor-based services.
- The integration of machine learning models demonstrated the potential for sophisticated peer-to-peer applications.

Advantages over Traditional Client–Server Architecture:

- Decentralization: Our peer-to-peer architecture eliminates the reliance on a central server, making it more resilient to server outages and potentially reducing infrastructure costs.
- Scalability: The architecture scales naturally as more devices join the network, as additional devices contribute to overall processing power and resource availability.

– Offline Functionality: Services can potentially function even in the absence of an Internet connection, as long as participating devices are within range and connected via Wi-Fi Direct.
– Reduced Latency: Direct communication between devices can minimize latency compared to client–server interactions, which may involve multiple hops and server processing.

6 Conclusion

The investigation highlights the viability of Wi-Fi Direct as a robust technology for establishing peer-to-peer connections without the need for centralized servers. The custom protocol enhances communication efficiency, opening avenues for diverse transient services. The demonstrated computation-based and sensor-based services emphasize the potential for real-world applications in decentralized environments.

6.1 Future Work

1. Expanded Service Offerings: Explore additional sensor-based and resource-based services to broaden the range of available transient services.
2. Framework Development: Develop a framework to enable developers to easily build and deploy their services on the peer-to-peer network.
3. Enhanced State Management: Implement state management mechanisms to seamlessly transition service provision between peers, ensuring continuity in case of disconnection.
4. Custom JSON Protocol Refinement: The current protocol will be further optimized with features like error handling, data validation, and potential security measures (e.g., encryption) to ensure continued robustness as services evolve.

In conclusion, the results affirm the feasibility and effectiveness of the proposed peer-to-peer transient service architecture. The contributions and insights gained from this study lay the groundwork for further advancements in decentralized and self-sustained networking paradigms.

For further exploration of the research, the codebase is available on GitHub Repository. Additionally, a demo video showcasing the prototype's functionalities can be found on Google Drive.

Acknowledgements We sincerely thank the Indian Institute of Information Technology Vadodara (IIIT Vadodara) for their support in this research. Their resources and guidance proved invaluable in the development of this research.

References

1. Sredojev B, Samardzija D, Posarac D (2015) WebRTC technology overview and signaling solution design and implementation. In: 38th International convention on information and communication technology, electronics and microelectronics (MIPRO). Opatija, Croatia, pp 1006–1009. https://doi.org/10.1109/MIPRO.2015.7160422
2. Wang Y, Xu S, Wang J, Xue Y, Fu J, Hu B (2015) Research of NAT traversal based on RTP relay server under mobile internet environment. In: 4th International conference on computer science and network technology (ICCSNT). Harbin, China, pp 1370–1374. https://doi.org/10.1109/ICCSNT.2015.7490983
3. Saloni S, Hegde A (2016) WiFi-aware as a connectivity solution for IoT pairing IoT with WiFi aware technology: enabling new proximity based services. In: International conference on internet of things and applications (IOTA). Pune, India, pp 137–142. https://doi.org/10.1109/IOTA.2016.7562710
4. Camps-Mur D, Garcia-Saavedra A, Serrano P (2013) Device-to-device communications with Wi-Fi Direct: overview and experimentation. IEEE Wirel Commun 20(3):96–104. https://doi.org/10.1109/MWC.2013.6549288
5. Khurana S, Deb N, Mistry S, Ghose A, Krishna A, Dam HK (2023) Egalitarian transient service composition in crowdsourced IoT environment. IEEE Trans Serv Comput 16(5):3305–3317. https://doi.org/10.1109/TSC.2023.3264581

Image Analysis

Comparison of the Effectiveness of Face Recognition Algorithms in Terms of Photo Distortion Level

Daniel Grabowski and Khalid Saeed

Abstract This research explores the impact of photo distortion on face recognition algorithms performance, focusing on convolutional neural networks (CNNs) due to their superior efficiency. Through a comprehensive review of current solutions and empirical experimentation with the LFW image database, this study investigates how compression, blurring, and noise affect algorithmic accuracy. Experiments utilized the Python-based DeepFace library for comparative analysis of photo similarity, optimizing algorithm parameters through density analysis and decision trees. The study confirms that while compression minimally affects performance, noise—especially at higher levels—significantly degrades accuracy. However, applying median filters to noisy images largely restores effectiveness. Blurring's impact varies with direction, with vertical blurring severely impairing recognition capabilities. Among the algorithms evaluated, FaceNet512 emerged as the most resilient in handling various types of distortions. The findings underscore the importance of distortion type and parameter optimization in maintaining recognition accuracy, offering valuable insights for enhancing algorithmic resilience against photo quality degradation.

Keywords Face recognition · Deep learning · Convolutional neural networks · Compression · Image distortion

D. Grabowski (✉) · K. Saeed
Faculty of Computer Science, Białystok University of Technology, Białystok, Poland
e-mail: daniel.grabowski@pb.edu.pl

K. Saeed
e-mail: k.saeed@pb.edu.pl

K. Saeed
Department of Computer Science and Electronics, Universidad de la Costa—CUC, Barranquilla, Colombia

© The Author(s), under exclusive license to Springer Nature Singapore Pte Ltd. 2025

N. Chaki et al. (eds.), *Applied Computing for Software and Smart Systems*, Lecture Notes in Networks and Systems 980, https://doi.org/10.1007/978-981-97-9762-2_7

1 Introduction

Human faces play a crucial role in communication and identification, embodying evolutionary significance by conveying essential information about others, such as gender, race, emotional state, and attractiveness. Facial expressions also serve as a vital source of non-verbal communication, with the ability to interpret faces to recognize family, friends, and foes starting from infancy. This innate skill is fundamental to daily interactions, enabling us to understand the emotions of those around us and identify acquaintances in a crowd. The history of biometric face recognition systems dates back to the 1960s, initially based on facial geometry, and has since evolved significantly with advancements in computational power, algorithm development, and the availability of high-quality image databases. These systems, now globally implemented for security purposes, have various applications, including identity verification at airports and identifying missing persons or bodies. Despite their widespread use and increasing integration into daily life, face recognition algorithms, especially those using deep neural networks, face challenges regarding data quality. Photo distortion negatively impacts algorithm precision, a hypothesis this research aims to explore further.

2 State of the Art

The first biometric systems for face recognition began to emerge in the 1960s. The earliest algorithms were primarily based on the geometry of the face, comparing distances between characteristic facial points, such as the corners of the mouth or the center of the eyes. However, these methods required the manual calculation of the locations of characteristic points, making them extremely labor-intensive. What distinguishes today's systems is the full automation of all stages of the face recognition system's operation, which are face detection, feature extraction, and face recognition. The most advanced algorithms have very high classification accuracy, reaching even above 99%; however, there are factors that negatively affect the classification level. These can be divided into internal factors, related to changes in a person's appearance, and external, related to the image [1].

2.1 FaceNet

FaceNet is a system developed by Google that learns to map directly from facial images to a compact Euclidean space, where distances directly correspond to a measure of facial similarity. In its basic version, the mapping occurs to a 128-dimensional Euclidean space, but there is also a variant of the algorithm where mapping occurs to a 512-dimensional space. In a study based on the LFW database, FaceNet achieved an accuracy of 99.63%.

In experiments, the system was trained using stochastic gradient descent (SGD) and the standard backpropagation algorithm on a cluster of computers for 1000 to 2000 h. The model used convolutional layers of sizes $1 \times 1 \times d$ between standard convolutional layers to increase efficiency, resulting in a model with a depth of 22 layers [2].

2.2 VGG-Face

The VGG-Face algorithm was created at the University of Oxford in 2015. The authors set themselves two goals, the first being to create a method for creating large datasets of photos for machine learning. The second goal was to explore different convolutional neural network architectures, including ways of normalizing photos and learning metrics. The authors analyzed currently used models for face recognition to select the best practices and sift through less significant details, achieving a simpler neural network structure while maintaining the results of the best solutions on the market. Compared to algorithms such as DeepFace and DeepID, two-dimensional face geometry recognition was used instead of three-dimensional [3].

2.3 ArcFace

ArcFace is a model of a deep convolutional neural network that is characterized by a small distance within decision classes and a large distance outside of decision classes. This was achieved by developing the Additive Angular Margin Loss (ArcFace) algorithm, which in this model replaces the often-used softmax algorithm. This significantly improves the discriminatory ability, which stabilizes the learning process of the model. In experiments, the ArcFace system was trained on a database of over 5 million objects. The ArcFace algorithm achieved an accuracy of 99.83% on the LFW database [4].

We have found the drawbacks of the known to us methods and tried to fill in the gap by our own approach. Hence, the authors have decided to improve the methodology and performance of deep learning algorithms for human face recognition. This way we could close the scientific gap concerning the impact of image distortion. Our proposed approach is given in Sect. 4.

3 Background

3.1 Introduction to Data Selection

The selection of data is a crucial step in any research project, particularly in studies involving face recognition technologies. It involves choosing datasets that are representative of the real-world scenarios in which the algorithms will be applied. This ensures that the findings of the research are valid and applicable outside of controlled experimental conditions.

3.2 Selected Dataset

One of the databases is the LFW Face Database, which was used in this study. The reason for this is that this database is used as a benchmark in many other studies to assess the effectiveness of face recognition algorithms. It contains over 13,000 cropped images of human faces. The database includes photos of more than 5,700 individuals, of which 1,680 have more than one photo.

3.3 Data Presentation

To facilitate the research, databases containing images altered with various distortions were developed. The effectiveness of the algorithms will be examined in terms of the level of image compression, blurring, and noise. Compression was performed at 3 levels: 50, 20, and 10. Blurring was divided into horizontal and vertical, each performed with filters of window size equal to 5 and 10. The last databases were created from images with applied Gaussian noise, salt-and-pepper noise, and from denoised images using a median filter (Fig. 1).

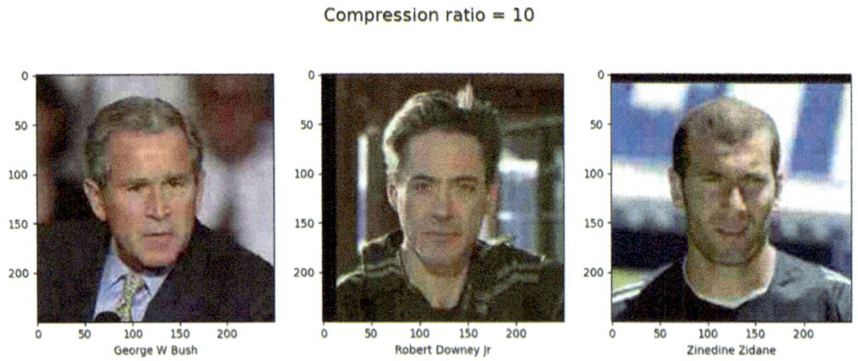

Fig. 1 Compressed photos, compression ratio = 10

4 Proposed Approach

For the execution of the experiments, the DeepFace library written in Python was chosen. It provides pre-built models of many face recognition algorithms available on the market and the possibility to download and load pretrained neural network weights, trained on millions of faces. This library also allows for the examination of attributes such as age, race, or mood of the person in the photo. To ensure the models work as well as possible, it is necessary to establish thresholds for all the algorithms.

4.1 Metric for Vector Distance

In the experiments, feature vectors returned by the studied algorithms will be compared. Each algorithm takes an input— a photo of a person—and returns a feature vector. Distances between vectors obtained from two images will be compared, and in this way, classification will be made whether the given photos depict the same person or not. The distance between feature vectors is calculated using the cosine method.

4.2 Setting Thresholds for Algorithms

To optimize the performance of the algorithms, it is necessary for each algorithm to set a threshold from which cases will be classified as negative or positive. To this end, a study of calculated distances on an undistorted database was conducted, and then density plots of the cases were drawn (Fig. 2).

From the chart, it is possible to see where positive and negative cases cluster. The greater the density distance of negative cases from positive ones, the greater the differences we can expect when comparing photos of two classes, which will positively affect the precision of the given model. It is also important for as many

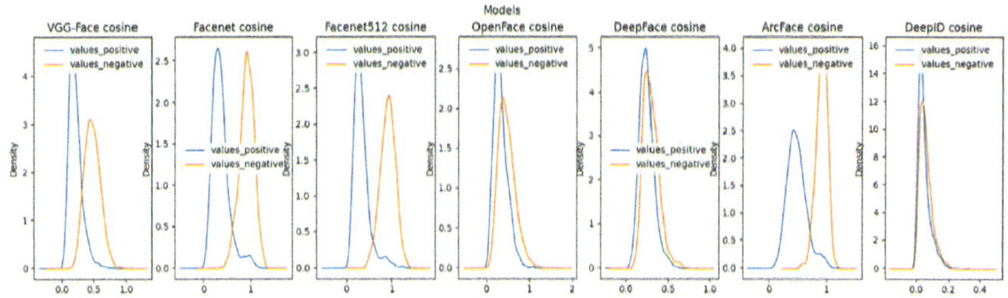

Fig. 2 Distribution of positive and negative cases of all models

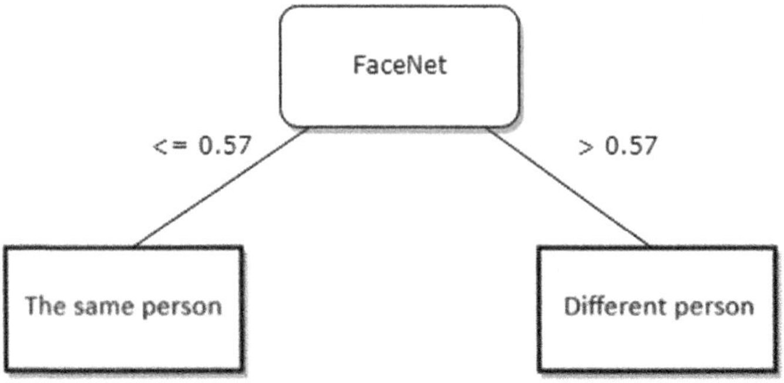

Fig. 3 An example of a decision tree created for the FaceNet algorithm

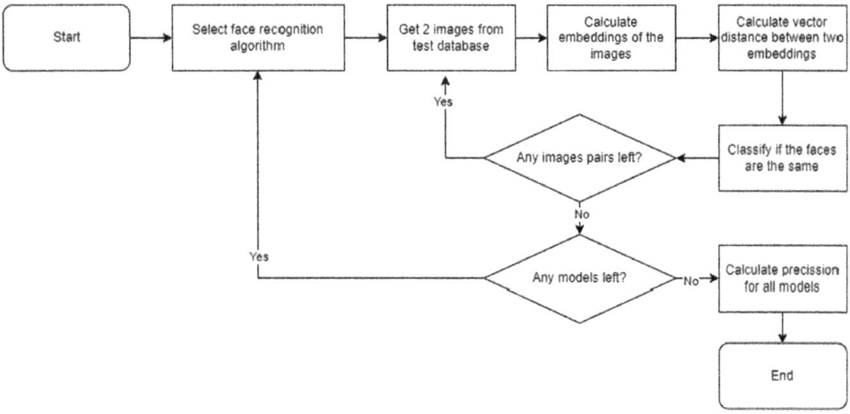

Fig. 4 Flowchart of the algorithm

cases as possible to cluster at one point, i.e., for the density plot to be as narrow as possible and not overlap with the plot of the opposite class (Fig. 3).

Subsequently, using this data, decision trees were created, aimed at finding the best-matched threshold. Using the example of a decision tree created for the FaceNet algorithm, if the cosine distance between two feature vectors is less than or equal to 0.57, the examined photos are recognized as photos of the same person. Otherwise, if the distance is greater than 0.57, the people in the photos are considered to be different (Fig. 4).

4.3 Face Detection Method

An important choice is also the algorithm for finding faces and cropping the photo. The DeepFace library implements algorithms such as OpenCV, Dlib, MTCNN, and

RetinaFace. The MTCNN and RetinaFace algorithms have the best effectiveness, however, they operate the slowest among the listed ones. To achieve the best results at the expense of execution time, the MTCNN algorithm was used [5–7].

5 Experiments

To determine how photo distortions affect the results, it is first necessary to check the algorithms' results on a database of undistorted photos.

5.1 Experiment 1: Compression

The first experiment examines how the level of compression affects the accuracy of the algorithms being studied. The photos were compressed to 50%, 20%, and 10%, respectively. It is also examined how the size of the photo database decreases relative to the undistorted base and the disk space savings this provides. Additionally, the time required for processing and comparing photos relative to their level of compression will be examined (Table 1).

Compression at 50% did not significantly affect the effectiveness of the algorithms tested. The ArcFace model achieved the same result as on the original base, and the other algorithms experienced a slight decrease in classification accuracy. A noticeable decrease in classification accuracy at 20% compression was observed for all algorithms, approximately 1.2%. Again, the ArcFace model registered the smallest decrease in accuracy compared to the official base. At 10% compression, the drop in classification quality for all algorithms is greater than in previous cases. The ArcFace algorithm, which had the smallest decreases in classification accuracy in previous experiments compared to the other algorithms, this time experienced the largest drop of all.

Table 1 Accuracy of face recognition models under different compression rates

Model	Normal (%)	50% compression (%)	20% compression (%)	10% compression (%)
VGG-Face	88.6	87.8	87.2	86.0
FaceNet512	95.5	94.7	94.1	92.3
FaceNet	93.8	93.3	92.5	91.2
ArcFace	93.7	93.7	92.8	89.6

5.2 Experiment 2: Noise

This experiment simulates noises that may appear on digital images, for example, during a sudden loss or significant weakening of the signal. Gaussian noise and "salt-and-pepper" noise, which replaces 5 and 10% of pixels on the photo, respectively, were applied to the images. Then, a median filter with a window size of 5 was applied to the images with S&P noise to smooth the noise and check how much the algorithms regain their effectiveness compared to undistorted photos.

Gaussian Noise Initially, the impact of applying Gaussian noise to the images was examined. All algorithms recorded a decrease in classification accuracy. The algorithms reacted in different ways to the application of noise; the DeepFace algorithm showed a significant tendency to classify photos as pictures of different people, thereby achieving a high recall level but low accuracy and very low precision, while the ArcFace method tended to classify images as images of the same person compared to the original base, thereby achieving high precision, but its overall accuracy significantly dropped (Table 2).

S&P Noise Initially, the effectiveness of the algorithms on a base of photos with applied S&P noise, where randomly 10% of pixels were set to a random value, is tested. Images with applied S&P noise caused a significant drop in classification accuracy of the algorithms tested. The tendencies of some algorithms to classify most photos into one class also remained. The ArcFace method, with S&P noise of 10%, recorded an excellent result in terms of precision, 97.8%, however, the algorithm's effectiveness dropped almost to 50%; the algorithm simply began to classify most pairs as pictures of the same person. The FaceNet and VGG-Face algorithms reacted similarly to the noise.

After testing the effectiveness of the algorithms on images distorted by noise, these images were subjected to filtering aimed at removing the noise. A median filter with a window size of 5 was used for this purpose. The experiment was then repeated on a database of filtered photos. After filtering the images, the algorithms achieved accuracy similar to that obtained on the original base. This proved to be an effective method for recovering the classification effectiveness of all algorithms (Tables 3 and 4).

Table 2 Precision of face recognition models with Gaussian noise

Model	Normal (%)	Gaussian noise (%)
ArcFace	93.7	74.6
FaceNet	93.8	85.2
FaceNet512	95.5	92.4
VGG-Face	88.6	83.3

Table 3 Accuracy of face recognition models under 5% salt-and-pepper noise and after median filter

Model	Normal (%)	S&P 5% noise (%)	After median filter (%)
VGG-Face	88.6	77.8	86.2
FaceNet512	95.5	87.4	92.2
FaceNet	93.8	80.0	90.3
ArcFace	93.7	74.2	91.9

Table 4 Accuracy of face recognition models under 10% salt-and-pepper noise and after median filter

Model	Normal (%)	S&P 10% noise (%)	After median filter (%)
VGG-Face	88.6	65.1	85.3
FaceNet512	95.5	77.1	93.3
FaceNet	93.8	60.8	89.5
ArcFace	93.7	53.8	89.9

5.3 *Experiment 3: Blurring*

The aim of the experiment is to check how the tested algorithms cope with images that have been blurred. The photos were blurred vertically and horizontally to check if it affects the results. Both blurrings were performed at two filter window sizes; a filter with a window size of 5 blurs less than a filter with a window size equal to 10.

Horizontal Blurring In the case of horizontal blurring, it had a slight impact on the quality of classification in the case of a filter window of 5 pixels. Using a filter size of 10 pixels had a slightly greater impact on accuracy, but the algorithms still recorded high accuracy.

Vertical Blurring Vertical blurring with a filter window size of 5 pixels had a similar impact on classification quality as horizontal blurring with a filter of the same size. Vertical blurring with a filter window size of 10 pixels caused a significant drop in accuracy for all algorithms, much greater than using a window of the same size in horizontal blurring (Tables 5 and 6).

Table 5 Impact of horizontal blur on the precision of face recognition algorithms

Model	Normal (%)	Horizontal blur = 5 (%)	Horizontal blur = 10 (%)
VGG-Face	88.6	87.1	85.4
FaceNet512	95.5	93.3	90.3
FaceNet	93.8	92.2	86.8
ArcFace	93.7	92.9	88.4

Table 6 Impact of vertical blur on the precision of face recognition algorithms

Model	Normal (%)	Vertical blur = 5 (%)	Vertical blur = 10 (%)
VGG-Face	88.6	87.2	77.5
FaceNet512	95.5	94.1	82.6
FaceNet	93.8	88.4	54.2
ArcFace	93.7	92.8	66.7

6 Result Analysis

In the conducted experiments, the impact of photo distortion on the effectiveness of face detection algorithms using convolutional neural networks was examined. The factors studied were compression, noise, and blurring of the photo.

Initially, the effectiveness of the studied algorithms on the LFW photo database was checked. Four algorithms achieved high effectiveness: VGG-Face, FaceNet, FaceNet512, and ArcFace. The algorithms OpenFace, DeepFace, and DeepID obtained significantly lower results. This could be due to poorly selected photos or network training on a different type of photos than those used in the experiments. Further analysis will focus on the algorithms with the highest level of classification effectiveness, as OpenFace, DeepFace, and DeepID do not contribute any research value due to their low effectiveness.

6.1 Compression

The experiments examined whether photo compression contributes to a decrease in the quality of classification of the studied algorithms. Compression was tested at three levels—50%, 20%, and 10%. Each of the studied algorithms noted a slight decrease in classification quality with a higher level of compression, which was expected, as the increase in the level of compression results in the loss of some features from the photo. The best results on each database were noted by the FaceNet512 algorithm, while the lowest result was noted by the VGG-Face algorithm. ArcFace achieved the lowest performance drops for the compression level of 50% and 20%, with no performance drop noted at the compression level of 50%, but it experienced the largest drop in performance of all algorithms at the compression level of 10%, showing a sharp drop in classification quality between the 10% and 20% levels. Other algorithms noted similar drops at each level of compression, up to 0.5% at 50% compression, up to 1.5% at 20% compression, and up to 3.2% at 10% compression. It is evident that the drops in classification quality at 50% and 20% compression levels are minor, but at the 10% level, it becomes noticeable (Table 7).

The size of the databases has significantly decreased after compression, from about 55% for 50% compression to about 80% for 10% compression. There is a

Table 7 Percentage drop in precision of face recognition models at different compression levels

Model	50% compression (%)	20% compression (%)	10% compression (%)
ArcFace	0	0.9	4.1
FaceNet	0.5	1.3	2.6
FaceNet512	0.8	1.4	3.2
VGG-Face	0.8	1.4	2.6

Table 8 Sizes of compressed image databases and the original database

Normal	50% compression	20% compression	10% compression
179 MB	79.7 MB	50.8 MB	37.5 MB

small difference in database size between 20% and 10% compression. Considering the previously discussed results of algorithm effectiveness, if we care about saving space, it would be best to apply 20% photo compression. This way, the loss in classification quality will be minimal, and we will save about 70% of disk space (Table 8).

6.2 Noise

Images were subjected to noise to check its impact on the quality of classification of algorithms. The application of noise simulates situations where the images processed by face recognition algorithms are distorted, for example, by a sudden signal loss or weakening during the transmission of a digital image. It was also checked whether it is possible to recover the classification quality by applying a median filter to the noisy images.

6.2.1 White Noise

All studied algorithms recorded a decrease in classification quality after applying Gaussian noise. This is expected behavior as a noisy photo loses some features, making classification difficult (Tables 9, 10, and 11).

6.3 Blurring

Images blurred vertically and horizontally with 5- and 10-pixel filters showed differing impacts on algorithm performance. Blurring of 5 pixels slightly affected accu-

Table 9 Percentage drop in precision of algorithms after Gaussian noise

Model	Precision drop after Gaussian noise (%)
ArcFace	20.38
FaceNet	9.17
FaceNet512	3.25
VGG-Face	5.98

Table 10 Percentage drop in precision of algorithms after 10% S&P noise

Model	Precision drop after 10% S&P noise (%)
ArcFace	42.58
FaceNet	35.18
FaceNet512	19.27
VGG-Face	26.52

Table 11 Percentage drop in precision of algorithms after 5% S&P noise

Model	Precision drop after 5% S&P noise (%)
ArcFace	20.81
FaceNet	14.71
FaceNet512	8.48
VGG-Face	12.19

racy without significant classification quality loss. However, 10-pixel blurring significantly reduced accuracy across all algorithms. Horizontal blurring at 10 pixels maintained classification capability with some accuracy loss, as precision and recall remained balanced. Unlike horizontal blurring, 10-pixel vertical blurring led to a more pronounced accuracy decline. Most algorithms, except FaceNet512, experienced a significant decrease in recall and an increase in precision, suggesting a higher likelihood of misclassifying images as the same person. These findings highlight that blurring's window size and direction crucially affect facial recognition accuracy. Specifically, while horizontal blurring somewhat preserves functionality despite accuracy dips, vertical blurring and larger window sizes increase misclassification risks, except in the FaceNet512 model, which showed resilience to these changes.

7 Conclusion

The objective of this study was achieved by examining the effectiveness of face recognition algorithms depending on the level of photo distortion, confirming the proposed research hypothesis. Distortions were applied in three ways: by compressing photos, by blurring photos, and by adding noise to the photos. Photo compression had the least impact on the classification quality of the algorithms among the three distortions tested. Despite using high compression, up to 10%, the algorithms lost little effectiveness. The experiments found that to save disk space, it is best to use compression at 20%; the loss in classification quality is minor, and the database size is significantly reduced. Applying noise had a greater impact on the quality of classification than the level of photo compression. The experiments tested how the performance of the algorithms would change after applying Gaussian noise and salt-and-pepper (S&P) noise at two levels—5% and 10%, that is, replacing 5 and 10 percent of the pixels with random other pixels. The drop in classification quality was similar between Gaussian noise and S&P 5% noise, with the ArcFace algorithm recording by far the performance drops. The application of S&P 10% noise resulted in a significant drop in classification quality across all algorithms. After applying the median filter to the noisy images, the algorithms regained most of their effectiveness and again successfully classified the images into the appropriate classes. In the case of blurring, the direction of the blur proved to play a significant role. The algorithms coped well with horizontal blurring, but in the case of vertical blurring, the ArcFace and FaceNet algorithms became useless, classifying most of the images as a photo of the same person, regardless of the input data.

The algorithm that performed by far the best in all experiments was FaceNet512. The OpenFace, DeepFace, and DeepID algorithms obtained very poor results on the tested data [8]. This could have been due to poorly selected data or incorrect model training. The ArcFace algorithm achieved high performance on the original database, but its classification quality declined most severely with higher distortions. This may be due to the algorithm used in the model, Additive Angular Margin Loss. It has been shown that compressing a photo can significantly reduce the database size, which can be important in the case of limited disk space and large numbers of photos, without causing significant losses in the accuracy of the algorithms. It was demonstrated that despite a significant decrease in classification quality when a photo is noised, it is possible to recover the accuracy of the algorithms by applying appropriate filters. In the case of blurring, the most important factor turned out to be its direction, since vertical blurring caused a much greater drop in accuracy, while causing a huge drop in the recall measure, indicating how many elements from a given class were correctly recognized.

References

1. Anwarul, S., Dahiya, S.: A Comprehensive Review on Face Recognition Methods and Factors Affecting Facial Recognition Accuracy. In: Singh, P., Kar, A., Singh, Y., Kolekar, M., Tanwar, S. (eds) Proceedings of ICRIC 2019 . Lecture Notes in Electrical Engineering, vol 597. Springer (2020)
2. Schroff, F., Kalenichenko, D., Philbin, J.: FaceNet: A unified embedding for face recognition and clustering. In: 2015 IEEE Conference on Computer Vision and Pattern Recognition (CVPR). Boston, MA, USA, pp. 815-823 (2015)
3. Parkhi, O., Vedaldi, A., Zisserman, A.: Deep Face Recognition. In: Proceedings of the 29th British Machine Vision Conference (BMVC). vol. 1, pp. 41.1-41.12. (2015)
4. Deng J, Guo J, Yang J, Xue N, Kotsia I, Zafeiriou S (2022) ArcFace: Additive Angular Margin Loss for Deep Face Recognition. IEEE Transactions on Pattern Analysis and Machine Intelligence 44(10):5962–5979
5. Akhtar Z, Rattani A (2017) A Face in any Form: New Challenges and Opportunities for Face Recognition Technology. Computer 50(4):80–90
6. Chauhan, H., Chauhan, A.: Implementation of Decision Tree Algorithm C4.5. In: International Journal of Scientific and Research Publications, 3(10) (2013)
7. Kumar, A., Kaur, A., Kumar, M.: Face Detection Techniques: A Review. In: Artificial Intelligence Review, 52(1) (2019)
8. Amos B, Ludwiczuk B, Satyanarayanan M (2016) OpenFace: A General-purpose Face Recognition Library with Mobile Applications. Carnegie Mellon University, Pittsburgh, School of Computer Science

Multi-class Histology Image Analysis Using Handcrafted Texture and Colour Features for Breast Abnormality Classification

Anindita Mohanta, Sourav Dey Roy, Niharika Nath, and Mrinal Kanti Bhowmik

Abstract Breast cancer is one of the most common cancers among women worldwide, affecting around 10% of all women. Its fatality rate is extraordinarily high when compared to other cancer types. Histopathological analysis is one of the reliable methods for identifying breast cancer with confidence. But it's a highly specialized, time-consuming procedure that depends on the pathologist's experience. An automatic computerized histopathological image classification model can assist doctors and pathologists for early diagnosis. But most of the work is based on the non-interpretable black box model for feature selection tasks which is not very encouraging in the medical field. To introduce features interpretability in histopathological imaging, handcrafted features analysis is important first. In this paper, a novel hybrid computer-aided multi-class classification framework is proposed for predicting breast abnormality using histopathological images and handcrafted features. The proposed framework uses traditional handcrafted features (i.e. texture features, colour features, and fusion of texture and colour features) and a deep neural network (DNN) for breast cancer abnormality classification in order to facilitate the screening process for early diagnosis. The DNN classifier is trained using the features that were extracted using the handcrafted methods for two-class and multi-class classifications. The experimental results show that the use of handcrafted texture features had a better prediction performance in histopathological image classification as compared to the other handcrafted features for breast abnormality prediction. Using texture features with a DNN

A. Mohanta · S. D. Roy · M. K. Bhowmik (✉)
Computer Science and Engineering Department, Tripura University (A Central University), Suryamaninagar, Agartala 799022, Tripura, India
e-mail: mrinalkantibhowmik@tripurauniv.ac.in

A. Mohanta
e-mail: aninditamohanta01@gmail.com

S. D. Roy
e-mail: souravdeyroy49@gmail.com

N. Nath
Biological and Chemical Sciences Department, New York Institute of Technology, New York, NY 10023, USA
e-mail: nnath@nyit.edu

© The Author(s), under exclusive license to Springer Nature Singapore Pte Ltd. 2025
N. Chaki et al. (eds.), *Applied Computing for Software and Smart Systems*, Lecture Notes in Networks and Systems 980, https://doi.org/10.1007/978-981-97-9762-2_8

classifier, we achieved average values for accuracy, recall, precision and F1-score of 73%, 79%, 81% and 82% respectively for multi-class classification.

Keywords Breast cancer · Histopathological imaging · Features · Deep neural network · Classification · Performance evaluation

1 Introduction

According to the World Health Organization (WHO) [1], breast cancer is one of the most occurring cancers among women worldwide. In breast cancer abnormal cells grow rapidly and form tumours. The tumours can spread throughout the body and become deadly. The early form of breast cancer (in situ) is treatable and not life-threatening. For breast cancer diagnosis, histopathological examination and imaging is considered as the gold standard as compared to the other pathological tests and imaging modalities [2]. It provides more reliable and comprehensive information on breast cancer diagnosis. However, analysis of histopathological images for diagnosis requires highly experienced and specialized pathologist which is very costly and time-consuming task. In addition, there is often a lack of agreement among the expert members regarding the diagnosis. Considering these facts, there is a pressing need for a computerized automatic system for diagnosis of breast cancer using histopathological imaging which serve as a supporting tool for pathologists and doctors for breast cancer diagnosis with confidence. Several research for histopathological images and two-class classification have been conducted in the literatures. However, very few works for histopathological images multi-class classification for breast cancer grading with high accuracy have been conducted. Moreover, most of the proposed approaches in the literature are based on only texture-based features which are not sufficient for highly accurate and confident classification. Also, in histopathological image classification and grading deep learning-based black box architecture is not very encouraging.

Objectives. Considering all these facts, in this paper, a novel hybrid computer-aided multi-class classification framework is proposed for predicting breast abnormality using histopathological images. The proposed framework uses conventional handcrafted features (i.e. texture features, colour features, and fusion of texture and colour features) and a deep neural network (DNN) for feature classification in order to facilitate the screening process for early diagnosis. The DNN classifier is trained using the features that were extracted using the handcrafted methods for two-class and multi-class classifications. The experimental results show that the use of handcrafted texture features had a better prediction performance in histopathological image classification as compared to the other handcrafted features for breast abnormality prediction.

Paper Outline. In Sect. 2, the existing methods adopted by the research community for breast histopathological image classification have been discussed. Section 3

elaborated our proposed hybrid framework for breast histopathological image classification using conventional handcrafted features and deep neural network. In Sect. 4, experimental results of our proposed framework for two-class and multi-class breast abnormality classification from histopathological images have been reported. Finally, Sect. 5 concludes the paper.

2 Related Works

There are a number of deep learning-based models for histopathological image classification tasks to improve early breast abnormality prediction and treatment. Gour et al. [3] proposed a residual learning-based convolutional neural network (CNN) for histopathological images two-class (benign and malignant) classification for breast abnormality detection. Hameed et al. [4] designed an ensemble deep learning framework for histopathological images for two-class classification of carcinoma and non-carcinoma breast cancer. Saxena et al. [5] proposed a hybrid machine learning model for histopathological images two-class (benign and malignant) classification using pre-trained ResNet50 and the kernelized weighted extreme learning. Senousy et al. [6] proposed a CNN-based dynamic deep learning ensemble model for breast histopathological images multi-class classification. Ahmad et al. [7] proposed transfer learning with patch-selection approach for breast cancer multi-class classification. Atban et al. [8] proposed optimized deep features-based framework using ResNet18 for feature extraction and meta-heuristic algorithms for more representative features selection for breast cancer histopathological image classification. Addo et al. [9] proposed a deep learning-based hybrid framework for breast cancer histopathological image two-class classification. Wang et al. [10] proposed a sample clustering domain-adaptation model to learn the semantic correlations for breast cancer histopathological image classification. Oyelade et al. [11] proposed twin CNN model with hybrid binary optimizer for multimodal breast cancer images (i.e. histopathology and mammography) multi-class classification. From our literature study, we conclude that there is a very small amount of work done on handcrafted features extraction for breast cancer histopathological images multi-class classification. Most of the work focuses on deep learning-based black box architecture for features extraction and selection tasks which is difficult to interpret, also the extracted features are not able to be validated by human experts. In this paper, we extract handcrafted features from histopathological images and train a DNN classifier using our extracted handcrafted features to analyse the effectiveness of handcrafted features selection in the medical domain for breast abnormality prediction with features interpretation by human experts.

3 Methodology

The overall diagrammatic representation of our proposed framework for breast histopathological image classification for cancer abnormality prediction is depicted in Fig. 1. The sequence of operations involved in our proposed framework for histopathological image classification task is described in the below subsections.

3.1 Handcrafted Feature Extraction

Handcrafted features also called as hand-engineered features are specific attributes or characteristics derived from given input data or images through manual design algorithm rather than automated machine learning-based algorithm. In machine learning-based feature extraction algorithms, features are the individual measurable properties of the input data that are used to make a prediction or classification result by the algorithm. In handcrafted feature extraction, features are extracted by human experts based on their understanding and knowledge of the input data and the given problem. These handcrafted features are designed to capture relevant information

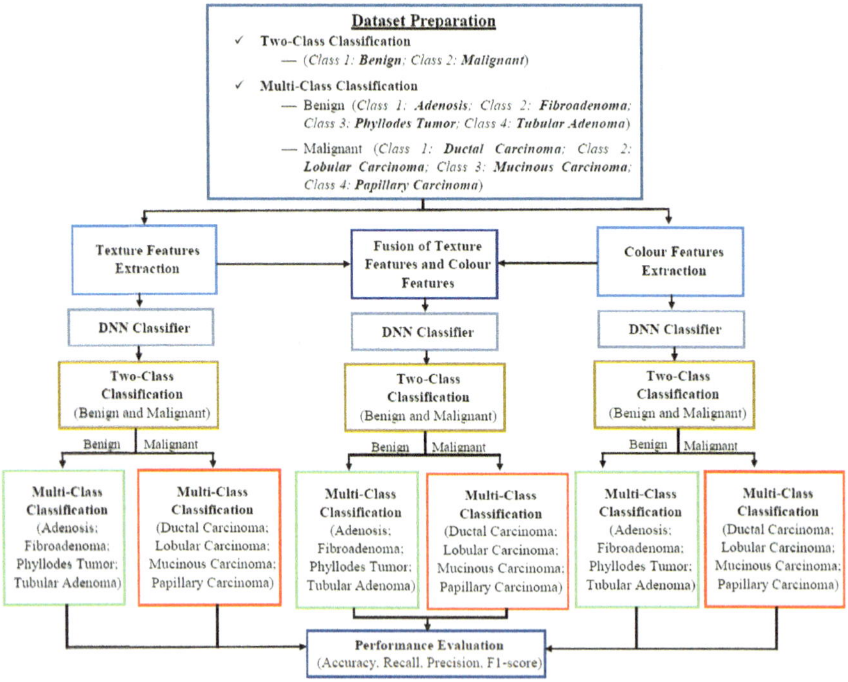

Fig. 1 Overview of the proposed framework for breast histopathological image classification for cancer abnormality prediction

to be able to make a prediction or classification decision. In our proposed work, we extract relevant handcrafted features from histopathological images and then classify extracted features using a DNN classifier. In handcrafted feature extraction, we extract different types of handcrafted features that are texture features, colour features, and fusion of texture and colour features from the histopathological image for two-class and multi-class classifications as described below.

Texture Feature. Texture is the visual pattern of an image, described by elements such as regularity, roughness or smoothness. It can be used in various image processing tasks such as image segmentation, detection and classification tasks. In this paper, we extract Haralick texture features [12] extracted from the Gray Level Co-occurrence Matrix (GLCM) of the histopathological images. These features include contrast, correlation, energy, homogeneity and entropy for classification task.

Contrast. Measure the intensity difference between neighbour pixels in an image. When there is no variation in pixel intensity values then the image contrast is 0. The definition of an image's contrast [13] feature is given by Eq. (1).

$$Contrast = \sum_{x}\sum_{y}(x - y)^2 I(x, y) \tag{1}$$

where $I(x, y)$ is the co-occurrence image matrix of relative frequencies with pixel x and y.

Correlation. Measure how the neighbour pixel is correlated over the whole image. The definition of an image's correlation [13] Feature is given by Eq. (2).

$$Correlation = \sum_{x}\sum_{y}\frac{(x - \mu_x)(y - \mu_y)I(x, y)}{\theta_x\theta_y} \tag{2}$$

where θ_x, θ_y and μ_x, μ_y are the means and standard deviations of an image $I(x, y)$.

Energy. Energy calculates the sum of squared pixel elements of an image. The degree of pixel pair repeats is measured by the energy [13] characteristic. It is a measurement of an image's textural disorder. The energy value of highly linked pixels is very high. The definition of an image's energy feature is given by Eq. (3).

$$Energy = \sum_{x}\sum_{y}[I(x, y)]^2 \tag{3}$$

where $I(x, y)$ is the co-occurrence image matrix of relative frequencies with pixel x and y.

Homogeneity. Inverse dissimilar movement is the usual term used to describe the homogeneity [13]. Greater values correspond to a lower grey tone difference between two objects when measuring image homogeneity. As contrast rises with constant energy, homogeneity falls. The definition of an image's homogeneity feature is given by Eq. (4).

$$Homogeneity = \sum_x \sum_y \frac{1}{1 + |x - y|^2} I(x, y) \tag{4}$$

where *I(x, y)* is the co-occurrence image matrix of relative frequencies with pixel *x* and *y*.

Entropy. The degree of regularity between pixels in an image is measured by the entropy [13] feature, whereas the texture of the image is described by randomness. Energy and entropy has a strong but negative relationship; images with higher grey levels have higher entropy. The definition of an image's entropy feature is given by Eq. (5).

$$Entropy = \sum_x \sum_y I(x, y) \log(I(x, y)) \tag{5}$$

where *I(x, y)* is the co-occurrence image matrix of relative frequencies with pixel *x* and *y*.

Colour Feature. Colour features are used to characterize the colour distribution of an image. It can be described in terms of 3 main attributes that are brightness, saturation and hue. In our work, we extract colour features for histopathological image two_class and multi-class classifications. In this paper, we extract histopathological images colour features that are brightness, saturation and hue for classification tasks.

Brightness. The average V channel of the HSV (i.e. Hue Saturation Value) colour space can be used to compute image brightness. By taking the mean of the V channel, which provides information on the image's intensity, the brightness of the image may be determined.

Saturation. The intensity or purity of a colour in an image is referred to as saturation. The colours are brighter and powerful when the saturation level is higher, and more muted and subdued when the saturation level is lower. The definition of an image's saturation feature is given by Eq. (6).

$$\begin{aligned} Saturation &= (\max - min)/(\max + min);\ When\, Luminosity < 0.5 \\ Saturation &= (\max - min)/(2 - \max - min);\ When\, Luminosity > 0.5 \end{aligned} \tag{6}$$

where Luminosity of a pixel is the range between the minimum (*min*) and maximum (*max*) values of Red, Green and Blue channels.

Hue. When Red, Green and Blue channels are equal then hue value is 0. Otherwise a value between 0 and 255, with the range of 0 to 255 being split into 3 strips for Green channel to Blue channel, Blue channel to Red channel and Red channel to Green channel gradients.

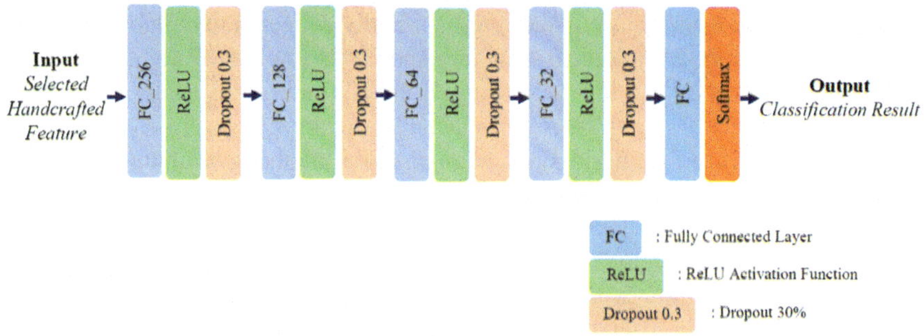

Fig. 2 Proposed DNN classifier for histopathological image extracted handcrafted features classification

3.2 Deep Neural Network for Classification

In recent years, Deep Neural Network (DNN) classifiers dominate all other traditional feature vector classification techniques [14]. In this paper, extracted handcrafted features (as described in Sect. 3.1) were classified using a DNN classifier which is shown in Fig. 2. Our proposed DNN classifier has an input layer, 4 dense layers (i.e., fully connected layers) have 256, 128, 64 and 32 numbers of neurons, respectively. Adam is used for optimization, each of the layers uses ReLU activation function. The last dense layer has 2 numbers and 4 numbers of neurons for two-class (i.e. benign and malignant) and multi-class (i.e. for individual benign and malignant subclasses) classification tasks respectively and a softmax activation function. Our DNN architecture used 30% dropout to prevent overfitting.

4 Experiments

4.1 Dataset Preparation

Used Dataset. For comparative analysis, we used openly accessible breast cancer histopathological image dataset, i.e. BreakHis [15] dataset which contains 7909 number of images in 4 different magnification factors (i.e. 40X, 100X, 200X and 400X) are in.PNG file format. These images mainly belong to 2 categories, namely, benign and malignant which are further sub-categories into 4 classes each namely adenosis (444 images), fibroadenoma (1014 images), phyllodes_tumour (453 images), tubu lar_adenoma (569 images), ductal_carcinoma (3451 images), lobular_ carcinoma (626 images), mucinous_carcinoma (792 images), papillary_carcinoma (560 images), respectively. All the images are in RGB colour format with 700 × 460 image resolution. For histopathological images collection in this dataset, breast tissues were collected from 82 patients which were processed and stained using

H&E staining method. These images were basically collected in collaboration with the P&D Laboratory Pathological Anatomy and Cytopathology, Parana, Brazil.

Data Augmentation and Pre-processing. BreakHis dataset has a very low number of histopathological images for multi-class classification task. To overcome this problem, we utilized 4 basic data augmentation methods without hampering histopathological image properties which are flipping (horizontal and vertical), rotation (45°, 75°, 110°, 145°, 180°) and random cropping to generate new images. After data augmentation, we generated 63,272 new images. For class balancing, we select the smallest class volume which is 444 number of images and add 3552 number of additional images to make its total image volume 3996 (444 randomly selected original images + 3552 augmented images). Similarly, we add require number of additional augmented images with original images in each class to make balance classes with image volume 3996. Thus, in our final dataset we have total of 31,968 images (i.e. 3996 images per class × 8), where 24,059 images are additional augmented images. From where 90% images are used for feature extraction and 10% images are used for testing and validation tasks.

4.2 *Parameter Configuration*

For two-class classification, we used our augmented BreakHis dataset into 2 classes, i.e. benign (non-cancerous tumour) and malignant (cancerous tumour) at 4 different magnification factors (i.e. 40X, 100X, 200X and 400X). For multi-class classification, we divided benign and malignant classes into 4 different sub-class, i.e. adenosis, fibroadenoma, phyllodes_tumour, tubular_adenoma for benign histopathological images and ductal_carcinoma, lobular_carcinoma, mucinous_carcinoma, papillary_carcinoma for malignant histopathological images at 4 different magnification factors (i.e. 40X, 100X, 200X and 400X). In our experimental analysis, we extract texture features and colour features from histopathological images for two-class and multi-class classifications. We also analyse the classification performance by using fusion of texture and colour features. After handcrafted features extraction, we classify the extracted handcrafted features by using a DNN classifier. For DNN classification using extracted handcrafted features, we used 30% dropout for regularization with batch sizes 20, Adam optimizer, 0.0001 learning rate and 1000 epochs. For performance measurement, we used 4 standard evaluation metrics for classification problem such as accuracy, recall, precision and F1-Score. The framework was implemented and tested by a single computing system (Model: DELL Precision Tower 5810) with MATLAB R2022a.

4.3 Results and Discussion

Table 1 shows experimental results of two-class classification using only the texture features. For two-class classification using extracted texture features, we achieved 82% accuracy, 88% recall, 94% precision and 93% F1-score. Table 2 shows experimental results of two-class classification using only the colour features. For two-class classification using extracted colour features, we achieved 32% accuracy, 41% recall, 39% precision and 43% F1-score which is very less compared to the texture features-based classification. We also analyse the two-class classification performance using fusion of texture and colour features as shown in Table 3. For two-class classification using fusion of texture and colour features, we achieved 35% accuracy, 35% recall, 38% precision and 39% F1-score. By fusion of texture and colour features which also shows decreasing classification performance. Similarly, we performed multi-class classification using extracted texture features, colour features, and fusion of texture and colour features. Table 4 shows experimental results of multi-class classification using only the texture features. For multi-class classification using extracted texture features, we achieved average of 73% accuracy, 79% recall, 81% precision and 82% F1-score. For multi-class classification using extracted colour features, we achieved average 21% accuracy, 32% recall, 26% precision and 30% F1-score. For multi-class classification using fusion of texture and colour features, we achieved average of 23% accuracy, 24% recall, 25% precision and 27% F1-score. In multi-class classification, fusion of texture and colour features also shows decreasing classification performance. Due to the brevity of space, we only show texture features based on multi-class classification in Table 4 which comparatively achieved better prediction performance as compared to colour features and fusion of texture and colour features-based classifications. Consequently, graphical illustrations of the experimental results for two-class and multi-class classifications are shown in Fig. 3. It is clearly observed from Fig. 3 that in both the cases of two-class and multi-class problems, individual texture features have been observed to perform better as compared to other considered individual and combined feature sets for breast abnormality prediction from the histopathological images.

5 Conclusion

Breast cancer is a type of cancer that forms in the cells of the breast and is mostly found to occur among women worldwide. Early diagnosis of the disease can reduce the mortality rate. Among various imaging modalities, histopathological images can well predict the abnormality with high true positive rate. In this paper, a hybrid framework for breast histopathological image classification is proposed towards predicting the breast abnormality. The proposed framework uses traditional handcrafted features (i.e. texture features, colour features, and fusion of texture and colour features) and a DNN classifier for classification in order to facilitate the screening process for

Table 1 Experimental results of histopathological image for two-class classification using texture features

Handcrafted feature name	MF	Class type	Accuracy	Recall	Precision	F1-score
Texture feature	40X	Benign	0.88	0.87	0.95	0.94
		Malignant	0.90	0.91	0.99	0.95
		Avg. Metrics	**0.89**	**0.89**	**0.97**	**0.94**
	100X	Benign	0.94	0.94	0.96	0.97
		Malignant	0.93	0.95	0.98	0.96
		Avg. Metrics	**0.93**	**0.94**	**0.97**	**0.96**
	200X	Benign	0.92	0.91	0.99	0.96
		Malignant	0.79	0.84	0.89	0.88
		Avg. Metrics	**0.85**	**0.87**	**0.94**	**0.92**
	400X	Benign	0.89	0.87	0.96	0.94
		Malignant	0.76	0.78	0.85	0.86
		Avg. Metrics	**0.82**	**0.82**	**0.90**	**0.90**
	Aggregate Avg. Metrics		**0.82**	**0.88**	**0.94**	**0.93**

* **MF:** Magnification Factor; **Avg. Metrics:** Average Metrics

Table 2 Experimental results of histopathological image for two-class classification using colour features

Handcrafted feature name	MF	Class type	Accuracy	Recall	Precision	f1-score
Colour feature	40X	Benign	0.25	0.35	0.46	0.47
		Malignant	0.31	0.39	0.43	0.46
		Avg. Metrics	**0.28**	**0.37**	**0.44**	**0.46**
	100X	Benign	0.38	0.54	0.49	0.56
		Malignant	0.35	0.53	0.42	0.43
		Avg. Metrics	**0.36**	**0.53**	**0.45**	**0.49**
	200X	Benign	0.28	0.45	0.32	0.40
		Malignant	0.34	0.42	0.41	0.45
		Avg. Metrics	**0.31**	**0.43**	**0.36**	**0.42**
	400X	Benign	0.35	0.34	0.33	0.37
		Malignant	0.31	0.28	0.34	0.35
		Avg. Metrics	**0.33**	**0.31**	**0.33**	**0.36**
	Aggregate Avg. Metrics		**0.32**	**0.41**	**0.39**	**0.43**

* **MF:** Magnification Factor; **Avg. Metrics:** Average Metrics

Table 3 Experimental results of histopathological image for two-class classification using fusion of texture and colour features

Handcrafted feature name	MF	Class type	Accuracy	Recall	Precision	F1-Score
Texture feature + Colour feature	40X	Benign	0.35	0.36	0.34	0.37
		Malignant	0.38	0.37	0.36	0.40
		Avg. Metrics	**0.36**	**0.36**	**0.35**	**0.38**
	100X	Benign	0.37	0.35	0.40	0.42
		Malignant	0.34	0.38	0.42	0.41
		Avg. Metrics	**0.35**	**0.36**	**0.41**	**0.41**
	200X	Benign	0.30	0.32	0.35	0.38
		Malignant	0.34	0.37	0.40	0.42
		Avg. Metrics	**0.32**	**0.34**	**0.37**	**0.40**
	400X	Benign	0.37	0.38	0.42	0.41
		Malignant	0.38	0.36	0.39	0.40
		Avg. Metrics	**0.37**	**0.37**	**0.40**	**0.40**
	Aggregate Avg. Metrics		**0.35**	**0.35**	**0.38**	**0.39**

* **MF:** Magnification Factor; **Avg. Metrics:** Average Metrics

early diagnosis with expert agreement in feature selection tasks. The DNN classifier is trained using the features that were extracted using the handcrafted methods for two-class and multi-class classifications. The experimental results show that the use of handcrafted texture features had a better prediction performance in histopathological image classification as compared to the other handcrafted features for breast abnormality prediction with an average accuracy of 82% and 73% for two-class and multi-class classification.

Table 4 Experimental results of histopathological image for multi-class classification using texture features

Handcrafted feature name	Category	MF	Class type	Accuracy	Recall	Precision	F1-Score
Texture feature	Benign	40X	Adenosis	0.69	0.84	0.86	0.87
			Fibroadenoma	0.52	0.56	0.68	0.64
			Phyllodes tumour	0.66	0.87	0.88	0.90
			Tubular adenoma	0.80	0.86	0.87	0.89
			Average Metrics	**0.66**	**0.78**	**0.82**	**0.82**
		100X	Adenosis	0.76	0.86	0.87	0.89
			Fibroadenoma	0.58	0.63	0.69	0.71
			Phyllodes tumour	0.71	0.76	0.78	0.79
			Tubular adenoma	0.87	0.88	0.87	0.89
			Average Metrics	**0.73**	**0.78**	**0.80**	**0.82**
		200X	Adenosis	0.78	0.87	0.89	0.88
			Fibroadenoma	0.64	0.68	0.70	0.74
			Phyllodes tumour	0.76	0.79	0.83	0.86
			Tubular adenoma	0.89	0.90	0.91	0.89
			Average Metrics	**0.76**	**0.81**	**0.83**	**0.84**
		400X	Adenosis	0.77	0.85	0.87	0.88
			Fibroadenoma	0.59	0.65	0.68	0.68
			Phyllodes tumour	0.73	0.78	0.83	0.85
			Tubular adenoma	0.85	0.87	0.89	0.90
			Average Metrics	**0.73**	**0.78**	**0.81**	**0.82**
		Aggregate Avg. metrics		**0.72**	**0.78**	**0.81**	**0.82**
	Malignant	40X	Ductal carcinoma	0.76	0.83	0.86	0.87
			Lobular carcinoma	0.89	0.86	0.87	0.88

(continued)

Table 4 (continued)

Handcrafted feature name	Category	MF	Class type	Accuracy	Recall	Precision	F1-Score
			Mucinous carcinoma	0.66	0.78	0.84	0.78
			Papillary carcinoma	0.53	0.67	0.72	0.70
			Average Metrics	**0.71**	**0.78**	**0.82**	**0.80**
		100X	Ductal carcinoma	0.78	0.84	0.87	0.87
			Lobular carcinoma	0.84	0.85	0.86	0.85
			Mucinous carcinoma	0.69	0.76	0.81	0.82
			Papillary carcinoma	0.72	0.83	0.85	0.86
			Average Metrics	**0.75**	**0.82**	**0.84**	**0.85**
		200X	Ductal carcinoma	0.75	0.81	0.82	0.84
			Lobular carcinoma	0.83	0.85	0.84	0.86
			Mucinous carcinoma	0.72	0.78	0.76	0.78
			Papillary carcinoma	0.76	0.84	0.85	0.80
			Average Metrics	**0.76**	**0.82**	**0.81**	**0.82**
		400X	Ductal carcinoma	0.80	0.86	0.85	0.85
			Lobular carcinoma	0.84	0.88	0.86	0.87
			Mucinous carcinoma	0.76	0.79	0.80	0.82
			Papillary carcinoma	0.82	0.83	0.87	0.84
			Average Metrics	**0.80**	**0.84**	**0.84**	**0.84**
		Aggregate Avg. Metrics		**0.75**	**0.81**	**0.82**	**0.82**

* **MF:** Magnification Factor; **Avg. Metrics:** Average Metrics

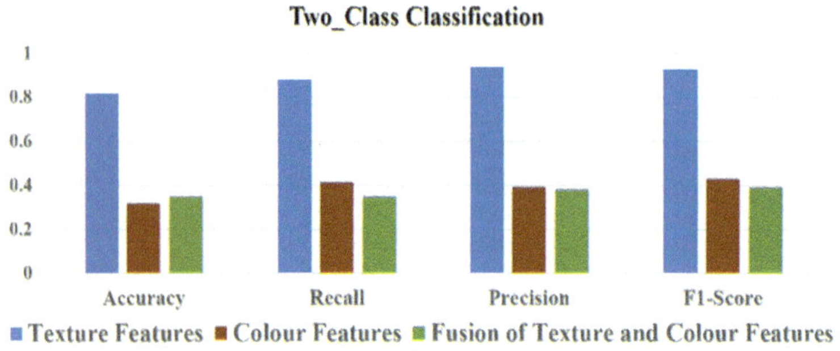

(a)

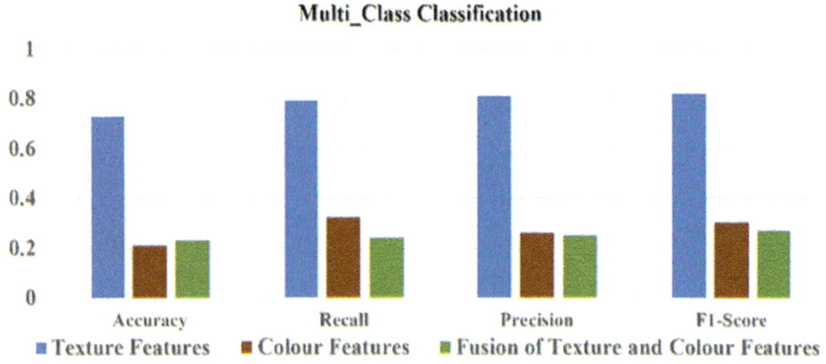

(b)

Fig. 3 Comparison of different types of features classification performance (i.e. texture features, colour features, and fusion of texture and colour features) **a** in two-class classification; and **b** in multi-class classification

Acknowledgments The work presented here is being conducted in the Bio-Medical Infrared Image Processing Laboratory of the Computer Science and Engineering Department, Tripura University, India. The first author is also grateful to All India Council for Technical Education (AICTE), Government of India for providing her AICTE Doctoral Fellowship (Application No. 1638385211866). Support to Mrinal Kanti Bhowmik as J1 Scholar was through Short Term ICMR-DHR International Fellowship, Government of India, Grant Number: INDO/FRC/452/S-69/2019-20-IHD, Dated: 09/12/2019.

References

1. https://www.who.int/news-room/fact-sheets/detail/breast-cancer. Accessed 8 Mar 2024
2. Shahidi F, Daud SM, Abas H, Ahmad NA, Maarop N (2020) Breast cancer classification using deep learning approaches and histopathology image: a comparison study. IEEE Access 8:187531–187552
3. Gour M, Jain S, Sunil Kumar T (2020) Residual learning based CNN for breast cancer histopathological image classification. Int J Imaging Syst Technol 30(3):621–635

4. Hameed Z, Zahia S, Garcia-Zapirain B, Javier Aguirre J, Maria Vanegas A (2020) Breast cancer histopathology image classification using an ensemble of deep learning models. Sensors 20(16):4373
5. Saxena S, Shukla S, Gyanchandani M (2021) Breast cancer histopathology image classification using kernelized weighted extreme learning machine. Int J Imaging Syst Technol 31(1):168–179
6. Senousy Z, Abdelsamea MM, Gaber MM, Abdar M, Acharya UR, Khosravi A, Nahavandi S (2021) MCUa: Multi-level context and uncertainty aware dynamic deep ensemble for breast cancer histology image classification. IEEE Trans Biomed Eng 69(2):818–829
7. Ahmad N, Asghar S, Gillani SA (2022) Transfer learning-assisted multi-resolution breast cancer histopathological images classification. Vis Comput 38(8):2751–2770
8. Atban F, Ekinci E, Garip Z (2023) Traditional machine learning algorithms for breast cancer image classification with optimized deep features. Biomed Signal Process Control 81:104534
9. Addo D, Zhou S, Sarpong K, Nartey OT, Abdullah MA, Ukwuoma CC, Al-antari MA (2024) A hybrid lightweight breast cancer classification framework using the histopathological images. Biocybern Biomed Eng 44(1):31–54
10. Wang P, Yang G, Li Y, Li P, Guo Y, Chen R (2024) Deep sample clustering domain adaptation for breast histopathology image classification. Biomed Signal Process Control 87:105500
11. Oyelade ON, Irunokhai EA, Wang H (2024) A twin convolutional neural network with hybrid binary optimizer for multimodal breast cancer digital image classification. Sci Rep 14(1):692
12. Haralick RM, Shanmugam K, Dinstein IH (1973) Textural features for image classification. IEEE Trans Syst Man Cybern 6:610–621
13. Mall PK, Singh PK Yadav D (2019) GLCM based feature extraction and medical X-RAY image classification using machine learning techniques. In: IEEE conference on information and communication technology. IEEE, pp 1–6
14. Grozdić ĐT, Jovičić ST, Subotić M (2017) Whispered speech recognition using deep denoising autoencoder. Eng Appl Artif Intell 59:15–22
15. Spanhol FA, Oliveira LS, Petitjean C, Heutte L (2015) A dataset for breast cancer histopathological image classification. IEEE Trans Biomed Eng 63(7):1455–1462

FvFc-Net: Forged Video Frame Classification Network

Santanu Das, Sourav Dey Roy, Priya Saha, and Mrinal Kanti Bhowmik

Abstract In the era of rampant digital manipulation, the need for robust forgery detection techniques for verifying the authenticity of the digital contents is imperative. Even though forgery detection in the images has received significant attention, the success in case of video de-forging (i.e., video manipulation detection techniques) has been less explored. This is because there exists a certain ghost phenomenon in the manipulated video that occurs when videos are altered and this makes the anti-forensic measures unsuitable for verifying their authenticity. Ghost phenomena in video manipulation refers to the artifacts created by multiple compression and decompression processes, resulting in faint, residual images, or trails of previous frames appearing in subsequent frames. These artifacts manifest as translucent duplicates or blurs, degrading the quality and fidelity of the video. Despite the lack of appropriate video forensic techniques, the paper highlights and analyzes the prevalence of ghost phenomenon associated with single video de-forging. Our objective is to provide insights into the ghost phenomenon associated with single video de-forging and thereafter propose a novel approach for forgery classification leveraging Deep Convolutional Neural Networks (DCNN). Our proposed method utilizes automatically learn discriminative features from manipulated images, enabling accurate classification of authentic and forged content. Evaluation on our own created dataset and available REWIND dataset showcases the superior performance of our approach, achieving an impressive accuracy of 96.73% and 95.31%, respectively. This indicates the effectiveness of our method in discerning subtle manipulations, making it a valuable contribution to the field of multimedia forensics.

Keywords Multimedia forensics · Video forgery · Splicing · Ghost phenomenon analysis · Convolutional neural network · Classification · Performance evaluation

S. Das · S. D. Roy · M. K. Bhowmik (✉)
Tripura University (A Central University), Suryamaninagar, Agartala 799022, India
e-mail: mrinalkantibhowmik@tripurauniv.ac.in

P. Saha
National Forensic Sciences University (Tripura Campus), Agartala 799006, India

© The Author(s), under exclusive license to Springer Nature Singapore Pte Ltd. 2025

N. Chaki et al. (eds.), *Applied Computing for Software and Smart Systems*, Lecture Notes in Networks and Systems 980, https://doi.org/10.1007/978-981-97-9762-2_9

1 Introduction

In the digital age, the proliferation of image editing software has made it increasingly challenging to distinguish between authentic and forged images [1]. This issue has significant implications across various domains, including journalism, forensics, and cyber security [2]. The emergence of Deep Convolutional Neural Networks (DCNNs) has provided a promising avenue for addressing this challenge by enabling automated image classification with remarkable accuracy and efficiency. Consequently, the ability to discern between authentic and manipulated images is crucial in numerous contexts. In journalism, the dissemination of fake news or manipulated images can mislead the public and erode trust in media outlets. Thus, developing reliable techniques for forged image classification holds immense societal and technological significance. In recent years, significant advancements have been made in forged image classification using DCNNs [3]. Researchers have explored innovative architectures, optimization techniques, and training methodologies to enhance model performance and generalization capabilities. Beyond technical advancements, the ethical and societal implications of forged image classification must also be carefully considered. The potential misuse of such technology for censorship, surveillance, or manipulation underscores the importance of responsible deployment and regulatory frameworks to safeguard against misuse and protect individual privacy and freedom of expression.

Even though the success of the verifying authenticity of the images using conventional detection methods [4–7] is being praised in forensic communities but success of these methods has been observed to be limited for video de-forging techniques (i.e., forensic methods for video manipulation detection). This is because the democratization of video manipulation, while empowering content creators, has also given rise to an alarming trend—the emergence of ghost phenomena within manipulated videos. Ghost phenomena refer to subtle yet perceptible artifacts and inconsistencies that manifest when videos are altered or manipulated. These anomalies, often overlooked by the untrained eye, can range from unnatural lighting and shadows to distortions in motion and background elements. Such imperfections pose a significant challenge, not only for forensic experts and digital media analysts but also for society at large, as they have the potential to mislead, deceive, and manipulate the truth. Depending upon this phenomenon, the primary contribution of this paper is:

1. Firstly, the paper investigates and analyzes the intricate realm of ghost phenomena during the creation of spliced videos. By addressing these challenges, our study contributes to the ongoing efforts in advancing the field of digital forensics and bolstering the reliability of multimedia content in an age where misinformation and manipulated media proliferate.
2. Secondly, to cope up with such challenging conditions of video-based forgery classification task, the paper proposes a novel framework for the classification of forged and authentic frames pertaining to the manipulated videos. The proposed framework combines the CNN-based features from the holistic frames and ELA (Error Level Analysis)-based CNN features for effective classification of forged

and authentic frames present in the video. For an effective forgery classification task, the paper further proposed a novel Forged Video Frame Classification Network (FvFc-Net) using the building blocks of convolutional neural networks (CNNs).

3. Thirdly, the experimental results on our own created dataset and publicly available REWIND dataset have been investigated and showcase the superior performance of our proposed framework for forgery classification task.

Paper Outline. In Sect. 2, review on the existing methods for forgery classification task was elaborated. Section 3 describes and analyzes the observable ghost phenomena of the manipulated videos, which deteriorates the performance of the conventional methods for forgery classification. In Sect. 4, our proposed network for forged frames classification from video has been described. The evaluation results of our proposed network are reported in Sect. 5. Finally, Sect. 6 concludes the paper.

2 Related Work

During the last few decades, various methodologies have been explored to accurately detect forgery within images. Here, we discuss the existing literature and approaches in this domain. Shen et al. (2017) introduced the Textural Features Based on the Grey Level Co-occurrence Matrices (TF-GLCM) method, focusing on texture features and utilizing the CASIA v2.0 dataset for evaluation. Similarly, Jaiswal et al. (2019) proposed a technique employing Resnet-50 with three distinct classifiers (Naïve Bayes, K-Nearest Neighbor, and Multi-Class Model using Support Vector Machine (SVM) Learner), demonstrating their effectiveness on the CASIA v2.0 dataset. Tiwari et al. (2019) adopted a CNN approach coupled with hierarchical agglomerative clustering (HAC) for forged frame classification, with experimentation conducted on the Dresden dataset. Furthermore, Hussien et al. (2020) introduced a method utilizing Deep belief network with Deep Neural Network (DBN-DNN) with Principal component analysis (PCA) on the Columbia Dataset, showcasing an alternative approach to tackle the classification task. In a recent study, Baleanu et al. (2023) proposed the generalized k- symbol Whittaker function (GKSWF) technique integrated with an SVM Classifier for forged frame classification, demonstrating promising results on the CASIA 2.0 dataset. While these existing methods have contributed significantly to the field, there remains room for improvement and exploration of novel techniques. However, the literature review presented in Table 1 indicates a notable deficiency in forgery detection methods, particularly in the domain of video, specifically within the context of forged frame classification. In this paper, to address the gap in the literature we present the Forged Frame Classification using Deep Convolution Neural Network (FvFc-Net), offering a unique approach to effectively detect forged frames within videos.

Table 1 Literature Review on State-of-the-Art Authentic/Forged Classification Methods

Author	Published in, years	Proposed method	Dataset used	Performance and discussion
Shen et al. [8]	IET Image Processing, 2017	TF-GLCM	CASIA v2.0	The technique demonstrated an impressive classification accuracy of 98.54%
Jaiswal et al. [9]	International Conference on Advanced Computing and Software Engineering (ICACSE), 2019	Resnet-50 with three different classifiers Naïve Bayes, K-Nearest Neighbor and Multiclass Model Using SVM Learner	CASIA v2.0	The proposed approach has attained classification accuracy across three distinct classifiers: Naïve Bayes (0.5991), Multiclass Model Using SVM Learner (0.7026), and K-Nearest Neighbor (0.5991)
Tiwari et al. [10]	International Conference, PReMI 2019	CNN with hierarchical agglomerative clustering (HAC)	Dresden	The proposed approach has attained an impressive accuracy rate of 94.90% in classifying forged images
Hussien et al. [11]	International Journal of Sociotechnology and Knowledge Development, 2020	DBN-DNN with PCA	Columbia Dataset	The proposed classifier is capable of determining whether an image is authentic or forged with an impressive classification accuracy of 98.20%
Baleanu et al. [12]	Computers, Materials & Continua, 2023	GKSWF with SVM classifier	CASIA 2.0	The proposed method aims to retain image information in smooth areas while detecting signs of tampering in textural details. The method achieved an impressive accuracy rate of 98.60%

3 Ghost Phenomena Analysis of Manipulated Video

In the realm of literature, numerous conventional forensic techniques [4–7] exist for detecting forgery in images. However, when it comes to videos, the task of forgery detection, especially identifying manipulated areas within video frames, poses a significant challenge. In this section, analysis of the ghost phenomena mostly responsible for deteriorating the performance of the splicing detection methods. The "**Ghost phenomena**" refers to the imperceptible loss or degradation of data that occurs during video editing and video-to-frame conversion processes. This phenomenon is characterized by the subtle disappearance of manipulated information within the multimedia contents, rendering it invisible to the naked eye due to certain level of editing. Despite the apparent visual fidelity, the compression algorithm selectively discards certain details, creating a spectral effect where the lost data becomes elusive or "**ghost-like**".

When we are trying to splice some objects in a video first we have to import two videos into a timeline of video editing tools where all the frames are situated by maintaining a sequence. The jagged diagrammatic representation of the creation of the manipulated video (i.e., object splicing) is demonstrated in Fig. 1. Here, V_1 and V_2 represent the two authentic videos and f_1, $f_2 \ldots f_n$ represent their corresponding frame sequences. Therefore, when manipulation of the video is performed, firstly two videos (V_1 and V_2) need to be imported into the timeline of the video editing tools where all the frames are located by maintaining the sequences as shown in Fig. 1. Then the object of interest from the video frames of V_2 is extracted and placed in the corresponding video frames of V_1. Thereby maintaining a frame sequence that helps to make forged video (S_1) more realistic. Secondly, to make the manipulated video more realistic, we have also adjusted the color balance of the spliced objects according to the background video where the spliced object has been pasted. After the completion of the all manipulation process, the forged video (S_1) is exported in a widely used common video format (i.e., MP4,.AVI, etc.). Therefore, video compression in the context of video editing and export involves two primary types, i.e., intra-frame compression and inter-frame compression [13]. However, to make the applicability of the image-based conventional blind detection methods on the manipulated video, the video is converted into frames. Therefore, multiple times of data loss happen in four different phases to make a spliced/tempered video as shown in Fig. 1, i.e., (1) during video to frame decompression in timeline of video editing tools before performing splicing, (2) intra-frame compression during exporting the manipulated video, (3) intra-frame compression during exporting the manipulated video, and (4) during extracting the frames from the manipulated video for performing the conventional forensic methods.

Due to these multiple compressions as shown in Fig. 1, all the artifacts of the image sequences or frames are suppressed and thereafter the digital forensic methods especially designed for image splicing detection fails to quantify the authenticity of a video. For effective understanding of the existence of the ghost phenomena in the manipulated video, Fig. 2 analyzes the DCT coefficient-based anomaly mask for different quality factors (QF) of manipulated JPEG image and a JPEG frame of a manipulated video thereby maintaining a certain timestamp (i.e., QF = 30, 60, 90, 100). It can be observed from Fig. 2a that in the case of the JPEG image with QF = 100, the anomaly mask (surrounded by respective yellow bounding box) of manipulated region within the holistic JPEG image is prominent and visible thereafter applying DCT. Also, it can be observed from Fig. 2a that as the quality factor of the JPEG compressed image decreases, the visibility of the manipulated region also gradually decreases. Therefore, from the analysis, it can be observed that when the compression is less with high-quality factor, the manipulated part of the image can be easily detected and therefore can generate a mask for spliced objects. Consequently, when the compression is high with low-quality factor, the manipulated part of the image cannot be well detected in the DCT coefficient-based anomaly mask. Conversely, for a JPEG frame extracted from manipulated video with different quality

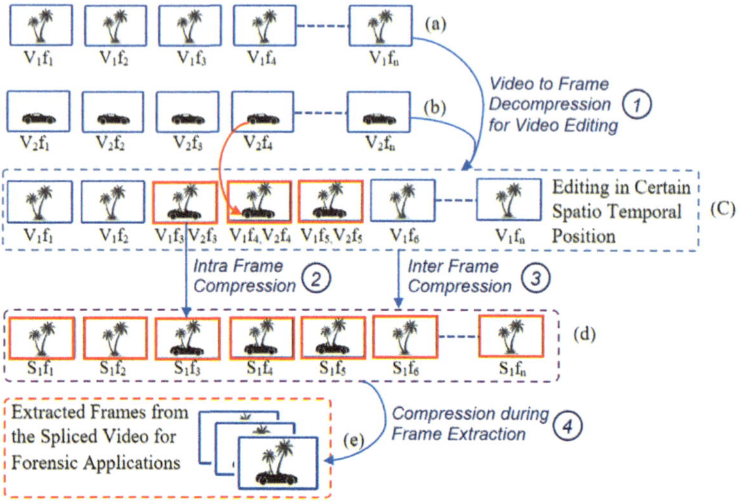

Fig. 1 Overall schematic diagram of creation of forged video (object spliced Video); **a** and **b** are the two different authentic videos; **c** Spliced the object of the second video into the first video; **d** Forged/spliced video; **e** Extracted frames of spliced video

factors as displayed in Fig. 2b, the anomaly mask for the manipulated regions is not prominent for QF = 100 and the same is observed for all the other considered quality factors. This is because multiple compression factors occurred in the spliced frames so as mentioned in Fig. 1. To mitigate this challenge, a novel forgery detection method is needed that can adapt to the semantic features representation of the spliced regions within the frames of the video clips. Depending upon this phenomenon, in this paper, a novel Forged Video Frame Classification Network (FvFc-Net) is proposed to classify the forged frames in the video so as elaborately described in the below sections.

4 Methodology

In this section, our proposed framework incorporating "FvFc-Net (Forged Video Frame Classification Network)" for forgery classification among frames present in the video is elaborately described. The overall diagrammatic representation of the proposed architecture for forgery classification in video is displayed in Fig. 3. The proposed architecture consists of five building blocks: Frame Extraction Layer, Error Level Analysis (ELA) [14], Feature Extraction Layer, Feature Fusion Module, and Classification Layer.

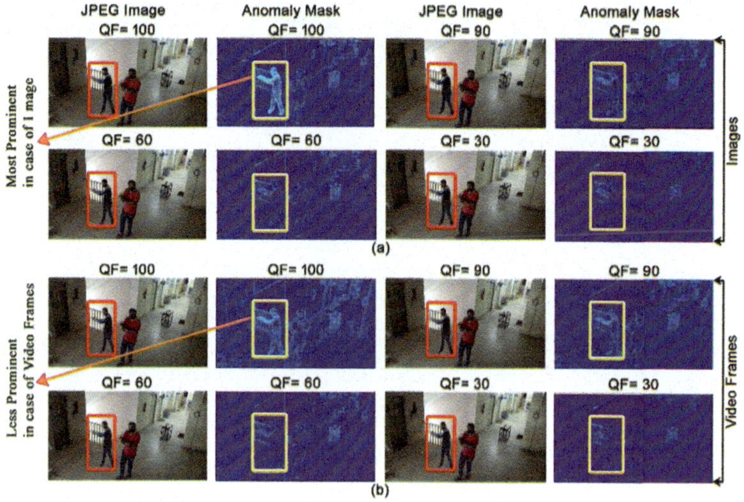

Fig. 2 Analysis of DCT coefficient-based anomaly mask; **a** Object spliced in image; **b** Object spliced in video frames

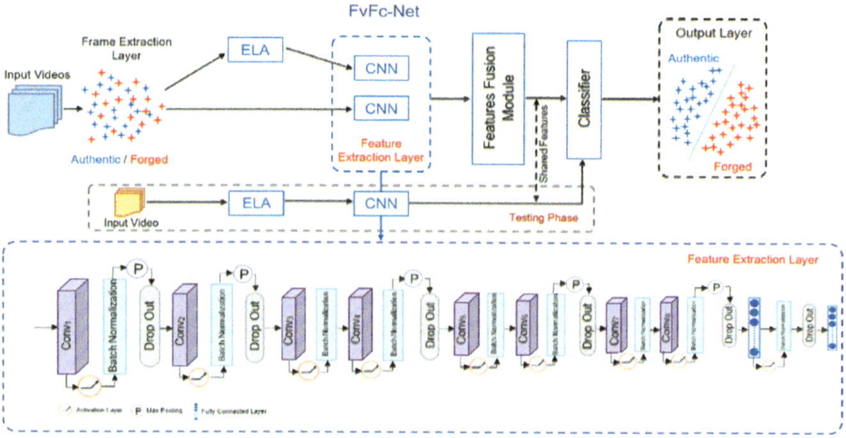

Fig. 3 Proposed framework for forged video frame classification network (FvFc-Net)

4.1 Frame Extraction Layer

In this layer, we have extracted frames I from the input video V to feed it into Deep Convolutional Network (DCNN). Here, I is the set of video Frames where $I = \{I_1, I_2, I_3, \ldots\ldots I_n\}$.

Fig. 4 Some of the output of our proposed FvFc-Net **a** and **b** are the frame sequences of our own designed dataset; **c** and **d** are the Frame sequences of the REWIND Dataset

4.2 *Error Level Analysis (ELA)*

Error Level Analysis (ELA) is a valuable technique for detecting digital image forgeries, particularly in the context of forged frame classification using Convolutional Neural Networks (CNNs). In the proposed work, we leverage the ELA (Error Level Analysis) technique [14], which has traditionally been utilized to identify noise in authentic images concerning their background. According to the literature, original images captured by digital cameras should have high ELA values. However, in the case of each resave (i.e., recompression) of the same image, the potential error rate will decrease. But in the case of tampered/modified (in the case of object/region splicing) images, the modified/spliced region will exhibit higher ELA values compared to the background or authentic regions [15]. When combined with CNNs, it can significantly enhance the accuracy of forgery detection. In the realm of forged frame classification, ELA serves as a preprocessing step to highlight potential areas of manipulation within video frames. By analyzing the error levels across different regions of a frame, ELA provides a visual cue to CNNs, guiding them towards areas that are likely to be tampered with. This approach enhances the CNN's ability to distinguish between authentic and forged frames by focusing its attention on regions where manipulation is more prevalent. Moreover, integrating ELA with CNNs in forged frame classification offers a robust solution capable of detecting various forms of forgery, especially in splicing. The synergy between ELA and CNNs empowers forensic investigators with a powerful toolset for accurately identifying forged frames within digital video content. In our proposed work, the ELA takes the input of I and generates I_e.

4.3 Error Feature Fusion Module (FFM)

In this module, both the feature maps $(F_M I_r, F_M I_e)$ have been concatenated, i.e.,$\prod FF_M = F_M I_r \oplus F_M I_e$. The feature level fusion facilitates the detection of correlated feature values extracted by the feature extraction module, thereby identifying a concise set of salient features that enhance classification accuracy.

4.4 Classification Layer

Finally, the $\prod FF_M$ is propagated through this layer, where the sigmoid function plays a crucial role in classifying each frame into one of two distinct classes: Authentic (Label 0) or Forged (Label 1).

4.5 Architecture Description of Our Proposed FvFc-Net

The Convolutional Neural Network (CNN) stands out as a remarkably efficient method for identification, incorporating an Artificial Neural Network (ANN) structured with multiple sequential layers: convolution layer (conv), pooling layer (pool), and fully connected layers (FC). The CNN takes in the entire video frame as input, and its output layer comprises numerous neurons, each representing a distinct class. In this paper, we have employed the ELA [14] technique to enhance classification accuracy. The block diagram illustrating the proposed DCNN architecture, dubbed "FvFc-NET (Forged Video Frame Classification Network)", for discerning between forged and authentic frames within videos, is depicted in Fig. 3. In designing the FvFc-NET, we adopted two-part approaches within the CNN network. Initially, we feed the input data in its raw form I_r into a CNN to extract features. Subsequently, we perform ELA on the input data I_e and pass it through another CNN to extract additional features. Finally, both sets of features $F_M I_r$, $F_M I_e$ obtained from the Feature Extraction Layer are passed through the Feature Fusion Module (FFM). The FFM will concatenate both the feature maps $F_M I_r$, $F_M I_e$ along the channel dimension, i.e.,$\prod FF_M$, where $\prod FF_M = F_M I_r \oplus F_M I_e$ and forwarded to the classification layer. Here, $\oplus$ represents element wise addition. While testing, $\prod FF_M$ will be shared for better classification. The classification layer evaluates all outputs from the FFM layer, providing a binary classification (i.e., Authentic or Forged). If a significant variance is detected, the frame is classified as forged O_f; otherwise, it is labeled as authentic O_a.

The description of each layers and number of learnable parameters associated with our proposed FvFc-NET architecture for classification of video frames on our created dataset are summarized in Table 2.

Table 2 Description of the FvFc-NET architecture for classifying forged frames in videos, presented layer by layer

Layer type	Input	Kernel size	Number of filters	Stride	Number of LPs in each layer
INPUT	224 × 224 × 3	–	–	–	–
CONV$_1$	224 × 224 × 3	3 × 3	32	1	896
RELU$_1$	224 × 224 × 32	–	–	–	0
BN$_1$	224 × 224 × 32	–	–	–	128
POOL$_1$	74 × 74 × 32	2 × 2	–	2	0
DROPOUT (25%)					0
CONV$_2$	74 × 74 × 32	3 × 3	32	1	9248
RELU$_2$	74 × 74 × 32	–	–	–	0
BN$_2$	74 × 74 × 32	–	–	–	128
POOL$_2$	24 × 24 × 32	2 × 2	–	2	0
DROPOUT (25%)					0
CONV$_3$	24 × 24 × 64	3 × 3	64	1	18,496
RELU$_3$	24 × 24 × 64	–	–	–	0
BN$_3$	24 × 24 × 64	–	–	–	128
CONV$_4$	24 × 24 × 64	3 × 3	64	1	36,928
RELU$_4$	24 × 24 × 64	–	–	–	0
BN$_4$	24 × 24 × 64	–	–	–	256
POOL$_3$	12 × 12 × 64	2 × 2	–	2	0
DROPOUT (25%)					
CONV$_5$	12 × 12 × 128	3 × 3	128	1	73,856
RELU$_5$	12 × 12 × 128	–	–	–	
BN$_5$	12 × 12 × 128	–	–	–	512
CONV$_6$	12 × 12 × 128	3 × 3	128	1	147,584
RELU$_6$	12 × 12 × 128	–	–	–	0
BN$_5$	12 × 12 × 128	–	–	–	512
POOL$_4$	6 × 6 × 128	2 × 2	–	2	0
DROPOUT (25%)					
CONV$_7$	6 × 6 × 128	3 × 3	128	1	147,584
RELU$_7$	6 × 6 × 128	–	–	–	0
BN$_6$	6 × 6 × 128				512
CONV$_8$	6 × 6 × 128	3 × 3	128	1	147,584
RELU$_8$	6 × 6 × 128	–	–	–	0
BN$_7$	6 × 6 × 128	–	–	–	512
POOL$_5$	3 × 3 × 128	2 × 2	-	2	0
DROPOUT (50%)					

(continued)

Table 2 (continued)

Layer type	Input	Kernel size	Number of filters	Stride	Number of LPs in each layer
FC_1	200 × 1	–	–	–	230,600
$RELU_9$	200 × 1	–	–	–	0
BN_8	200 × 1	–	–	–	800
DROPOUT (50%)					0
FC_2	2 × 1	–	–	–	402
$RELU_{10}$	2 × 1	–	–	–	0

SIGMOID activation with two classes (2 × 1)
$Conv_n$—nth convolution layer; $RELU_n$—nth activation layer; $Pool_n$—nth pooling layers, FC_n—nth fully connected layer; BN_n—nth batch normalization, LPs—learnable parameters

4.6 Training of Proposed FvFc-Net Architecture

We have implemented our proposed architecture using the Python platform, specifically leveraging the TensorFlow and Keras libraries. This implementation was conducted on a system equipped with an Nvidia TITAN XP GPU and 64 GB of installed RAM. To train the FvFc-Net for forged frame classification, 110,000 frame sequences are arbitrarily selected from own created forged video dataset. In order to produce the forged videos, we have captured videos using a variety of camera models, such as the Nikon D5100 and the FLIR T650sc. These videos contain indoor and outdoor scenes. To bolster our dataset, we've captured videos under both daylight and nighttime conditions, ensuring a more comprehensive and robust collection. The spliced video clips are generated through a multi-step process. Initially, the background is eliminated from the authentic videos. The salient objects are then extracted using the background subtraction method facilitated by the Rotobrush tool [16] within Adobe After Effects 2022 [17]. Finally, these salient objects are seamlessly targeted or "spliced" into other authentic videos, resulting in the creation of the manipulated video clips. To enhance learning, training frames are randomly shuffled before inputting them into the network. The frame sequences are split into training and validation sets in a 4:1 ratio. The learning rate is set at 0.001 and weight decay at 0.0002 based on experimentation. Additionally, random weight initialization is performed. We conducted empirical comparisons of the training and validation accuracy of our proposed architecture, utilizing categorical cross-entropy loss [18], with three commonly employed optimizers: Adam [19], RMSprop (root mean square propagation) [20], and SGD (stochastic gradient descent) [18]. Observations indicate that the Adam optimizer yields the highest learning accuracy. In our proposed approach, we employ a total of 100 epochs with a batch size 32 to minimize the softmax loss. Figure 5a displays the convergence plot of the training process depicting the training and validation loss over epochs of our proposed network.

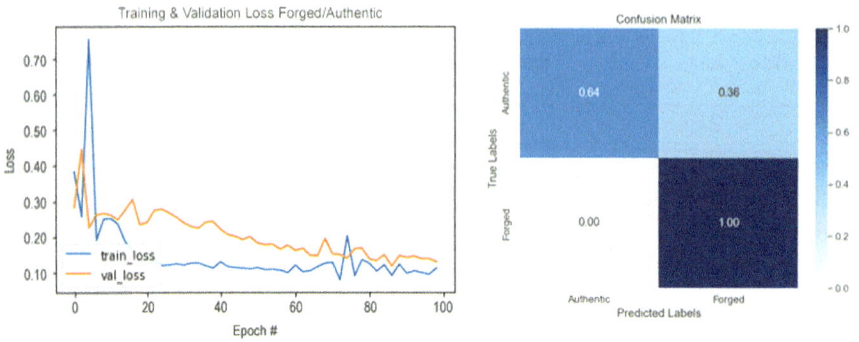

(a) (b)

Fig. 5 **a** Confusion matrix of FvFc-Net; **b** Convergence graph of the proposed FvFc-Net

5 Experimental Results and Discussion

In this section, we demonstrate the performance evaluation of FvFc-NET, for classifying video frame sequences into authentic and forged frames. For quantitative evaluation of our proposed model, our model has been tested with another set of frame sequences from our own designed dataset those are not used for training our model. Consequently, we also tested our proposed model (FvFc-NET) on publically available REWIND dataset [21]. Figure 5b displays the confusion matrix of the proposed model in terms of classification accuracy for forgery classification task on the used datasets. Therefore, this plot offers valuable insights into the model's ability to correctly classify instances across different classes, aiding in the evaluation of its overall performance. To assess the efficacy of our proposed model, we conducted a comparative analysis with state-of-the-art pre-trained CNN models. Numerous pre-trained CNN models have been released to the public, accompanied by their learned kernels and weights, specifically designed for the ImageNet challenge dataset [22]. In our study, we utilize ten renowned and extensively employed pre-trained CNN models sourced from literature: VGG-16 [23], VGG-19 [23], AlexNet [24], Inception-V3 [25], GoogleNet [26], Resnet-101 [27], Resnet-50 [27], Resnet-18 [27], Densenet-201 [28] and Inception-Resnet-V2 [29]. To evaluate the effectiveness of our proposed model (FvFc-NET) with state of the art models, we employed the pre-trained CNNs mentioned earlier, utilizing them as both a fixed feature extraction module (CNN_{FFE}) and a transfer learning module (CNN_{TLM}). In our work, we employ a fixed feature extraction approach by removing the fully connected layers from the CNNs while preserving the rest of the network architecture. Following the feature extraction process using these pre-trained networks, we integrate support vector machines (SVM) with various kernels and employ k-fold cross-validation on the fixed feature extractor. This results in the classification of authentic or forged frames within our own designed dataset. For evaluating the performance of (CNN_{FFE}), similar 10,000 frames of our dataset used for testing our proposed CNN models are used. The performance of the state of the art CNNs against the best-performing kernel (i.e., SVM-Linear) is provided in Fig. 6a. In our comparative study, we also have used transfer learning module on the pre-trained CNN networks (as shown in Fig. 6b). Here, we initially train the base networks using a base dataset, such as the ImageNet challenge dataset. Subsequently, the generic features acquired from this extensive dataset, represented by pre-defined weights, are transferred to the target network while preserving the entirety of the network structure. This enables training these models on our own dataset. For transfer learning with pre-trained CNN models, we leveraged a subset of 100,000 frames from our dataset, which corresponds to the data used in training our proposed model. Meanwhile, for evaluating the performance of (CNN_{TLM}), we have used similar test set of 10,000 frames from our dataset used for testing our proposed model. The effectiveness of both the pre-trained models and our novel model is assessed by examining accuracy, sensitivity, and specificity, as depicted in Fig. 6. In case of CNN as a fixed feature extraction module (CNN_{FFE}), it can be observed that VGG-19 [23] outperforms remaining nine pre-trained CNN models with an average accuracy, specificity, and sensitivity of 82.70%, 81.34%, and 83.43%. respectively. Another interesting observation can be clearly observed from Fig. 5 that the use of pre-trained CNN models as a transfer learning module (CNN_{TLM}) enhances the Forged Frame classification performance in all of the specific cases as compared to the pre-trained CNN models as a fixed feature extraction module (CNN_{FFE}). In this case, VGG-16 [23] outperforms with an average accuracy, specificity, and sensitivity of 84.85%, 86.16%, and 83.22%, respectively. In comparison, we can observe from Fig. 6 that our proposed CNN model enhances the classification performance of forged frame classification with an average testing accuracy, specificity, and sensitivity of 96.73%, 89.58%, and 87.31%, respectively. Moreover, in order to assess the efficacy of the proposed network's performance, we conducted a comparative analysis of its classification capabilities against those of state-of-the-art methods employed for similar tasks on our dataset. The compared methods are TF-GLCM [8], HAC [10], DBN-DNN + PCA [11], and GKSWF + SVM [12]. The prediction performance of the proposed network and the existing methods on our own dataset is provided in Table 3 in terms of average accuracy. It can be observed from Table 3 that as compared to the state-of-the-art techniques, the method GKSWF + SVM [12] performed well on our created dataset with an average accuracy of 95.87%. Moreover, it can also be concluded that FvFc-Net on our proposed framework has superior performance for classification of forged and authentic frames with an average accuracy of 96.73%. Additionally, our proposed CNN model is also tested on REWIND Dataset [21]. The proposed framework is able to predict the class of video frames with an average accuracy of 95.31% on REWIND dataset.

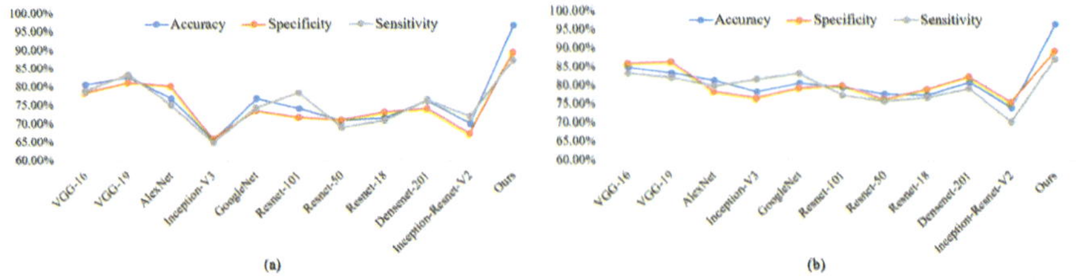

Fig. 6 Performance of our proposed (FvFc-Net) in comparison to the state-of-the-art pre-trained CNNs for classification of forged video frames on our own created Dataset. **a** Pre-Trained CNN as a fixed feature extraction (CNN_{FFE}); **b** Pre-trained CNN with transfer learning module (CNN_{TLM})

Table 3 Accuracy comparison with other state-of-the-art techniques

State-of-the-art technique	Accuracy (in %)
Our proposed framework	**96.73**
TF-GLCM [8]	95.21
HAC [10]	93.51
DBN-DNN + PCA [11]	94.52
GKSWF + SVM [12]	95.87

6 Conclusion

In this paper, we delve into the occurrence of ghost artifacts in manipulated videos and assess their impact on the reliability of conventional forensic techniques for verifying video authenticity. Additionally, we introduce a novel Forged Video Frame Classification Network (FvFc-NET) for classifying forged and authentic frames within videos. This framework integrates a backbone DCNN network with both a Feature Fusion Module (FFM) and an ELA Analysis module. Through rigorous experimentation on our own designed dataset and publically available REWIND dataset, we demonstrate the efficacy of our proposed FvFc-NET method with respect to the existing state-of-the-art methodologies. In the future, we will refine or fine-tune the proposed model to perform well on forged images or videos generated using generative models. Moreover, a key limitation of our approach is the forgery classification task in multiple compression settings which occurs in the videos compressed under various social media platforms and is an example of social media laundering in the domain of multimedia forensics. Therefore, generalization of the proposed model with respect to different compression settings will play a significant role to meet up the forgery classification task and make it applicable for various real-world forensics applications and will be extended in our future work.

Acknowledgements The work is supported by DST SERB International Research Experience (SIRE) Fellowship awarded to Mrinal Kanti Bhowmik for the year 2022–2023 under the Grant Number SIR/2022/000387, Dated: 12/05/2022 from Science and Engineering Research Board (SERB), Department of Science and Technology, Government of India as a visiting Research scholar

(Post Ph.D. Experience) at the NYU Center for Cybersecurity (CCS), Department of Computer Science and Engineering, Tandon School of Engineering, New York University, New York City, USA. Mrinal Kanti Bhowmik would also like to thank his mentor/supervisor, Prof. Nasir Memon, Emeritus Professor of New York University, USA for providing him infrastructural and research support during the Post-Ph.D. SERB International Research Experience.

References

1. Sharma P, Kumar M, Sharma H (2023) Comprehensive analyses of image forgery detection methods from traditional to deep learning approaches: an evaluation. Multimed Tools Appl 82(12):18117–18150
2. Alghamdi MI (2021) Digital forensics in cyber security—recent trends, threats, and opportunities. In: Cybersecurity threats with new perspectives
3. Rawat W, Wang Z (2017) Deep convolutional neural networks for image classification: a comprehensive review. Neural Comput 29(9):2352–2449
4. Hakimi F, Hariri M, GharehBaghi F (2015) Image splicing forgery detection using local binary pattern and discrete wavelet transform. In: 2nd International conference on knowledge-based engineering and innovation (KBEI). IEEE, pp 1074–1077
5. Kaur G, Khehra BS (2016) An efficient approach for digital image splicing detection using adaptive SVM. Int J Comput Sci Inf Secur 14(6):168
6. Liu Q, Sung AH (2009) A new approach for JPEG resize and image splicing detection. In: Proceedings of the First ACM workshop on multimedia in forensics. ACM, New York, pp 43–48
7. He Z, Lu W, Sun W, Huang J (2012) Digital image splicing detection based on Markov features in DCT and DWT domain. Pattern Recogn 45(12):4292–4299
8. Shen X, Shi Z, Chen H (2017) Splicing image forgery detection using textural features based on the grey level co-occurrence matrices. IET Image Proc 11(1):44–53
9. Jaiswal AK, Srivastava R (2019) Image splicing detection using deep residual network. In: Proceedings of 2nd international conference on advanced computing and software engineering (ICACSE)
10. Tiwari SK, Mazumdar A, Bora PK (2019) Detection of splicing forgery using CNN-extracted camera-specific features. In: Pattern recognition and machine intelligence: 8th international conference, PReMI 2019, Tezpur, India, 17–20 Dec 2019, Proceedings, Part I. Springer International Publishing, pp 473–481
11. Hussien NY, Mahmoud RO, Zayed HH (2020) Deep learning on digital image splicing detection using CFA artifacts. Int J Sociotechnol Knowl Dev (IJSKD) 12(2):31–44
12. Baleanu D, Al-Shamayleh AS, Ibrahim RW (2023) Image splicing detection using generalized whittaker function descriptor. Comput Mater Contin 75(2):1–2
13. Chen W, Verbist F, Deligiannis N, Schelkens P, Munteanu A (2013) Efficient intra-frame video coding for low resolution wireless visual sensors. In: 18th International conference on digital signal processing (DSP). IEEE, pp 1–6
14. Abd Warif NB, Idris MYI, Wahab AWA, Salleh R (2015) An evaluation of error level analysis in image forensics. In: 5th IEEE international conference on system engineering and technology (ICSET). IEEE, pp 23–28
15. Sudiatmika IBK, Rahman F, Trisno T, Suyoto S (2019) Image forgery detection using error level analysis and deep learning. TELKOMNIKA (Telecommun Comput Electron Control) 17(2):653–659
16. Rotobrush. https://cmsc426.github.io/rotobrush/#ref. Accessed 21 Jan 2013
17. Adobe: Stand out with after effects. https://www.adobe.com/in/products/aftereffects.html. Accessed 6 Sept 2022

18. Sutskever I, Martens J, Dahl G, Hinton G (2013) On the importance of initialization and momentum in deep learning. In: Proceedings of the international conference on machine learning, pp 1139–1147
19. Kingma DP, Ba J (2014) Adam: a method for stochastic optimization. Preprint at arXiv:1412.6980
20. Khandelwal R (2019) Overview of different optimizers for neural networks. https://medium.com/datadriveninvestor/overview-of-different-optimizers-for-neural-networks-e0ed119440c3. Accessed 12 Mar 2023
21. REWIND—Video (2017) Copy-move forgeries dataset. https://sites.google.com/site/rewindpolimi/downloads/datasets/videocopy-move-forgeries-dataset, Accessed 12 Jan 2024
22. Deng J, Dong W, Socher R, Li L, Li K, Fei-Fei L (2009) Imagenet: a large-scale hierarchical image database. In: Proceedings of the IEEE conference on computer vision and pattern recognition. IEEE, pp 248–255
23. Russakovsky O, Deng J, Su H, Krause J, Satheesh S, Ma S, Huang Z, Karpathy A, Khosla A, Bernstein M, Berg AC (2015) Imagenet large scale visual recognition challenge. Int J Comput Vision 115(3):211–252
24. Krizhevsky A, Sutskever I, Hinton GE (2012) ImageNet classification with deep convolutional neural networks. In: Advances in neural information processing systems. ACM, pp 1097–1105
25. Szegedy C, Vanhoucke V, Ioffe S, Shlens J, Wojna Z (2016) Rethinking the inception architecture for computer vision. In: Proceedings of the IEEE conference on computer vision and pattern recognition. IEEE, pp 2818–2826
26. Szegedy C, Liu W, Jia Y, Sermanet P, Reed S, Anguelov D, Erhan D, Vanhoucke V, Rabinovich A (2015) Going deeper with convolutions. In: Proceedings of the IEEE conference on computer vision and pattern recognition. IEEE, pp 1–9
27. He K, Zhang X, Ren S, Sun J (2016) Deep residual learning for image recognition. In: Proceedings of the IEEE conference on computer vision and pattern recognition. IEEE, pp 770–778
28. Huang G, Liu Z, Van Der Maaten L, Weinberger KQ (2017) Densely connected convolutional networks. In: Proceedings of the IEEE conference on computer vision and pattern recognition. IEEE, pp 4700–4708
29. Szegedy C, Ioffe S, Vanhoucke V, Alemi AA (2017) Inception-v4, inception-resnet and the impact of residual connections on learning. In: Thirty-first AAAI conference on artificial intelligence proceedings, ACM, vol 31(1)

Semi-automated Ground Truth Generation System for Bangla Offline Handwritten Text Documents

Rejoy Chakraborty, Payel Rakshit, Chayan Halder, and Kaushik Roy

Abstract In the field of image processing, document-level image processing is one of the trending topics. Various researchers have proposed works on Bangla handwritten documents, and one of them is segmentation and ground truth generation. Segmentation schemes for lines, words, and characters are a crucial preprocessing task in many applications. Segmentation algorithms for Bangla handwritten offline documents that have been proposed so far were unable to achieve significantly high accuracy due to various problems like overlapping of components, cut-through, etc. Fully manual segmentation can produce highly accurate results but it is not feasible for any user. In this paper, a semi-automated system has been proposed for Bangla handwritten documents which can potentially produce fully accurate segmentation results. Furthermore, ground truth generation for words has also been added to the system using the isolated character recognition CNN model.

Keywords Bangla offline handwritten document · Semi-automated system · Segmentation · CNN

1 Introduction

Ground truth data are crucial components for handwritten document recognition like any other recognition task. When it comes to line, word, or character recognition, a preprocessing stage for performing segmentation is required to extract the individual components from the handwritten documents. Hence, the requirement of developing

---

R. Chakraborty
Indian Institute of Technology Ropar, Rupnagar, Punjab, India

P. Rakshit
Maheshtala College, Kolkata, WB, India

C. Halder
Ramakrishna Mission Vivekananda Centenary College, Kolkata, WB, India

K. Roy (✉)
West Bengal State University, Barasat, WB, India
e-mail: kaushik.mrg@gmail.com
© The Author(s), under exclusive license to Springer Nature Singapore Pte Ltd. 2025

N. Chaki et al. (eds.), *Applied Computing for Software and Smart Systems*, Lecture Notes in Networks and Systems 980, https://doi.org/10.1007/978-981-97-9762-2_10

an efficient segmentation algorithm comes to the place. The segmentation algorithms may act differently based on the pattern of the dataset. A dataset that can be curated either online or in an offline approach. In the offline approach, the digitization of the documents is required, and hence scanning of the document is required before performing any operation. But in the case of an online approach, the generated document is itself digital and contains some additional features like pixel coordinates, timestamps of the strokes [13], etc. which in terms leads to provide easy and better results for post-processing like segmentation.

Before digging into the problem, a brief information about Bangla scripts is given. The origin of Bangla is from the Brahmi script. There are in total 11 vowels and 39 consonants. A character may consist of a horizontal line in the upper region, called "Matra" and a consonant can be merged with a vowel in a modified shape. Two compound characters also get merged between them to form a compound character.

To the best of our knowledge, no research work has produced significantly high segmentation accuracy. The major challenge the researchers face in handwritten document processing is the variation of style in writing. However, in some applications, finely segmented components are desirable, like converting handwritten text to digital text. Furthermore, there is no standard ready-to-use ground truth generation system for offline Bangla handwritten text documents.

This paper presents a GUI system that can perform segmentation in a semi-automated fashion by which lines, words, and character components from Bangla offline handwritten text documents can be extracted and can produce ground truth using the characters. The proposed system is easy to use, and robust for any kind of Bangla document. The paper has been divided into several parts. Section 2 provides some of the key literature works proposed so far. Section 3 elaborates on the proposed methodologies of the ground truth generation system and Sect. 4 demonstrates extensive results with analysis. Finally, Sect. 5 concludes the research work by mentioning possible future scopes.

2 Literature Review

Numerous researchers have proposed their work in the field of Bangla handwritten offline text document processing. Furthermore, some researchers have also presented datasets on handwritten documents.

Shapiro et al. [14] proposed a vertical light projection scheme on the text documents for line extraction, but it is limited to constrained handwritten documents only. Extending the segmentation work from ICDAR 2013 [15], Rakshit et al. [10] have presented a connected component-analysis-based approach which is followed by Hough transformation for text line detection in Bangla text documents. Inspired by the projection scheme, Rakshit et al. [9] have presented a generalized line segmentation algorithm that can perform in multi-script handwritten text documents.

They used an approach of horizontal light projection to the line followed by a vertical smoothing technique. They then traced each line to extract it from the document. The authors have also proposed WBSUBNdb_text, a Bangla text dataset.

Louloudis et al. [6] have presented a word segmentation scheme in their research work where inter- or intra-word distances of overlapped components have been calculated in a Gaussian mixture modeling framework. To address the issues with unconstrained Bangla handwritten text documents, Pal et al. [7] proposed a water reservoir-based principle that is forwarded by character segmentation from the word. Sarkar et al. [12] have presented research work for word segmentation and baseline detection for Bangla handwritten text documents. Rakshit et al. have achieved 90.06% accuracy in word segmentation in their proposed work [10] based on connected component analysis and extraction using bounding boxes.

Character segmentation technique for Bangla words has been presented by Roy et al. [11] with a novel approach for skew detection. Bag et al. [2] have presented a character segmentation technique using vertex characterization of isolethic covers for handwritten words. Ahmed et al. [1] have proposed a CNN-based character recognition approach preprocessed by line and character segmentation.

From the above discussion and the best knowledge of ours, no research work has reached to a significant accuracy score in the text line, word, or character segmentation. Hence, a semi-automated system can come to the picture to address this issue. Though Sen et al. [13] have proposed a semi-automated web-based segmentation and ground truth generation system for Bangla online handwritten documents, there is no such semi-automated system for segmentation and ground truth generation for Bangla handwritten offline documents. To fill this gap, a semi-automated approach for offline Bangla handwritten documents has been proposed in this research work.

3 Methodology

3.1 Line Segmentation

For automated line segmentation, a light projection scheme is found to be efficient for the extraction of line components which can be implemented in numerous ways. Motivated by the research works of Rakshit et al. [9, 10], a light projection scheme forwarded by connected component analysis is applied in our proposed system to segment the lines from the documents. Figure 1a shows the result on a sample document on performing connected component analysis.

If a user finds unsatisfactory results in the automated line segmentation algorithm, the user may rectify the result with the optimal number of mouse clicks to generate the desired outcome. An example of an incorrect result using an automated line segmentation algorithm is presented in Fig. 1b. In this instance, it can be observed that the segmentation algorithm failed to split the first two lines and as a consequence

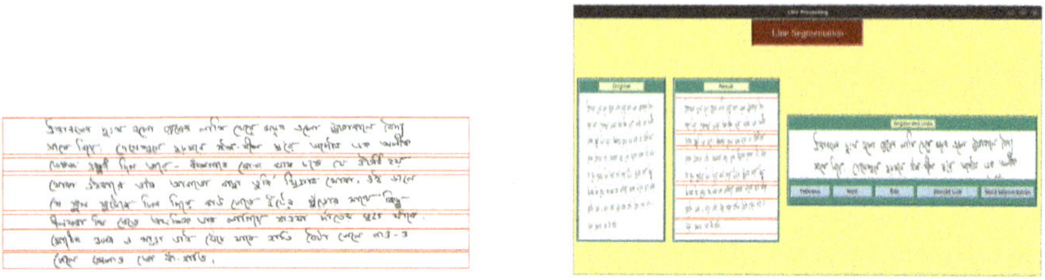

(a) (b)

Fig. 1 Extracted line components after line segmentation

combined two lines as one single line. So, the user may click the "Edit" button of the interface holding the undesired image in Fig. 2.

Once the user reaches the editable window, the user has to pinpoint the places from where the separator needs to be passed. Users may get the desired outcome with an optimal number of clicks only. Figure 2 demonstrates the working process. After a satisfactory number of clicks, the user may split the two lines by clicking "q" and two lines will be separated based on the separator line passing through the selected points. At any point in time, if a user finds any unsatisfactory point selection, pressing "u" can undo the selection of the point. After the rectification, the result will be directly reflected on the interface including updating of the image showing connected components. It will also handle local files and will store them accordingly. The view of the interface after the updation has been shown in Fig. 3.

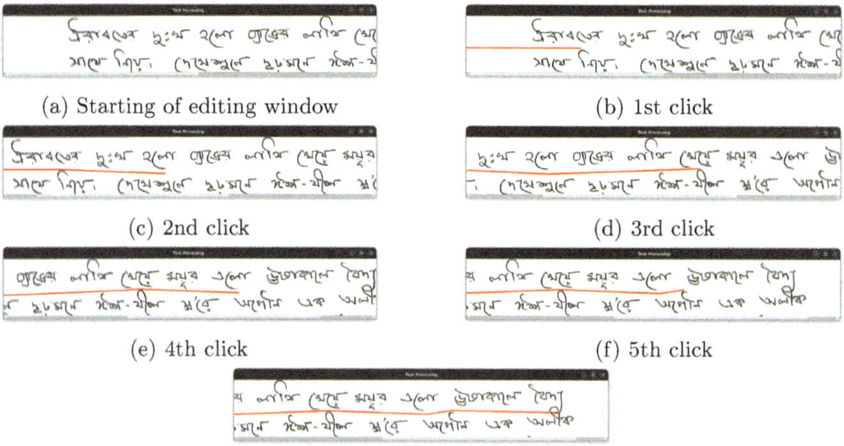

(a) Starting of editing window (b) 1st click

(c) 2nd click (d) 3rd click

(e) 4th click (f) 5th click

(g) Final click

Fig. 2 Demonstration of editing the line-segmented image

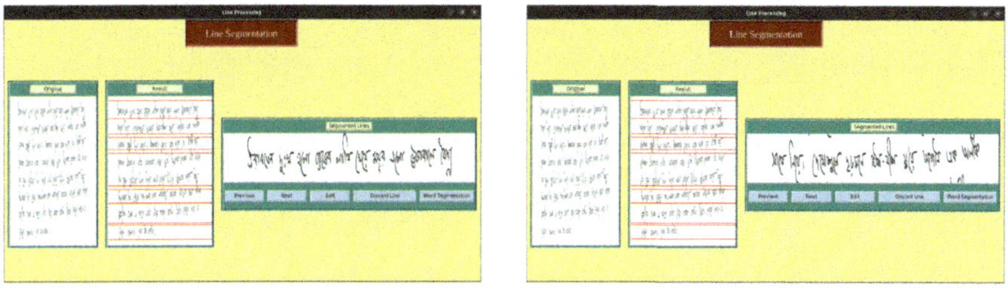

(a) Displaying modified line 1 (b) Displaying modified line 2

Fig. 3 Demonstration of line segmentation window after editing the incorrect image

3.2 Word Segmentation

After the line segmentation, the word segmentation can be performed on the segmented components. Motivated by the research work of Rakshit et al. [10], a word segmentation technique based on connected component analysis has been performed. To handle the inter-gap and intra-gap issues, "Matras" have been identified and removed, by extracting morphological horizontal lines from the upper zone of the components. The 35%–55% of the height of the line components has been experimentally considered as the upper region of the component based on the density of the black pixels. Figure 4 demonstrates the result of the matra removal process. Matra removal is forwarded by finding the connected components using contour tracing. As a preprocessing step, erosion and dilation have been performed for noise removal. In Fig. 5, the result of recognized components has been presented.

For instance, Fig. 6 shows where a user might find an unsatisfactory result of the automated word segmentation algorithm, where the "matra" is slightly truncated, which occurred due to the "matra removal" process.

The user may proceed for rectification to gain satisfactory output. The window displays the whole line component and the user has to make a polygon which will be extracted as a word region. To form a polygon, the user must click a sufficient amount number of times where points will be treated as vertices of the polygon around the word. The same procedure can be applied to add a word component too. Users can hit the "q" key on the keyboard to extract the region using the polygon masking method. There is also a facility to undo a placed vertex of the polygon using the "u" key on

(a) Line before matra removal (b) Line image after matra removal

Fig. 4 Demonstration of matra removal process in a sample line image

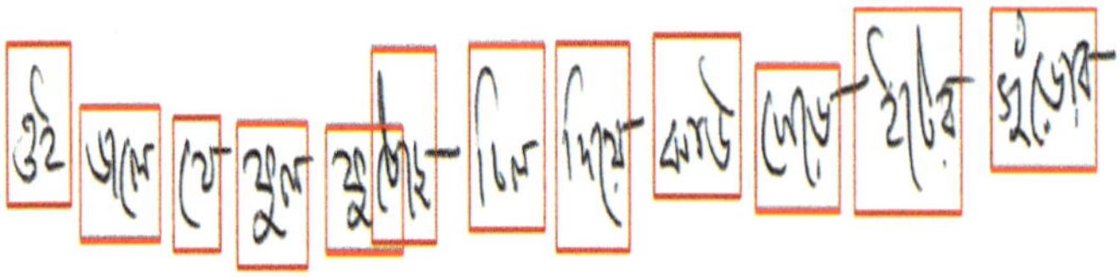

Fig. 5 Bounding boxes of each word component

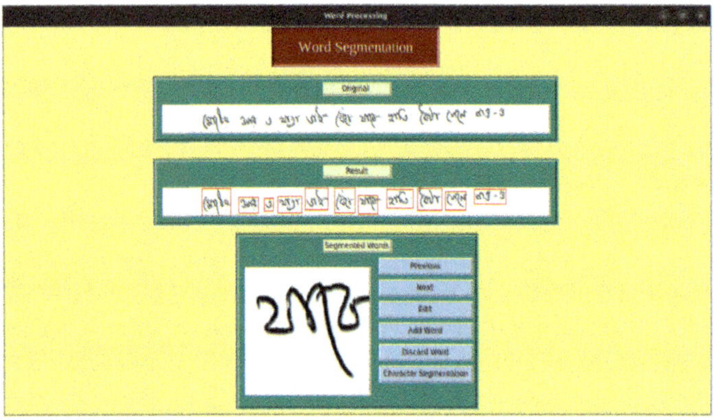

Fig. 6 Example of unsatisfactory word segmentation result using automated algorithm

the keyboard. Figure 7 demonstrates the polygon masking technique to extract the word.

Like the line segmentation scheme, the updated results will also be reflected in the word segmentation window. Figure 8 shows a view of the window after updation. Furthermore, the image showing connected component words will also be updated and actions like add/discard/delete will also act according to the updations.

(a) Starting of editing/adding window

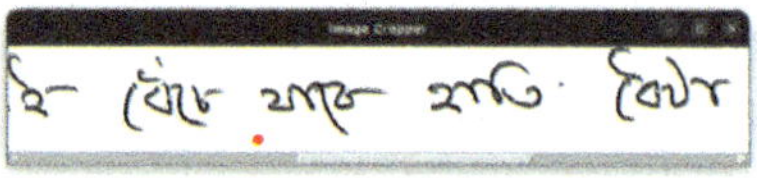

(b) First click

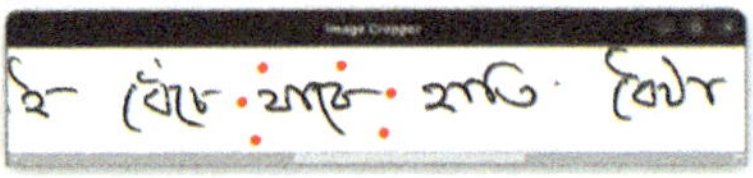

(c) After a few clicks

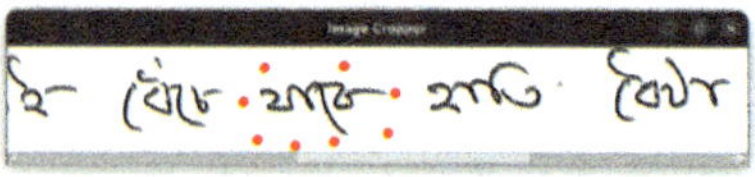

(d) Final click

Fig. 7 Demonstration of editing/adding word segmentation image

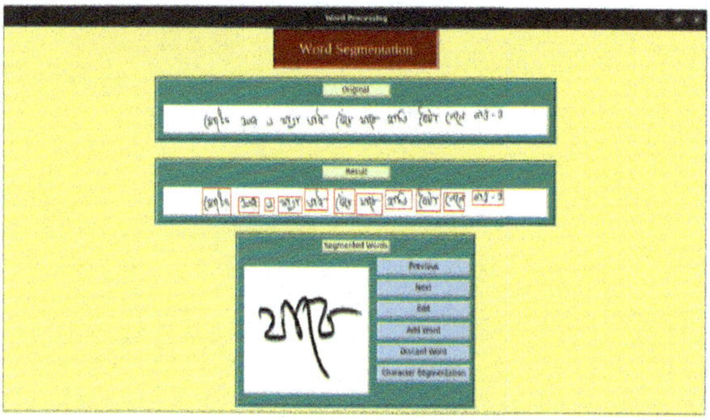

Fig. 8 Demonstration of word segmentation window after editing the unsatisfactory image

3.3 Character Segmentation

In the final phase of segmenting the document into small components, characters/strokes have been extracted from the previously extracted word components. Motivated by the research work of Rakshit et al. [10], a character segmentation algorithm has been implemented. Each word component has been divided into upper, lower, and middle regions and the area of the regions has been computed experimentally based on the black pixel density and height. After the zone calculation, the components are separated. The lower region used to have a higher black pixel density due to the presence of the vowel characters in that region. The characters have been isolated by utilizing the intra-gap between the characters. Figure 9 demonstrates the result of the character segmentation.

The procedure of correction of character segmentation follows the same as the polygon masking technique of rectifying the word components. Figure 10 shows an instance where rectification of the segmented character component might be needed and Fig. 11 demonstrates the polygon masking technique for extracting the character component. The same approach can also be applied to add a new character component. Once the editing/adding of a character is finished, the result will be reflected on the character segmentation window, shown in Fig. 12.

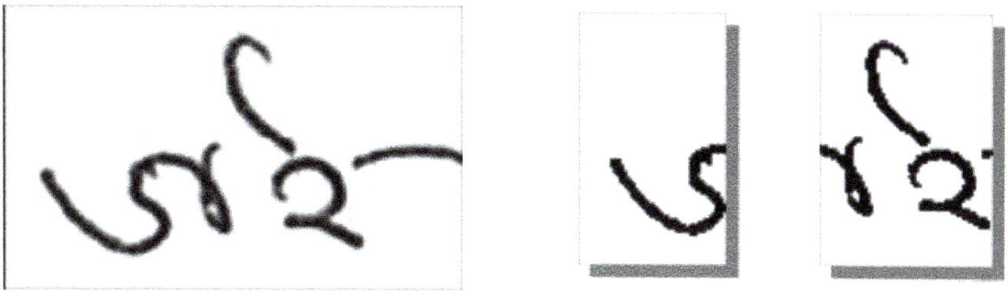

(a) Sample word for segmentation

(b) Segmented characters

Fig. 9 Demonstration of character segmentation

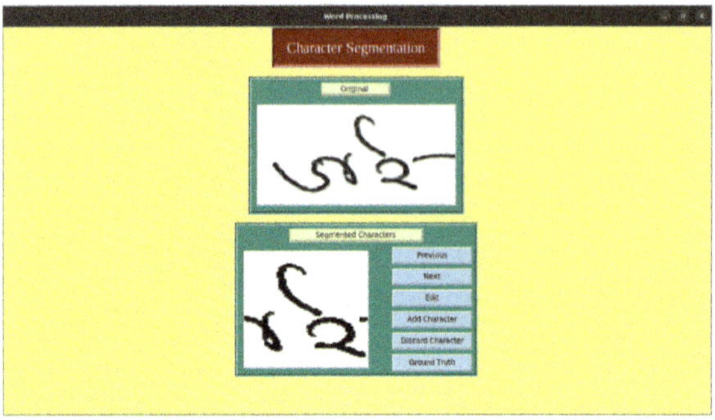

Fig. 10 Example of unsatisfactory character segmentation result using automated algorithm

(a) Starting of editing/adding window

(b) First click

(c) After a few clicks

(d) Final click

Fig. 11 Demonstration of editing/adding character segmentation image

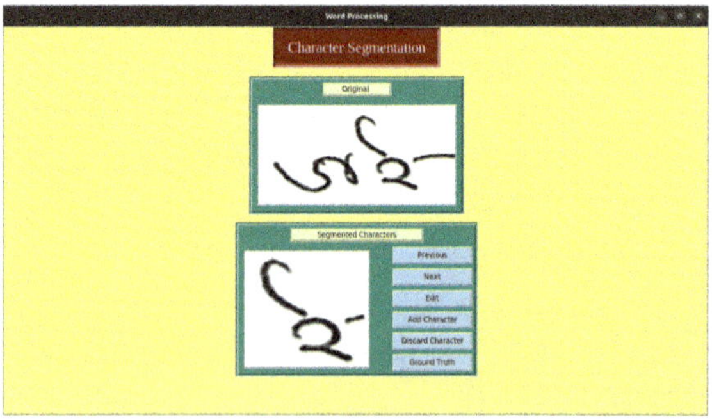

Fig. 12 Demonstration of character segmentation window after editing the unsatisfactory image

3.4 *Ground Truth Generation*

A CNN model has been deployed by considering the standard MobileNet [5] as the base architecture. The model has been trained with a transfer learning approach using the "Ekush" Bangla character dataset [3, 8] and the "MatrrivaSha" Bangla compound character dataset [4]. Just to note, the deployment of the character recognition model is still in progress.

4 Results

The proposed semi-automated system including the segmentation algorithms was evaluated with 50 sample Bangla handwritten document images of WBSUBNdb_text dataset [9]. These evaluations have been done on a system with a specification of 4GB DDR4 RAM and an Intel Dual Core processor.

In Table 1, the result of the automated line segmentation technique has been illustrated with 50 document images. Hence from Table 1, it can be observed that 51 lines need rectification. By experimental observation, if we go for manual segmentation then on average 5 clicks are required for each line. Hence, if we perform fully manual segmentation over the whole document for all these 50 documents then all these 429 lines have to be manually segmented whereas in the semi-automated system, only 51 line needs to go through the handcrafted segmentation. In Table 2, the performance gain on clicks for the sample set of documents has been presented.

Now from all the 429 segmented lines (either processed with fully automatic or semi-automatic mode), a subset of 70 lines has been taken for further evaluation. In Table 3, the performance of the automated word segmentation algorithm on the taken subset of 70 line components has been mentioned.

Table 1 Result of automated line segmentation

# Documents	# Lines	# Lines segmented	Accuracy (%)
50	429	378	88.11

Table 2 Performance gain in semi-automated line segmentation over fully handcrafted line segmentation

Fully manual	Semi-automated	Performance gain
2145	270	7.94

Table 3 Automated word segmentation result

# Lines	# Words	# Words segmented	Accuracy (%)
70	651	595	91.39

Table 4 Performance gain in semi-automated word segmentation over fully manual word segmentation

Fully manual	Semi-automated	Performance gain
6510	560	11.62

Table 5 Result of automated character segmentation

# Words	# Characters	# Characters segmented	Accuracy (%)
100	600	362	60.33

Table 6 Performance gain in semi-automated character segmentation over fully manual character segmentation

Fully manual	Semi-automated	Performance gain
9000	3570	2.52

From Table 3, it can be measured that 56 words require handcrafted segmentation additionally after automated segmentation. It has been experimentally observed that on average 10 clicks are needed to draw the polygon for applying the polygon masking technique to extract the desired word region. Hence, if we perform fully manual handcrafted segmentation to all these 70 lines which contain 651 words then the number of clicks will be massive compared to the semi-automated system. The performance gain on clicks for the sample set of line images is described in Table 4.

From the set of all these 651 segmented words (either processed with fully automatic or semi-automatic mode), a subset of randomly selected 100 words has been taken for evaluation of the automated character segmentation algorithm, and the result of the evaluation has been tabulated in Table 5.

As a result of slightly low accuracy, 238 characters require additional handcrafted segmentation after the automated segmentation. In this scenario, an average of 15 clicks are needed to extract the desired character region using the polygon masking technique. Hence, if we perform fully manual segmentation of all these 100 words that contain 600 characters then the number of clicks will be higher than the semi-automated system, since here we need to take care of those 238 characters. The performance gain on clicks for the sample set of word images is illustrated in Table 6.

The CNN architecture which has been deployed with a base architecture of MobileNet [5] for isolated character recognition has been trained for 100 epochs using the "Adam" optimizer with an initial learning rate of 1.0000^{-9}. After 100

Table 7 Metrics of the CNN model

Training accuracy	Validation accuracy	Training loss	Validation loss
0.9986	0.9639	0.0077	0.9007

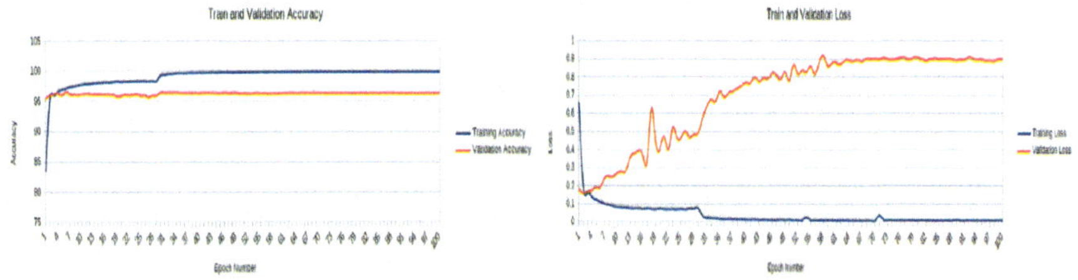

(a) Accuracy (b) Loss

Fig. 13 Accuracy and loss of character recognition CNN model

epochs, the metrics of the model are mentioned in Table 7 mentioning the accuracy and the loss during training and validation. Figure 13 presents the training, validation accuracy, and loss, respectively, corresponding to each epoch. All the evaluations have been done in the "Google Colab" environment utilizing the T4 GPU.

5 Conclusion

In this proposed work, a semi-automated system has been presented that can potentially generate significantly high accuracy in the cost of handcrafted segmentation only for unsatisfactory components. The performance gain and the accuracies indeed imply the effectiveness of the system. Furthermore, the recognition task completes the system's utility and goal to generate ground truth information of a handwritten Bangla document. For future work, the development of a language-agnostic system could be pursued, capable of functioning regardless of the language of the document. Deploying a system like that requires generalized segmentation techniques which can be applicable with any script documents. Our work offers a framework to address this area and assistance to the research community working on Bangla handwritten document images. A request can be made to access the system through the link: https://forms.gle/2jFRNj57f51dXkbT7.

References

1. Ahmed T, Uddin M, Khan M, Rahman A, Hasan A (2022) Offline handwritten character recognition including compound character from scanned document. Asian J Res Comput Sci 14(4):119–129
2. Bag S, Bhowmick P, Harit G, Biswas A (2011) Character segmentation of handwritten bangla text by vertex characterization of isothetic covers. In: 2011 Third national conference on computer vision, pattern recognition, image processing and graphics. IEEE, pp 21–24
3. Ekush: a multipurpose and multitype comprehensive database for online off-line bangla handwritten characters. https://rabby.dev/ekush/#home
4. Ferdous J, Karmaker S, Rabby ASA, Hossain SA (2021) Matrivasha: a multipurpose comprehensive database for bangla handwritten compound characters. In: Emerging technologies in data mining and information security: proceedings of IEMIS 2020, vol 3. Springer, pp 813–821
5. Howard AG, Zhu M, Chen B, Kalenichenko D, Wang W, Weyand T, Andreetto M, Adam H (2017) Mobilenets: efficient convolutional neural networks for mobile vision applications. arXiv preprint arXiv:1704.04861
6. Louloudis G, Gatos B, Pratikakis I, Halatsis C (2009) Text line and word segmentation of handwritten documents. Pattern Recogn 42(12):3169–3183
7. Pal U, Datta S (2003) Segmentation of bangla unconstrained handwritten text. In: Seventh international conference on document analysis and recognition, proceedings vol 3. Citeseer, pp 1128–1128
8. Rabby ASA, Haque S, Islam MS, Abujar S, Hossain SA (2019) Ekush: a multipurpose and multitype comprehensive database for online off-line bangla handwritten characters. In: Recent trends in image processing and pattern recognition: second international conference, RTIP2R 2018, Solapur, India, 21–22 Dec 2018, Revised selected papers, Part III 2. Springer, pp 149–158
9. Rakshit P, Halder C, Sk MO, Roy K (2023) A generalized line segmentation method for multi-script handwritten text documents. Expert Syst Appl 212:118498
10. Rakshit P, Halder C, Ghosh S, Roy K (2018) Line, word, and character segmentation from bangla handwritten text-a precursor toward bangla hocr. In: Advanced computing and systems for security, vol 5, pp 109–120
11. Roy A, Bhowmik TK, Parui SK, Roy U (2005) A novel approach to skew detection and character segmentation for handwritten bangla words. In: Digital image computing: techniques and applications (DICTA'05). IEEE, pp 30–30
12. Sarkar A, Biswas A, Bhowmick P, Bhattacharya BB (2010) Word segmentation and baseline detection in handwritten documents using isothetic covers. In: 12th International conference on frontiers in handwriting recognition. IEEE, pp 445–450
13. Sen S, Bhattacharyya A, Sarkar R, Roy K (2021) Byanjon: a ground truth preparation system for online handwritten bangla documents. Trans Asian Low-Resour Lang Inf Process 20(6):1–16
14. Shapiro V, Gluhchev G, Sgurev V (1993) Handwritten document image segmentation and analysis. Pattern Recogn Lett 14(1):71–78
15. Stamatopoulos N, Gatos B, Louloudis G, Pal U, Alaei A (2013) Icdar 2013 handwriting segmentation contest. In: 12th International conference on document analysis and recognition. IEEE, pp 1402–1406

SWFEM: Sparse Weighted Fine-Tuned Ensemble Model for Leaf Disease Detection

Rakesh Paul, Susobhan Ghosh, Mridul Ghosh, Sk Md Obaidullah, and Kaushik Roy

Abstract Globally, the greatest concerns of farmers are plant diseases. A considerable loss of yield has an immediate impact on the economy. Machine learning models exhibited capabilities to detect plant diseases. To enhance the generalization, flexibility, and adaptability of machine learning models, an ensemble model was introduced to combine many models and produce an appropriate and optimal predictive solution. To achieve better accuracy in deep learning models, it is necessary to have a significant amount of training time and memory resources. This is frequently utilized in engineering fields. Nowadays, the computational burden of training multiple independent networks increases with the growing number of deep learning models. To address these computational challenges, we introduced pruning and tuning-based fast, cost-effective sparse ensemble networks. The popularity of sparse architecture has increased because of its low memory requirements and faster inference rates. To identify an appropriate subset of ensemble members to build a smaller ensemble model, pruning is used to refrain from retraining the model. To achieve this, we first trained a base network. After that, we clone the parent network and drastically reduce each child's features to generate an ensemble of members with various topologies. We soon trained each successor network for a smaller number of epochs. An empirical investigation demonstrated how well our method works on the pruned ensemble model.

R. Paul (✉)
Department of Computer Science, Debra Thana S. K. S Mahavidyalaya, Debra, India
e-mail: rakeshpaul470@gmail.com

S. Ghosh
Department of Computer Science, Dinabandhu Mahavidyalaya, Bongaon, India

M. Ghosh
Department of Computer Science, Shyampur Siddheswari Mahavidyalaya, Howrah, India

S. M. Obaidullah
Department of Computer Science and Engineering, Aliah University, Kolkata, India

K. Roy
Department of Computer Science, West Bengal State University, Barasat, India

© The Author(s), under exclusive license to Springer Nature Singapore Pte Ltd. 2025

N. Chaki et al. (eds.), *Applied Computing for Software and Smart Systems*, Lecture Notes in Networks and Systems 980, https://doi.org/10.1007/978-981-97-9762-2_11

Keywords Deep learning · Sparse fine-tuned ensemble · Agriculture · Leaf disease

1 Introduction

Due to technological advancements, Deep Learning (DL) [1] has achieved great success with automatically extracting features. Deep Neural Networks (DNNs) [2], which include hundreds of parameters, have an extremely intricate, multi-layered architecture. Nevertheless, a large number of the neural designs that have been presented have substantial computational and memory needs. Because of this, the final predictions tend to be regarded as being incomprehensible. Even though DL has shown great promise in several technological domains, there are still certain concerns that need to be fixed. The computational cost is one of them. The computational cost for training multiple individual networks increases with the number of state-of-the-art neural networks; furthermore, the accuracy of the individual base model's categorization is poor. Despite the benefits of DL models derived from deep architecture, several obstacles, such as vanishing or exploding gradients and degradation problems, stand in the way of achieving this objective. Yet, building deep ensemble [3] models involves a steep learning curve, considering the computational expense. To improve standardized achievement, ensemble learning [4] integrates several distinct models. The benefits of ensemble learning and DL are combined with deep ensemble learning models to improve the generalization performance of the resulting model. Before training the ensemble model [5], it is crucial to tune hyperparameters owing to the inter-dependencies among neural networks [6]. Nevertheless, determining the hyperparameters and their values beforehand poses a significant challenge in the pre-training phase. There is a work shortage as a direct consequence. We emphasized the computational parameter before training a model to find a solution to that issue. We chose the agricultural field [7] as our experiment because, by the time the world's population reaches 9.1 billion by the end of 2050, it will have gradually increased, and 70% more food will be needed [8]. About 600 million tons of wheat are produced worldwide every year. Wheat is regarded as one of the "big three" grains, but 95% of wheat produced globally is impacted by fungus, bacteria, and viruses [9]. Younis et al. [10] considered VGG-16 and CNN for ensemble learning, taking a dataset of brain images, and achieving 98.14% accuracy. Lasker et al. [3] examined lung images considering MobileNetV2, EfficientNetB0, and InceptionResNetV2 and designed a stacked skeleton of ensemble and refined their work over U-Net model 3.43% IoU experimentation and 3.02% on Dice coefficient. Mariappan et al. [11] employed MRI images to assess 10 CNN models individually, and among them, 4 CNNs were chosen for the ensemble and obtained 95% accuracy. Wang et al. [12] employed pruning the channel using regularization; they assessed seven CNN models using multiple datasets and derived 42.84% deduction of parameter with minimal accuracy loss. Chen et al. [13] proposed CCPrune to assess the impact of channels on the CIFAR-10 dataset; they revealed 85.50% and 78.31% FLOPs of model VGG-19 and ResNet50, respectively. Blakeney et al. [14] proposed an SVCCA analysis tool to represent the pruning layer of VGG, ResNet, and ConvNet and observed that the pruned model preserved the same accuracy and also explored that the sparse model accomplished reasonable accuracy compared to the pruned model. Mishra et al. [15] revealed sparse tensor accomplished 50% sparsity on accuracy maintaining extensive areas of tasks. Liu et al. [16] spotted that arbitrary pruning, in the beginning, can be highly efficient for modern neural networks' sparse training. The key contribution of our work is as follows:

- In this work, a deep ensemble model is proposed to detect and classify wheat leaf diseases.
- To make the model efficient and robust in terms of computational cost, sparse-enabled pruning is considered for tuning hyperparameters.

In Sect. 2, the proposed method has been elaborated. Section 3 introduces the investigation, whereas Sect. 4 presents the ensuing work's findings and discussion. The paper concludes in Sect. 5.

2 Proposed Method

In recent years, ensemble learning has gained prominence as a research trend, which has increased the number of studies utilizing ensemble learning across a range of application domains [17]. Conversely, ensemble learning is a methodology that builds a larger single model that is more powerful than its individual components by combining multiple baseline models.

Training an ensemble model has a substantial time and computational cost penalty. The training and integration process of numerous base models leads to an increase in time and resource overhead. In particular, the training phase becomes computationally intensive with modern DNNs, which reduces the effectiveness of ensemble learning for applications requiring strict time or resource constraints, especially on mobile devices.

2.1 Ensemble Model

As we are trying to deal with computational complexity, on the basis of that, we considered multiple-base learning models based on the parameters used, named VGG-19 [18], ResNet50 [19], Xception [20], DenseNet201 [21], and NasNetMobile [22], including a customized CNN [23]. The overall structure of any ensemble model using an independent technique, the aggregation function F to merge a 'n' number baseline classifier set denoted as $m_1, m_2, ..., m_n$, predicts individual output. Considering dataset size x and feature dimension y, D=(a_i, b_i), where $1 \leq i \leq x$, $a_i \in R^y$ predicted output can be explained by Eq. (1):

Table 1 Performance metrics of CNN model and customized DNN models along with parameters used

Sl. no.	Model	Parameter	Accuracy
1	CNN	397539	90.02
2	**Customized VGG-19**	**28446531**	**93.77**
3	Customized ResNet50	57175683	92.55
4	**Customized Xception**	**40710691**	**97.66**
5	Customized InceptionV3	40710691	92.12
6	**Customized DenseNet201**	**49812803**	**97.92**
7	Customized NasNetMobile	21604759	91.97

$$b_i = \phi(a_i) = F(m_1, m_2, ..., m_k) \quad (1)$$

where $b_i \in Z$ indicates the final classification. We first revealed the pre-trained model mentioned above. We extended a base model by adding additional layers for classification. To prevent overfitting, we first flattened the output into a one-dimensional vector. Then, we used the ReLU activation function and regularization with a coefficient of 0.01 to create a dense layer with 256 neurons. During training, we used dropout regularization to randomly drop 50% of the connections from the previous layer. Again, dropout regularization is applied to the second dense layer, the final output layer, with 3 neurons suggesting a 3-class classification problem. In the same manner, except for CNN, we trained each DL model and selected the top three models among them for ensembling based on accuracy. The results of CNN and DL models are described in Table 1.

2.2 *Proposed Sparse Weighted Fine-Tuned Ensemble*

DNNs often experience overfitting since they are overparameterized models. Scientific research indicates that sub-networks within randomly initiated DNNs can enhance generalization in comparison to dense networks. Sparse neural networks are sub-networks that have a much lower number of non-zero trainable parameters than their dense equivalent. Their usefulness can also be observed in terms of improved training and inference computing performance, as well as improved storage and energy efficiency. Such sparsification is typically very helpful for large-scale models requiring millions to billions of parameters and arithmetic operations. In summary, the sparsification of complicated models will result in simpler models that are easier to train, analyze, and generalize. The proposed methodology structure is shown in Fig. 1. There are numerous methods and techniques available to train sparse networks. We selected an individual block pruning methodology. Block sparsity optimizes software for all hardware types by reducing indices and controlling

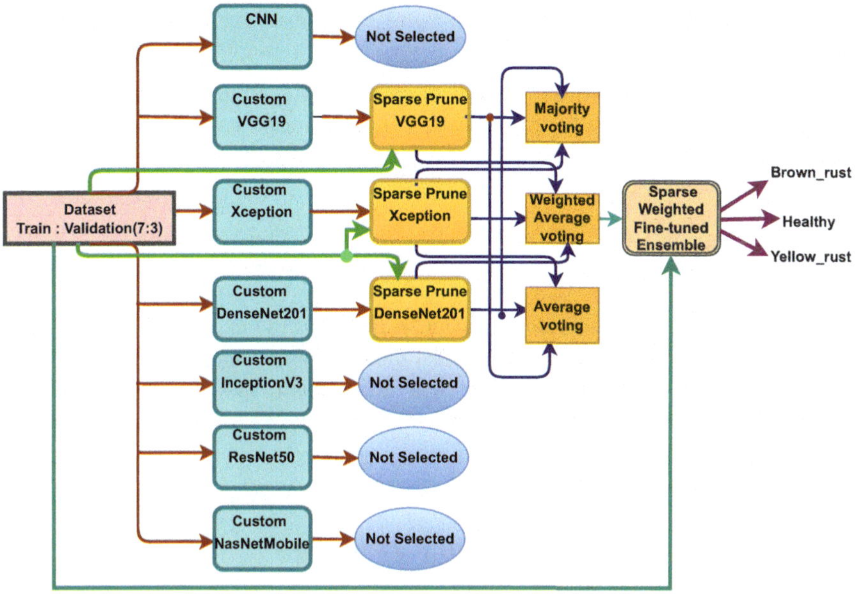

Fig. 1 Block diagram of the proposed SWFEM model

logic overhead. Instead of combining filters, Hoefler et al. [24] deconstruct sparse convolutions into both beneficial and detrimental ones, then translate the beneficial ones to a dense version for efficient processing. Utilizing mechanism training and sparsification, the model is re-trained for improved accuracy. This plan tries to enhance throughput and scalability during infertility. The sparsified model is the ideal baseline to assess model quality, as it can be easily compared to the initial dense model. From several ensemble approaches, the best method for our system of multi-class classification was achieved through a weighted stacking ensemble. Assuming we've got an 'n' number of base models denoted as $m_1, m_2, ..., m_n$ and a meta-model M_{meta} for merging the base models, each base model's weight is designated by $w_1, ..., w_n$, where $\sum_{i=1}^{n} w_i = 1$ to verify that the weights total up to one. The stacked ensemble prediction equation can be represented as a weighted sum of the basic model predictions:

$$y_{ensemble} = \sum_{i=1}^{n} w_i.M_i(input) \quad (2)$$

3 Experiment

It is crucial to follow a consistent and scientific approach to confirm suitable and feasible solutions for any experiment. Below are some key ways of assessing and summarizing experimental research.

3.1 *Dataset*

We gathered a public wheat dataset from Kaggle for the experiment, focusing on mainly two diseases, namely Brown Rust caused by the fungus Puccinia triticina Eriks and Yellow Rust caused by Puccinia striiformis f. sp. tritic, which primarily attacks leaves [25]. The dataset includes 1395 healthy, 1128 brown rust, and 1156 yellow rust-detected wheat leaves [26].

3.2 *Hardware Configuration and Training*

We conducted our experiment using an Intel Core i7-7820HQ processor with a clock speed of 2.90GHz, 1 TB of SSD, 32 GB of RAM, and a Nvidia Quadro P3000 graphic processing unit (GPU) with 6 GB of graphics memory to assist training performance and enable CUDA cores.

3.3 *Evaluation*

We considered a sparse weighted pruning ensemble model with batch size 32, dividing the dataset into a 7:3 ratio to get the best accuracy. To legitimize the outcome, accuracy is not enough. The classification report is also an integral component. To evaluate this, we assessed True Positive (A), True Negative (B), False Positive (C), and False Negative (D). Precision (P), recall (R), and F1-score (F1) will be calculated as follows:

$$Accuracy = \frac{A + B}{A + B + C + D} \tag{3}$$

$$Precision\,(P) = \frac{A}{A + C} \tag{4}$$

$$Recall\,(R) = \frac{A}{A + D} \tag{5}$$

$$f_1 Score\,(F1) = 2 * \left(\frac{P * R}{P + R}\right) \tag{6}$$

4 Result

SWFEM model reforming the benefits of model pruning with ensemble learning. It provides a novel way to attain both increased accuracy and decreased computing expenses.

We generate offspring networks by cloning the parent and drastically reducing their parameters, resulting in an ensemble of members with distinct topologies. We train each child network for a total of 32 epochs, resulting in faster convergence.

During training, the model is sparsely pruned to remove excess weights and connections. After the models are retrained, each of these blocks of layers with the desired sparsity is pruned to remove any extra weights or connections. These prune model results are described in Table 2. Pruned models minimize individual model parameters and boost efficiency. We named these sparse-enabled, fine-tuned, pruned models. The model with greater performance on the test set is used for further analysis. We performed majority voting, average voting, and weighted average voting. In majority voting, a multi-class classification task is used to determine the best number of ensemble pruning depending on ranking. In average voting, we simply assigned dummies to create replicas of current prediction arrays, most commonly for clarity or to prevent changing the original data. It adds the associated prediction arrays unit-wise, essentially aggregating the results for every input point. Weighted average voting is an expansion of the average approach. All models are given varying weights, which indicate the relevance of each model for prediction. This method uses weighted averaging to aggregate predictions from various models, giving more weight to models regarded as more dependable based on previous performance. This method has the potential to enhance accuracy over simple averaging since it better utilizes the strengths of diverse models. Weight assignments are critical and must be carefully considered or adjusted depending on empirical facts. We attain the best results by employing weighted average voting results. Consider the stacked ensemble model when utilizing neural networks as a meta-learner. Specifically, the sub-networks could be integrated with a larger multi-headed neural network, which in turn trains how it can optimally integrate the projections from each input sub-network. It allows stacking ensembles to be handled as a single, gigantic model. The advantage of this

Table 2 Performance metrics of the sparse pruned DNN models along with reduction in parameter

Sl. no.	Pruning model	Parameter after pruning	Accuracy	Approx reduction of parameter (%)
1	Customized VGG-19	23758542	96.56	16.48
2	Customized Xception	25004506	97.48	38.58
3	Customized DenseNet201	28796781	97.85	42.19

Table 3 Detailed result of wheat leaves disease identification using proposed SWFEM model

	Precision	Recall	F1-score	Support
Class 0 (Brown_rust)	0.9735	0.9735	0.9735	339
Class 1 (Healthy)	0.9812	0.9952	0.9882	419
Class 2 (Yellow_rust)	0.9912	0.9741	0.9826	347
Accuracy	0.9819			1105
Macro average	0.9819	0.9809	0.9814	1105
Weighted average	0.9820	0.9819	0.9819	1105

strategy is that the sub-network findings are sent directly to the meta-learner. It is additionally possible to change the parameter values of the sub-networks using the meta-learner technique if necessary. We carried out our experiment considering the top 3 accuracies of the sparsely weighted, fine-tuned DNN model. A detailed result is given in Table 3.

Figure 2 shows the confusion matrix for our presented approaches.

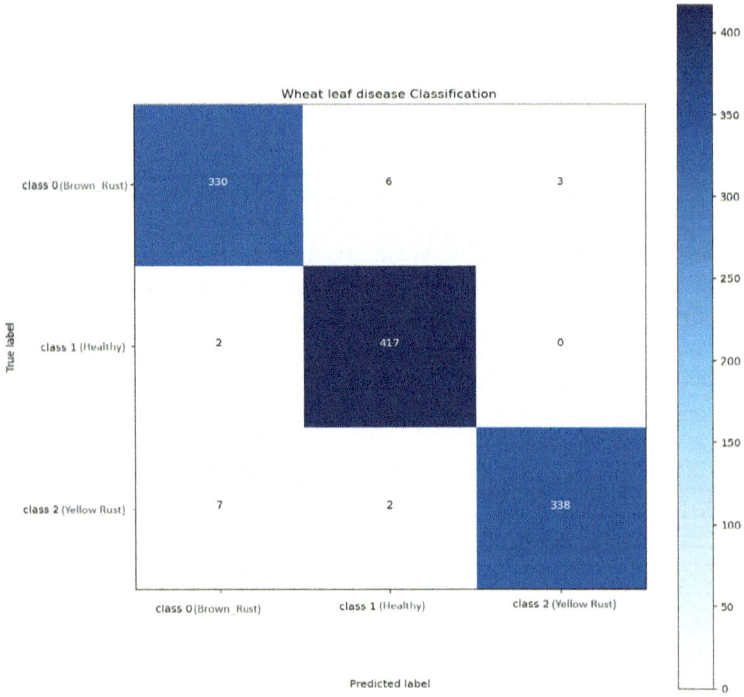

Fig. 2 Confusion matrix of SWFEM on wheat leaves dataset

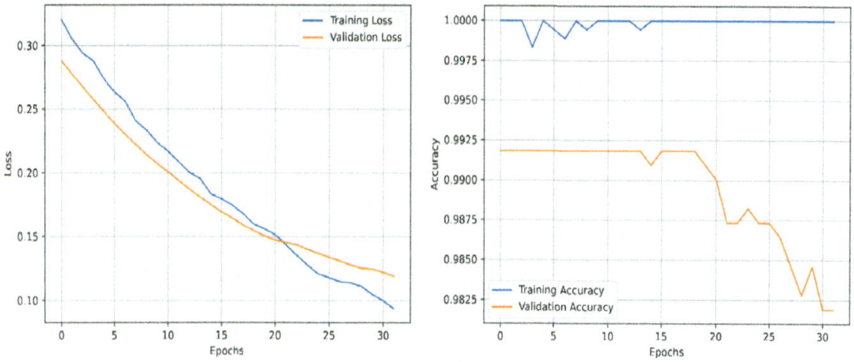

Fig. 3 Wheat model accuracy versus loss plot using the proposed SWFEM model

To achieve improved performance on unseen data, a generalizable model requires reduced training and validation losses and higher training and validation accuracy. Figure 3 depicts detailed results.

4.1 Performance Evaluation on Wheat Leaf Disease Detection

High-level feature extraction is a crucial task in DL. The task of fitting extracted features into a model to achieve better accuracy is a major challenge while minimizing computation time and resources. The recommended approach is effectively demonstrated in Table 2 as a reduction of parameters while maintaining accuracy, which directly affects the computational parameters. Besides, sparse pruning diminishes the use of tensors while maintaining accuracy. The AUC measure has the smallest error margin and provides a more exact estimate than other ensemble methods (95% confidence interval). The results are depicted in Fig. 4.

The off-diagonal of the confusion matrix shown in Fig. 2 embodies inaccurate estimations; however, the diagonal illustrates the number of properly detected wheat-diseased leaves. It has been observed that out of 339 images, 330 were correctly classified as Brown Rust leaves, 6 as Healthy, and only 3 as Yellow rust. In the case of Healthy, 417 instances were correctly classified, and only 2 were classified as Brown Rust. Out of 347 images, 338 were correctly classified as Yellow rust, 7 were misclassified as Brown Rust, and only 2 were Healthy leaves.

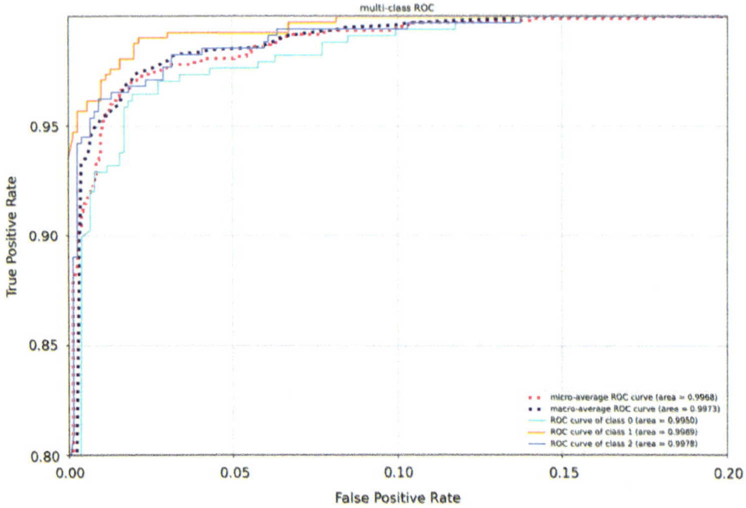

Fig. 4 ROC curves for micro/macro-averaged and class-specific AUC generated from the weighted-average pruned ensemble

Table 4 Comparison SWFEM with available DL method

Sl. no.	Work	Method	Accuracy (%)
1	VGG-19	Pruning	96.56
2	Xception	Pruning	97.48
3	DenseNet201	Pruning	97.85
4	Nakata et al. [27]	Pruning	78.3
5	Whitaker et al. [28]	Prune and tune	96.48
6	Jiang et al. [29]	ER-Stacking	88.1
7	**Our proposed approach SWFEM**	**Sparse prune and tune**	**98.19**

4.2 Comparative Analysis

Our main focus was on the impact of a sparse ensemble pruning model. The current experiment aims to establish the efficacy of our suggested strategy. To accomplish this goal, we compared our approach using our own individual networks to show the difference between our proposed ensemble model and others' efforts as well. The comparative results are described in Table 4. It has been observed that our proposed methodology improves accuracy by 1.69%, 0.73%, 0.35%, 25.40%, 1.77%, and 11.42% compared to VGG-19, Xception, DenseNet201, Nakata et al. [27], Whitaker et al. [28], and Jiang et al. [29] respectively.

5 Conclusion

Combining base models to establish an ensemble network is not sufficient for the classification task. It is also important to keep in mind about the computational power. Our proposed SWFEM approach efficiently illustrates how the number of extracted parameters affects the accuracy of the ensemble model. Our idea merges the concepts of both DL and ensemble. Sparse pruning in each block of a DL model shrinks a model's dimensions and computation complexity by getting rid of replicated or worthless weights. Detecting exact pruning at each block plays an important role before the final assembly. We eminently completed that task. Besides, the sparsity of an ensemble also diminishes the reduction of the tensor, so we also kept our attention on that area. Brown rust and yellow rust mainly occur on leaves. It can be controlled biologically, chemically, using a fungicide, or by monitoring the pathogen.

Early crop monitoring is an effective approach to increasing agricultural productivity. In remote areas, it can be challenging to maintain continuous monitoring and have access to agricultural professionals who can detect issues by examining leaves using their expertise. The ensemble model's versatility in vision approaches renders it a valuable solution to this situation. The proposed technique is unique and has demonstrated promising results.

This research aims to inspire and encourage researchers in the agricultural domain to employ DL, especially for disease diagnosis. This advancement in smart farming will assist farmers who previously struggled with disease management due to poor technological support. The proposed strategy holds enormous potential to empower the agricultural business. Our goal is to generalize and broaden the scope of this technique by applying it to numerous agricultural datasets.

References

1. Ghosh M, Baidya G, Mukherjee H, Obaidullah SM, Roy K (2022) A deep learning-based approach to single/mixed script-type identification. Adv Comput Syst Secur 13:121–132
2. Zdimalova M, Ghosh M, Lasker A, Obaidullah S, Poornima R, Shvydka S, Boratkova K, Kopani M (2023) Traditional and deep learning-oriented medical and biological image analysis. Bratisl Med J/Bratislavské Lekárske Listy 124(9)
3. Lasker A, Ghosh M, Obaidullah SM, Chakraborty C, Goncalves T, Roy K (2022) Ensemble stack architecture for lungs segmentation from x-ray images. In: International conference on intelligent data engineering and automated learning. Springer, pp 3–11
4. Obaidullah SM, Ghosh M, Mukherjee H, Roy K, Pal U (2022) Sen: stack ensemble shallow convolution neural network for signature-based writer identification. In: 26th International conference on pattern recognition (ICPR). IEEE, pp 1414–1420
5. Lasker A, Ghosh M, Das S, Obaidullah SM, Chakraborty C, Goncalves T, Roy K (2023) Segmented-based and segmented-free approach for covid-19 detection. In: International conference on computational intelligence in communications and business analytics. Springer, pp 332–342
6. Lasker A, Ghosh M, Obaidullah SM, Chakraborty C, Roy K (2023) Lwsnet-a novel deep-learning architecture to segregate covid-19 and pneumonia from x-ray imagery. Multimed Tools Appl 82(14):21 801–21 823

7. Ghosh M, Lasker A, Banerjee P, Manna A, Obaidullah SM, Gonçalves T, Roy K (2023) Plant disease detection and classification using a deep learning-based framework. In: International conference on intelligent data engineering and automated learning. Springer, pp 39–50
8. Sharma A, Jain A, Gupta P, Chowdary V (2020) Machine learning applications for precision agriculture: a comprehensive review. IEEE Access 9:4843–4873
9. Shewry PR (2009) Wheat. J Exp Bot 60(6):1537–1553
10. Younis A, Qiang L, Nyatega CO, Adamu MJ, Kawuwa HB (2022) Brain tumor analysis using deep learning and VGG-16 ensembling learning approaches. Appl Sci 12(14):7282
11. Mariappan P, Valarmathi R (2023) Deep learning techniques in classification of stages in dementia: an ensemble approach. J Adv Res Appl Sci Eng Technol 36(1):131–146
12. Wang Z, Li F, Shi G, Xie X, Wang F (2020) Network pruning using sparse learning and genetic algorithm. Neurocomputing 404:247–256
13. Chen Y, Wen X, Zhang Y, Shi W (2021) Ccprune: collaborative channel pruning for learning compact convolutional networks. Neurocomputing 451:35–45
14. Blakeney C, Yan Y, Zong Z (2020) Is pruning compression? Investigating pruning via network layer similarity. In: Proceedings of the IEEE/CVF winter conference on applications of computer vision, pp 914–922
15. Mishra A, Latorre JA, Pool J, Stosic D, Stosic D, Venkatesh G, Yu C, Micikevicius P (2021) Accelerating sparse deep neural networks. Preprint at arXiv:2104.08378
16. Liu S, Chen T, Chen X, Shen L, Mocanu DC, Wang Z, Pechenizkiy M (2022) The unreasonable effectiveness of random pruning: return of the most Naive baseline for sparse training. Preprint at arXiv:2202.02643
17. Mohammed A, Kora R (2023) A comprehensive review on ensemble deep learning: opportunities and challenges. J King Saud Univ-Comput Inf Sci
18. Simonyan K, Zisserman A (2014) Very deep convolutional networks for large-scale image recognition. Preprint at arXiv:1409.1556
19. He K, Zhang X, Ren S, Sun J (2016) Deep residual learning for image recognition. In: Proceedings of the IEEE conference on computer vision and pattern recognition, pp 770–778
20. Chollet F (2017) Xception: deep learning with depthwise separable convolutions. In: Proceedings of the IEEE conference on computer vision and pattern recognition, pp 1251–1258
21. Huang G, Liu Z, Van Der Maaten L, Weinberger KQ (2017) Densely connected convolutional networks. In: Proceedings of the IEEE conference on computer vision and pattern recognition, pp 4700–4708
22. Sutaji D, Yildiz O (2023) Weldonnet: weighted ensemble Densenet121, Mobilenetv2, and NasNetMobile to improve skin lesion diseases classification
23. Ghosh M, Mukherjee H, Obaidullah SM, Roy K (2021) STDNet: a CNN-based approach to single-/mixed-script detection. Innov Syst Softw Eng 17(3):277–288
24. Hoefler T, Alistarh D, Ben-Nun T, Dryden N, Peste A (2021) Sparsity in deep learning: pruning and growth for efficient inference and training in neural networks. J Mach Learn Res 22(1):10 882–11 005
25. Mylonas I, Stavrakoudis D, Katsantonis D, Korpetis E (2020) Better farming practices to combat climate change. In: Climate change and food security with emphasis on wheat. Elsevier, pp 1–29
26. Wheat Dataset. https://www.kaggle.com/datasets/sinadunk23/behzad-safari-jalal. Accessed 20 Jan 2024
27. Nakata N, Siina T (2023) Ensemble learning of multiple models using deep learning for multiclass classification of ultrasound images of hepatic masses. Bioengineering 10(1):69
28. Whitaker T, Whitley D (2022) Prune and tune ensembles: low-cost ensemble learning with sparse independent subnetworks. In: Proceedings of the AAAI conference on artificial intelligence, vol 36(8), pp 8638–8646
29. Jiang H, Zhang S, Yang Z, Zhao L, Zhou Y, Zhou D (2023) Quality classification of stored wheat based on evidence reasoning rule and stacking ensemble learning. Comput Electron Agric 214:108339

Machine Learning

A Comparative Study of Hospital Length of Stay Prediction of Indoor Patients for Admission, Post Admission, and Discharge Stage Data Using Machine Learning Algorithms

Somnath Ghosh and Debotosh Bhattacharjee

Abstract Globally, healthcare administrators are working to reduce healthcare costs without sacrificing the quality of service provided. The biggest portion of health expenses is related to hospitalization. Length of stay has a direct impact on the hospitalization cost and resource management. Researchers for the last few years are trying to find the relevant factors to accurately identify LOS. In this research, a strategy to forecast the number of hospitalization days in a community was developed using USA SPARCS 2017 de-identified data repository. The data is not individually identifiable. The dataset consists of 2,343,569 instances and 34 features. In this study, we are proposing a three stage model: admission, post admission, and discharge stage for an improved hospital length of stay prediction. These stage-wise models will help both the patient and hospital management to correctly predict LOS from the beginning of the hospitalization to the time of discharge. This study suggests a paradigm for predicting patient length of stay (LOS) using various machine learning (ML) and ensemble techniques. This stage-wise prediction will be performed using a weighted probability averaging ensemble model. The weights of the individual classifiers will be selected based on their performance through a grid search method. The proposed ensemble model has outperformed the individual base ML models. The proposed ensemble model has improved balanced classification performance and it improves across stages: admission to discharge. The discharge stage has the highest accuracy 0.9065 followed by the post admission stage with 0.8921 and the admission stage has the lowest among all 0.6824.

Keywords Health care · Hospitalization · Multi stage · Data mining · Machine learning · Ensemble learning

S. Ghosh (✉) · D. Bhattacharjee
Jadavpur University, 188, Raja Subodh Chandra Mallick Rd, Jadavpur, Kolkata, West Bengal 700032, India
e-mail: ghoshsomnath791@gmail.com

© The Author(s), under exclusive license to Springer Nature Singapore Pte Ltd. 2025
N. Chaki et al. (eds.), *Applied Computing for Software and Smart Systems*, Lecture Notes in Networks and Systems 980, https://doi.org/10.1007/978-981-97-9762-2_12

1 Introduction

Healthcare systems have started putting more of an emphasis on efficient resource management and forecasting in order to assure optimal levels of care, lower associated costs, and enhance patient care. The main goal of hospital managers is to create an effective healthcare plan by distributing the resources both human and physical necessary for a successful hospital operation. The most crucial feature is to accurately estimate how long a patient who is admitted to the hospital will stay. Many hospitalization lengths of prediction analysis have been already performed; those are mostly based on ICU [1, 2] patients or for specific categories. For ICU patients some predictions used conventional techniques such as SAPS, APACHE II, and APACHE IV. Most of the existing analysis has been limited to a specific study either on hospital, department, or disease [3–14]. This piece of work is different from the existing length of stay predictions models; this prediction will be performed based on USA SPARCS de-identified 2017 dataset. In this study, we are proposing a unique three stage model: admission, post admission and discharge stages to predict the hospital length of stay. The model is new generation model that is not stick to a particular phase it explores all possibilities of LOS prediction in a hospital stay. Three stages will have an increasing number of features based on the available data on each phase. The feature selection of each stage is performed based on the business knowledge gathered from the metadata of SPARCS [15]. We have utilized various machine learning models such as Logistic regression, KNN, ANN, Random Forest, Adaboost, XGBoost, LightGBM, Catboost, and ensemble learning for our analysis. Final prediction was achieved by a weighted probability average hybrid ensemble [16–19] model by combining the top performing base ML models across three stages.

This is how the rest of the paper is organized. A review of earlier studies in the same field is included in Sect. 2. In Sect. 3, the dataset that was studied and the methodologies employed to assess the initial state are described. The supervised machine learning algorithms and the methodology for stage-wise length of stay prediction is then covered. Results are presented in Sects. 4 and 5, along with a description of which stage is better suited to forecast length of stay. Section 6 concludes with some thoughts and recommendations.

2 Literature Survey

Predicting length of stay is an important factor for hospitalization cases. The length of stay is the direct factor of hospitalization elements such as cost of hospitalization or the indicator of the patient's condition. The prediction of length of stay not only helps the hospital to manage its resources but also indicates the well-being of the patient. For a long length of stay, engagement of the hospital authority toward the patient also increases. Several studies on predicting hospital length of stay have been carried out by researchers. Some multi-stage LOS predictive models are being developed. They used different techniques such as a Markov process model, some used CART to predict LOS [20], some used a linear regression and neural network [21]. The USA SPARCS 2017 de-identified that data repository is an open source data available for research purpose. Few LOS studies have been already performed on this dataset, and the studies are presented in Table 1. The studies performed in this dataset is limited to single stage, mainly pre admission/admission stage. For other studies on LOS please refer Table 2.

In this study, LOS prediction will be based on classification. Regression prediction gives an exact value of predicted LOS but it might not be accurate always. A small error can hugely impact the hospital authority to manage the resources. The prediction using classification will be much more informative with small groups of LOS. Multiclass and binary both have been explored in LOS classification task. For our model, we have identified 7 small groups for LOS; those are 0–5 days, 6–10 days, 11–15 days, 16–30 days, 31–60 days, 61–90 days, and > 90 days. Most of the studies performed except SPARCS data are limited to a single area only, e.g., Department, hospital, and disease. Few studies proposed two/multi stage models but they are lacking with diverse data category. The studies performed on SPARCS data are of single stage only. Previous studies predict LOS on SPARCS data at admission stage [22, 23, 25] or at pre admission stage [22]. The studies identified features captured only at admission and discharge stage [23, 25] some were identified at pre admission and admission stage [22]. Our idea is different from these existing studies and we are introducing a new form of Los prediction using a three stage model for the SPARCS data utilizing the fact that the data set has certain features that cannot be captured only at pre admission, admission stage, and discharge stage. Few of those are Procedure code, Risk of mortality, Severity of illness [33], and birth weight. We have identified potential stages for prediction are admission, post admission and discharge stages. The prediction will be performed in each stage based on varying size of features captured at that stage. This stage-wise prediction model will benefit

Table 1 LOS prediction referenced articles for SPARCS data

Article	Model	Keywords	Features used
Zeng [22]	LightGBM	Pre admission/admission model, hybrid encoding, regression, single stage	All
Jain et al. [23]	XGBoost RF, DT	Admission stage, regression, distribution dependent target encoding, single stage	Removed total cost, total charges, patient disposition
Rahman et al. [24]	LR, Lasso, Ridge	Federated learning, 10 clients (hospitals), flower framework, single stage	All
Alnsour et al. [25]	XGBoost, ANN, DT	Regression (only for short stays), classification (all LOS groups), SMOTE, admission stage, single stage	Removed total charges, cost, disposition, Birth weight etc. used 16 features

Table 2 LOS prediction on other referenced articles

Article	Model	Keywords
Caetano et al. [26]	RF	Hot deck method, Portuguese study, admission stage
Jaotombo et al. [27]	GB	Outlier removal
Chen et al. [28]	6 Models, KNN, XGBoost	Sepsis, outlier removal model
Abd-elrazek et al. [1]	Fuzzy logic	ICU, bias variance adjustment, general admission features , admission stage
Shahid Ansari et al. [29]	LR with EN	Dengue, regression imputation, 24hrs. model, Include lab tests until discharge
Ebinger et al. [30]	Meta learners	Covid-19, ENET blender, advanced AVG blender, day1-day2-day3 model
Tsai et al. [21]	NN, LR	Two stage model, coronary atherosclerosis, heart failure and acute myocardial, preadmission and predischarge model
Kumar et al. [20]	CART, RF	Surgical patients, two stage model, stage I (before surgery) ,stage II (after surgery)
Rouzabahman et al. [2]	LR, K-Means	Clustering, supervised learning, death and Los analysis
Li et al. [31]	Lasso regression	Box cox transformation, skewed data, icu, survivor and non-survivor Los analysis
Gutiérrez et al. [32]	DT, RF	Multiple departments LOS prediction

Caption: GB = Gradient Boosting, LR with EN = Logistic regression with elastic net, RF = Random Forest, NN = neural Network, CART = Classification and regression tree, DT = Decision Tree, LOS = Length of stay

the hospital authority to verify patient discharge status time by time. The stage-wise predictions will be obtained using a weighted probability average ensemble model by utilizing several machine learning models. This stage-wise hybrid ensemble method is unexplored in the area of length of stay prediction with potential actual small LOS groups. Some studies cut down the extreme LOS [27, 28] groups into an average group but our study also focuses on the extreme cases. Class imbalance is an important factor for predictive task. Some studies [25] use SMOTE technique to encounter this. In this study no sampling method was incorporated using the fact that synthetic data might influence the actual prediction instead we have given importance to the performances of different evaluation metrics to better understand the precision of the model. This study also utilizes charlson comorbidity factors for Los prediction.

3 Methods

The purpose of the study is to predict hospitalization length of stay using an incremental model called three stage model. The stage-wise: admission to discharge time features will be utilized in length of stay prediction. The dataset, pre-processing, and proposed model will be discussed in the later sections.

3.1 Data Repository

In 1979, the government and the healthcare sector collaborated to create SPARCS, a comprehensive all payer data reporting system. The New York State Department of Health contributed to the Hospital Inpatient Discharges (SPARCS De-Identified) 2017 dataset [15]. This is an open dataset with de-identified data the file is not individually identifiable. It contains basic record level detail. This repository contains 2.3 Million records. It features patient's demographic data, diagnosis data, hospital level data, and expenses. Patient's lab test related parameters and patients private information are excluded and are not part of our prediction. The target feature is Length of Stay. In the dataset, the number of female patient is higher compared to male. The hospitalization case is higher for older age groups. The dataset contains multiple hospitals, diseases in the US based on different regions. This dataset includes all patients visited both in emergency and general admissions.

3.2 Data Pre-processing

Data pre-processing improves the data quality for ML models. Our data has some inconsistency: the Length of stay has value '120+', and we have modified it to make it 120. Gender has value 'unknown', APR Medical Surgical Description' has value 'Not Applicable' and birth weight has value 'unknown' we have removed these entries from dataset. Other pre-processing will be discussed in the later sections:

Missing Values handling. We have several columns in our data having missing values. The detailed information is presented in Table 3:

Populate with Default value. For Facility name = 'Redacted for Confidentiality', some attribute values are intentionality being removed for confidentiality. This entries are belongs to abortion patients. We have populated default value '999' for numeric features and 'Not provided' for categorical features. Impacted Features are Hospital Service Area, Hospital County, Operating Certificate Number, Permanent Facility Id, and Zip Code—3 digits.

Table 3 Missing data overview

Feature	Missing value count	Percentage
Hospital service area	5155	0.219964
Hospital county	5155	0.219964
Operating certificate number	5155	0.219964
Permanent facility id	5155	0.219964
Zip code—3 digits	39,019	1.664939
APR severity of illness description	240	0.010241
APR risk of mortality	240	0.010241
Payment typology 2	878,722	37.495034
Payment typology 3	1,737,244	74.128135
Birth weight	2,115,685	90.276198

Removal of features. Few columns had very high missing percentages. Payment Typology 2 and Payment Typology 3 having missing percentages of 37% and 74%.We have removed these two features from the dataset. Birth weight is having missing values more than 90%. Birth weight is important feature for newborns. So, we have kept birth weight for analysis of newborn length of stay.

Removal of missing rows. APR Severity of Illness Description and APR Risk of Mortality have around 240 missing data. Valid values for APR Severity of Illness Description and APR Risk of Mortality are 'Minor', 'Moderate', 'Major', and 'Extreme'. As APR Severity of Illness Description and APR Risk of Mortality features cannot be blank, we have removed the 240 rows from the dataset. These missing rows belongs to the Live Born or pregnancy related hospital cases.

Populate missing ZIP code. Zip code has a total 1.66% of missing rows. Patient zip code is limited to the first three digits. However, in cases where the population size for the zip code is less than 20,000, if the cell size on patient classification strata is less than 10, or if the record is an abortion then the zip code is blank. For the redacted facility/abortion cases of the 5155 entries, it has been updated with default value as '999'. For the rest of the 33,864 entries, we have implemented a Zip code population logic based on Zip code related features. Zip code is assigned based on features Hospital Service Area, Hospital County, and Facility Name. We have identified valid zip codes for those missing entries utilizing the above-mentioned features. We have randomly assigned zip codes from valid zip codes group to the missing entries.

Feature Engineering. We have created few new features from the existing 34 features. We have CCS (Clinical Classification code) diagnosis code in our feature set which is ICD-9 standard code. Using CCS diagnosis code [34], we have generated Comorbidity [34] related features based on Charlson comorbidity index. It added 18 new features. Patient disposition is the final destination of patients after release from hospital we have created a feature Expired Ind utilizing patient disposition 'Expired'. It informs that the patient is died in hospital. The birth weight feature is explicitly for the newborn. We have created a new feature birth weight Indicator utilizing birth weight. This field holds binary values '1' or '0' based on availability of birth weight. We have also generated few features based on estimated Total Cost utilizing features CCS procedure code, Emergency Indicator and Payment Typology 1. We have calculated the Min, Max, Mean, Median, and Standard Deviation for the features CCS procedure code, Emergency Indicator, and Payment Typology 1 based on the estimated total Costs. These newly generated 35 features will be utilized for prediction along with the existing features.

Feature Selection. In our dataset 5 duplicate features are present. Please refer Table 4 for the details:

We will delete one of the features from Table 4. Discharge year has constant value of '2017'. We have discarded the field from the feature set. Mutual Information and Chi-Square test also has been applied for feature selection.

Feature Transformation and Normalization. Total estimated cost and total charge features distribution are highly skewed. To make them balanced normally distributed, we have performed Log transformation. This final features set will be normalized. Certain distance or weight based models cannot produce good results without proper scaling. We have applied Min max normalization to the independent features.

Feature Encoding. We have used a mixed encoding technique. For few categorical features, we have a large number of unique categories one hot encoding cannot be applied on those features; it will unnecessarily increase the feature dimension. For large number categories features (e.g. Procedure code has 200+ unique categories), we have opted Label encoding. Please refer to Table 5 for the encoding details.

Table 4 Duplicate features details

Feature name	Similar feature (Will be deleted)
CCS diagnosis code	CCS diagnosis description
CCS procedure code	CCS procedure description
APR DRG code	APR DRG description
APR MDC code	APR MDC description
APR severity of illness code	APR severity of illness description

Table 5 Encoding mapping details

Encoding method	Features used
One hot encoding	Health service area, race, ethnicity, type of admission, payment typology 1, APR medical surgical description, expired Ind
Ordinal encoding	APR risk of mortality, age group, emergency department indicator
Label encoding	Hospital county, facility name, zip code—3 digits, gender, patient disposition, abortion edit indicator, operating certificate number, facility id, CCS diagnosis code, CCS procedure code, APR DRG code, APR MDC code

Relationship between features. After pre-processing, feature selection and engineering total 57 features are left with us. We have performed statistical Pearson correlation coefficient test between target and independent features and the few top-performing results are published in Table 6. We have observed that Total cost and Total charge have very high positive correlation 0.616172266 and 0.576748042 respectively with LOS. APR Severity of Illness Code, APR Risk of Mortality, and Patient Disposition also have high positive correlation with LOS with co-efficient as 0.387247695, 0.339236665, and 0.209421147 respectively. Age Group, Expired_Ind, Emergency Department Indicator, and Charlson Index Comorbidity Score also have a significant amount of positive correlation with the LOS. We can see that Birth Weight Ind has a significant amount of negative correlation with the target feature that is identified as one of the valuable predictors of the LOS. The new cost-related features have a significant amount of both positive and negative correlation with the target feature; few top performing cost-related features are presented in the table. A few Comorbidity factors have less impact toward the prediction; those are Rheumatologic Disease and Moderate or Severe Liver Disease. As part of general processing, we are not going to remove any of the features. We will perform a feature selection based on Mutual information and Chi-Square test to cross verify the performance of the model using the reduced feature set.

Table 6 Correlation of LOS with independent variables

Features	Correlation
Total costs	0.616172266
Total charges	0.576748042
APR severity of illness code	0.387247695
APR risk of mortality	0.339236665
Patient disposition	0.209421147
Total costs_procdre_med	0.256985534
Age group	0.148518559
Total cCosts_pay_typ_med	0.127150528
Expired_Ind	0.101501961
Emergency department indicator	0.087134496
Charlson index comorbidity score	0.069836535
Gender	0.069332564
Any malignancy, including leukemia and lymphoma	0.059219741
CCS procedure code	0.054908829
Cerebrovascular disease	0.045864754
Zip code—3 digits	−0.022035095
Race	−0.036714269
CCS diagnosis code	−0.038629603
Total costs_Emer_ind_max	−0.087134496
Birth_Weight_Ind	−0.092799894

3.3 Proposed Model

This study introduces a three stage model: admission, post admission, and final discharge model. These three phase predictive models will give hospital authority and patient parties an estimated stay in hospital in an incremental way. The model will have three predictions, first of all at the time of admission, next is at post admission phase that is generally after 1–2 days (48hrs.) of hospitalization and the final is at the time of discharge. This multi-stage predictions will be performed based on the features collected at each stage. The features will be incremented from the admission to discharge phase this will impact the LOS prediction proportionally. This phase-wise prediction is dependent on the predictive power of the features added in each stage. Different feature sets for the phase-wise prediction model have been identified. The identification of features is performed based on reference given in SPARCS data dictionary [15], business understanding, and based on few study [33]. Please refer to Table 7 for the stages and their corresponding features. At the admission stage, patient's demographic and preliminary diagnosis details can be extracted. After admission some diagnoses and procedures will be performed many features related to the diagnosis and condition of the patient will be added and the same will be utilized for post admission stage prediction. The final or discharge stage consists of features that will only be available at the time of discharge. The features Total cost and Total charges are very important feature of LOS prediction. Total cost is the estimated cost provided by hospital authority or by physician after admission. We have included Total cost in the post admission stage prediction and the Total charge is the actual cost of the hospitalization which can only be available at the time of discharge.

Table 7 Three stage model feature mapping

Stage	Features used for prediction
Admission	Health service area, hospital county, operating certificate number, permanent facility id, facility name, age group, zip code—3 digits, gender, race, ethnicity, ethnicity, CCS diagnosis code, payment typology 1, emergency department indicator, charlson comorbidities
Post admission	CCS procedure code, APR DRG code, APR MDC code, APR severity of illness code, APR risk of mortality, APR medical surgical description, birth weight, abortion edit indicator, total costs
Discharge time	Patient disposition, total charges, expired Ind

Discharge disposition is the final destination of patient after discharge. We have included this feature in the final discharge stage only. The admission stage model prediction will be performed based on the features mentioned in the table and the subsequent stages will use the features mentioned in the table along with the prior stages features not vice versa. The final discharge stage will have all the features mentioned in table.

In the proceeding section, we will discuss the different machine learning models utilized in our prediction model and at last our proposed hybrid model will be presented.

Logistic regression. This is one simple model for classification tasks which mainly deals with linear data. Multiclass Softmax function will be used for our prediction. We will be using the default setting for logistic regression models such as penalty as 'L2' and iteration = 100 across the stages.

Random Forest Classifier. Among the supervised learning methods is the well-known machine learning algorithm Random Forest. Its foundation is the idea of tree based ensemble learning. Rather than depending on a single decision tree, because there are more trees in the forest, accuracy is higher and over fitting is avoided. For our prediction, the parameters used is estimator as 150 and max tree depth of 25 irrespective of the stages.

Adaboost. By combining several weak classifiers, the ensemble modeling technique known as 'boosting' aims to create a powerful classifier. It is accomplished by using weak models in series to develop a model. We used the number of estimators = 500 for Adaptive boosting ensemble learning irrespective of the prediction stage.

XGBOOST. A distributed gradient boosting library optimized for efficiency and scalability in machine learning model training is called XGBoost. It is an ensemble learning technique that generates a stronger prediction by aggregating the predictions of several weak models. It can handle large datasets and achieve state-of-the-art performance in many machine learning tasks. In this study, the parameters used are estimators as 1000 and max tree depth of 8 across the stages.

LightGBM. Microsoft created LightGBM, an open-source, distributed, high-performance gradient boosting framework. Its design prioritizes accuracy, scalability, and efficiency. It is predicated on decision trees, which are intended to decrease memory utilization and increase model efficiency We have used the Number of estimators = 500 and used other default configurations across all the stages.

KNN. Because of its simplicity and ease of use, the KNN method is a popular and adaptable machine learning technique. No presumptions on the distribution of the underlying data are necessary. The KNN algorithm locates the K closest neighbors, using a distance metric like Euclidean distance, to a given data point. The majority vote or the average of the K neighbors are then used to establish the class or value of the data item. We have taken the number of neighbors = 10 across all the stages for our prediction.

Artificial Neural Network. Artificial neural networks are based on the architecture and functions of human neurons. Neural nets or neural networks are other names for it. An artificial neural network's first layer, known as the input layer, transfers data from outside sources to the second layer, known as the hidden layer. Each neuron in the hidden layer takes in information from the neurons in the layer above, calculates the weighted sum, and then relays it to the neurons in the layer below. The ANN structure we have used has 1 input layer, 4 hidden layers and the final output layer across all the stages. Maximum hidden neurons used is 80 and minimum 24.

Catboost. It is an open-source boosting library known as Catboost, sometimes known as Categorical Boosting. It is intended for use with very large numbers of independent features in regression and classification tasks. A gradient boosting variation that can handle both numerical and categorical features is called catboost. For our prediction, m we have used Max Depth = 8 and Iterations = 2000 across all the stages.

Ensemble Learning. Through the method of ensemble learning, several machine learning models are trained to produce improvised predictions that improve the performance of individual machine learning models. The predictors that have been trained to compute the predictions are essentially referred to as the 'ensemble'. The majority of the time, decision trees which are useful for solving quantitative problems are used to facilitate ensemble learning. Using a variety of machine learning models, including Random Forest Classifier, XGBoost, ANN, Catboost, among others, one such ensemble learning implementation trains and then displays the combined output of these models to make final judgements. Two different voting classifier types are available. The soft voting classifier and the hard voting classifier are used to generate the findings. Several soft Ensemble approaches are available: average probability, weighted average probability, max Rule, sum rule, and product rule.

Our Architecture. In our study, we are introducing a weighted average probability ensemble-based LOS model. This model will take different well-tuned model's predicted probabilities and averages the probabilities based on the weight assigned to each classifier. Classifier's weights will be selected based on the contribution of the each model toward the generation of an optimal ensemble model through a robust grid search method. We are proposing a 3WP ensemble model. The 3WP ensemble model will be selected based on top three performing base models. The algorithm for the prediction is discussed as below.

Algorithm:

Input: Training dataset and testing dataset.
Procedure: Below steps will be repeated for three stages: Admission, Post admission and discharge.

1. Train different base models.
2. Test the base models and compare the accuracy, precision, recall and f1 score.
3. Calibrate the individual base models to obtain optimum precision.
4. Select top models based on descending order of classification metrics considering accuracy, precision, recall and f1 score.
5. Top performing models selected are Random Forest, Xgboost, ANN, Catboost and LightGBM.
6. Select top three models based on balanced classification metrics.
7. Obtain an initial weighted average ensemble model using equal weights.
8. Calculate the weights for each model using grid search method and assign optimum weights to each classifier.
9. Apply weighted probability average ensemble for prediction using optimized weights.
10. Generate probability weighted voting with a new set of classes.
11. Compare and evaluate the precision of the new ensemble model with the base models.

The proposed model's architecture is shown in Fig. 1.

3.4 Evaluation Criteria

For classification problems classic evaluation metrics are accuracy, precision, recall, F1-score, and AUC-ROC value. We will be using these metrics for validation of our multi-stage model. These metrics define different aspects of the model's performance. Our aim is to produce maximum efficiency from these metrics through combining the top-performing base models. Our LOS groups are imbalanced so, we will also give importance to precision and recall apart from the accuracy.

4 Result

Initially the USA SPARCS dataset had 2,343,569 records with 34 features. After data pre-processing, duplicate removal, feature selection, and feature engineering, we have 2,308,391 instances left in our dataset. The feature dimension got increased from 34 to 82. We have utilized the online data science and ML platform Kaggle for our coding and analysis. Kaggle is an open source online platform for machine learning and data science research work. The platform uses Python 3.10 with the

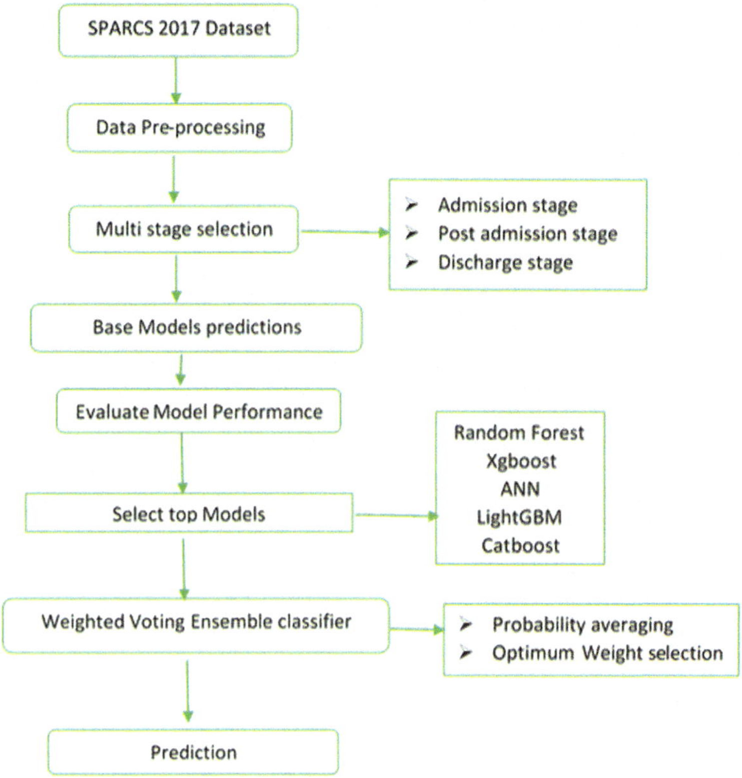

Fig. 1 Schematic diagram of the LOS prediction model

latest scikit learn, tensorflow, and Keras ML packages. The models are trained using 80% data and the rest 20% data is used for testing. The comparative study of different stages of the proposed model is illustrated in Table 8. The performance of base models and the weighted average model is also presented in the table. The final optimum weights of the 3WPE are presented in the table. The optimum weight search space is used between 0.1 and 0.8. The number of features that are utilized for different stages are 55, 79, and 82 for admission, post admission, and discharge stages respectively. Two different evaluations are performed. One is based on full feature set and another one with a reduced feature. Reduced feature set evaluation is only applicable for post admission phase.

4.1 Admission Phase Model

The accuracy of admission stage models lies between 0.64 and 0.68. Most of the model's performance is similar according to accuracy, but varies in precision, recall,

Table 8 Result across all the stages

Stage	Models	Accuracy	Precision	recall	f1-score	Auc	Weights
Admission	LR	0.6721	0.14	0.14	0.14	0.5237	X
	RF	0.672	0.27	0.18	0.19	0.7687	X
	Adaboost	0.6746	0.2	0.15	0.14	0.6906	X
	XGBoost	0.6769	0.28	0.19	0.2	0.7853	X
	LightGBM	0.6806	0.29	0.18	0.18	0.7911	X
	Catboost	0.6821	0.34	0.18	0.18	0.803	X
	ANN	0.6762	0.3	0.16	0.15	0.7619	X
	KNN	0.6563	0.22	0.16	0.16	0.6141	X
	3WPE	**0.6824**	**0.34**	**0.18**	**0.18**	**0.8042**	**0.6,0.1,0.1**
Post admission	LR	0.7324	0.14	0.14	0.13	0.608	X
	RF	0.8854	0.74	0.68	0.71	0.9835	X
	Adaboost	0.7475	0.45	0.48	0.46	0.7485	X
	XGBoost	0.8909	0.7	0.66	0.68	0.9858	X
	LightGBM	0.8493	0.53	0.51	0.51	0.8737	X
	Catboost	0.8833	0.7	0.64	0.67	0.9837	X
	ANN	0.8692	0.69	0.62	0.64	0.9803	X
	KNN	0.7501	0.37	0.22	0.24	0.7380	X
	3WPE	**0.8921**	**0.74**	**0.69**	**0.71**	**0.9863**	**0.8,0.6,0.3**
Discharge	LR	0.7320	0.14	0.14	0.13	0.6105	X
	RF	0.8995	0.77	0.72	0.74	0.9868	X
	Adaboost	0.7387	0.48	0.52	0.49	0.7387	X
	XGBoost	0.9053	0.74	0.71	0.72	0.9891	X
	LightGBM	0.8633	0.52	0.5	0.51	0.8483	X
	Catboost	0.8966	0.74	0.69	0.71	0.9873	X
	ANN	0.8812	0.71	0.68	0.69	0.9831	X
	KNN	0.7519	0.4	0..21	0.24	0.7446	X
	3WPE	**0.9065**	**0.77**	**0.73**	**0.75**	**0.9892**	**0.8,0.5,0.2**
Post admission reduced features (55)	RF	0.8853	0.74	0.68	0.71	0.9837	X
	XGBoost	0.8908	0.7	0.66	0.68	0.9856	X
	ANN	0.8693	0.69	0.62	0.64	0.9805	X
	Catboost	0.8833	0.7	0.64	0.67	0.9837	X
	3WPE	**0.8920**	**0.74**	**0.69**	**0.71**	**0.9861**	**0.8,0.6,0.3**

f1-score, and Auc scores. LightGBM, Catboost, and XGBoost are performing well not only with respect to accuracy but the other parameters are also performing well compared to other models. For AUC, LightGBM tops the table with value 0.7911 and XGBoost is in second position with value 0.7853. Catboost is the winner for precision with value 0.34 XGBoost has the highest recall and f1-score 0.19 and 0.2 respectively. For accuracy, Catboost is the winner with value 0.6821. The performance of random forests and ANN is not much significant at this stage. The 3WPE ensemble classifier combines Catboost, XGBoost, and LightGBM in admission phase. The ensemble accuracy is 0.6824 and precision, recall, f1-score, and AUC are 0.34, 0.18, 0.18, and 0.8042 respectively which perform decently well with respect to top base models. The weights used for 3WPE are 0.6, 0.1, and 0.1.

4.2 *Post Admission Phase Model*

In post admission phase, logistic regression, KNN, and AdaBoost models are not performing well. Top performing models are Random forest, XGBoost, ANN, and catboost. Random forest model's accuracy, AUC, precision, recall, and f1-score are 0.8854, 0.9835, 0.74, 0.68, and 0.71 respectively. For the XGBoost model, accuracy, AUC, precision, recall, and f1-score are 0.8909, 0.9858, 0.7, 0.66, and 0.68 respectively. For the Catboost model, accuracy, AUC, precision, recall, and f1-score are 8833, 0.9858, 0.7, 0.64, and 0.67 respectively. For Artificial neural network, accuracy, AUC, precision, recall, and f1-score are 0.8692, 0.9803, 0.69, 0.62, and 0.64 respectively. In this phase, XGBoost has the highest accuracy and AUC but random forest has the highest precision, recall, and f1-score. For post admission phase 3WPE ensemble model combines Random Forest, XGBoost, and Artificial Neural network model. The combined model's accuracy, AUC, precision, recall, and f1-score are 0.8921, 0.9863, 0.74, 0.69, and 0.71 respectively. The ensemble model outperformed its base models. The weights used for 3WPE are 0.8, 0.6, 0.3. The 3WPE ensemble model is also evaluated with reduced optimal feature set of 55 and the results are similar compared to its original features.

4.3 *Discharge Phase Model*

For the discharge model, logistic regression, KNN, and Adaboost models are not performing well. The top-performing models are XGBoost, Catboost, ANN, and Random Forest. For XGBoost, accuracy, AUC, precision, recall, and f1-score are 0.9053, 0.9891, 0.74, 0.71, and 0.72 respectively. For Random Forest, accuracy, AUC, precision, recall, and f1-score are 0.8995, 0.9868, 0.77, 0.72, and 0.74 respectively. For Artificial Neural network, accuracy, AUC, precision, recall, and f1-score are 0.8812, 0.9831, 0.71, 0.68, and 0.69 respectively. In discharge phase, XGBoost has the highest accuracy and AUC but random forest has the highest precision, recall, and

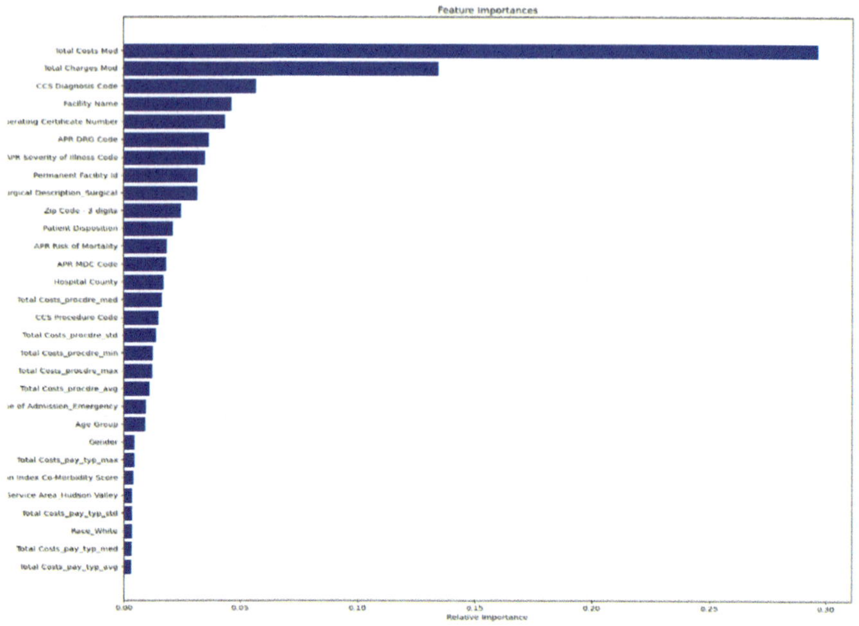

Fig. 2 Feature importance of Random Forest for top 30 features at discharge stage

f1-score. For discharge phase 3WPE ensemble model combines XGBoost, ANN, and Random forest models. The combined model's accuracy, AUC, precision, recall, and f1-score are 0.9065, 0.9892, 0.77, 0.73, and 0.75 respectively. The ensemble model outperformed its base models. The weights used for 3WPE are 0.8, 0.5, and 0.2. The feature importance of the final discharge stage for top 30 features of RF model is shown in Fig. 2. Total cost, charges, diagnosis code, APR DRG code, facility name, operating certificate number, severity of illness are the top predictors of LOS.

5 Discussion

The performance of the stage-wise models got improvised from admission to discharge phase. The accuracy, AUC, precision, recall, f1-score obtained in admission phase through the 3WPE model is 0.6824, 0.8042, 0.34, 0.18, and 0.18 respectively which is improved version of the base models. The performance in this phase is quite less due to lack of important features. The accuracy, AUC, precision, recall, and f1-score obtained in post admission phase through the 3WPE model are 0.8921, 0.9863, 0.74, 0.69, and 0.71 respectively which is quite improved from the admission phase. The performance of the ensemble model is quite good in this phase as the features included in this phase have significant importance in length of stay prediction. The accuracy, AUC, precision, recall, and f1-score obtained in discharge stage through the 3WPE model is 0.9065, 0.9892, 0.77, 0.73, and 0.75 respectively which again improved compared to post admission phase. Features selected for this phase have significant impact on LOS prediction. The 3WPE model has outperformed its base ML models in each phase. The admission phase has limited features related to demographic and diagnosis. The performance got improved in post admission phase due to availability of important features like estimated cost, procedures performed, Severity score, etc. The discharge phase has performing best in this three stage predictive model. Features like actual hospitalization cost, patient disposition, expired Ind have been included in this phase. We have also included Charlson comorbidity score [34] in our prediction which also has significant correlation with the LOS. Charlson comorbidities are important for this study, as patient's lab tests or vital signs are not included in this study. Prior studies on sparcs data did not consider these comorbidities for Los prediction. Compared to other studies [25] they used sparcs data to predict Los only at admission phase. They identified discharge features as total charge, cost, patient disposition, birth weight and abortion Ind and removed them from the feature set. In earlier studies on sparcs [23] they excluded total cost, total charge, and patient disposition from the feature set with the assumption that those are discharge related features though total cost is estimated cost for hospitalization which will be provided at post admission phase. Some study uses sparcs data as admission time prediction considering all the features [22]. Our model is differing from these existing approaches and we have introduced a three phase model to give proper importance to features collected in each stages. Many studies used sampling technique [25] to resolve the class imbalance problem. We have not applied any sampling technique in our study rather we have given importance to the performance of all classification metrics. The synthetic data might break the purity of the data. Short stays have better accuracy compared to the long stays due to lacking in number of samples in long stays. Admission phase model can predict only short stays whereas post admission and discharge phase model predicts both the short and long stays well. This study using a weighted average ensemble model across three stages showed promising results, competing with the existing length of stay studies [22, 23, 25].

However this study has some limitations. Firstly, SPARCS do not share the lab test-related parameters for the patients. Those parameters might be impactful to accurately predict the length of stay. It also lacks the features related to number of available beds in the hospital, number of outpatient visits performed by the patient in that year, whether patients have any existing comorbidities like high blood pressure, diabetes. These information might have helped to predict the length of stay more accurately.

6 Conclusion

Forecasting how long a patient will stay in hospital is a potentially valuable strategy for healthcare resource management systems. Earlier studies on SPARCS dataset have considered only single phase models [22, 23, 25]. We have introduced a three stage model to predict length of stay to give flexibility to the patient and hospital party to get an estimation from admission to discharge. The accuracy got increased from admission to discharge phase with the addition of new features. The first prediction will be performed at admission with quite a few features. The performance at this stage is not that significant as availability of less information. In post admission stage, prediction will be performed with an increased number of features accumulated after hospitalization. This model is more robust and accurate compared to the admission model. The final prediction will be performed by the discharge model. This final stage model can be used as a reference point for the hospital authority and insurance companies to verify how well their models are contributing and will be utilized for analysis purposes. This three stage model will help in betterment of hospital admission procedures. These three phase predictions have been achieved by an ensemble model called 3WPE. For future scope perspective, there is a scope of improvement of the model's performance by adding a few other models into our analysis and build a more improved ensemble model.

Acknowledgements I am grateful to the HOD of computer science and engineering department and Vice chancellor of Jadavpur University for their help.

Disclosure of Interests We do not have any known competing financial interests or personal relationships that could have appeared to influence the work reported in this paper.

References

1. Abd-Elrazek MA et al (2021) Predicting length of stay in hospitals intensive care unit using general admission features. Ain Shams Eng J 12(2):1561–1570. https://doi.org/10.1016/j.asej.2021.02.018
2. Rouzabahman M et al (2017) Can cluster-boosted regression improve prediction of death and length of stay in the ICU? IEEE J Biomed Health Inform 21(3):851–858. https://doi.org/10.1109/JBHI.2016.2525731
3. Xie Y et al (2014) Predicting the number of hospitalization days based on health insurance claims data using bagged regression trees. In: Proceedings of the Annual International Conference of the IEEE Engineering in Medicine and Biology Society (EMBC), pp. 3314–3317. https://doi.org/10.1109/EMBC.2014.6944181
4. Xie Y et al (2016) Analyzing health insurance claims on different timescales to predict days in hospital. Journal of Biomedical Informatics 60:276–286. https://doi.org/10.1016/j.jbi.2016.01.002
5. Jabeen R et al (2016) Predicting days in hospital using health insurance claims. International Journal of Scientific Engineering and Technology Research (IJSETR) 5(29):6081–6086 [Online]. Available: http://ijsetr.com/uploads/5/3/2/9/53299371/ijsetr_vol_5_issue_29.pdf

6. Srimannarayanaa G et al (2021) Study on days of hospitalization of insured with the claims data. Turkish Journal of Computer and Mathematics Education (TURCOMAT), 12(3):5230–5238 [Online]. Available: https://turcomat.org/article/download.
7. Wang C et al (2017) Prediction of days in hospital for children using random forest. ResearchGate [Online]. Available: https://www.researchgate.net/
8. Bertsimas D et al (2020) Predicting inpatient flow at a major hospital using interpretable analytics. https://doi.org/10.1101/2020.05.12.20098848
9. Ortiz-Barrios M et al (2021) Predicting 15-day unplanned readmissions in hospitalization departments: an application of logistic regression, Ingeniare. Revista chilena de ingeniería 29(2):378–398
10. Prayaga L et al (2021) Predicting inpatient status for the next 30/60/90 days with machine learning. Int J Big Data Anal Healthcare 6(2):1–18. https://doi.org/10.4018/IJBDAH.20210701.oa9.
11. Chrusciel J et al (2021) The prediction of hospital length of stay using unstructured data. J Med Internet Res 23(12):e01722. https://doi.org/10.1186/s12911-021-01722-4.
12. Klein B et al (2023) Forecasting hospital-level COVID-19 admissions using real-time mobility data. Communications in Medicine 3(1):25. https://doi.org/10.1038/s43856-023-00253-5
13. Levin S et al (2020) Machine-learning-based hospital discharge predictions can support multidisciplinary rounds and decrease hospital length-of-stay. BMJ Innov 7(2). https://doi.org/10.1136/bmjinnov-2020-000420
14. Wilk M et al (2020) Predicting length of stay in hospital using electronic records available at the time of admission. Stud Health Technol Inform 270:191–195. https://doi.org/10.3233/SHTI200186
15. Hospital Inpatient Discharges—SPARCS De-Identified. New York State Department of Health, Health Data NY. https://health.data.ny.gov/dataset/Hospital-Inpatient-Discharges-SPARCS-De-Identified/22g3-z7e7
16. Reddy GT et al (2020) An ensemble based machine learning model for diabetic retinopathy classification. In: 2020 International Conference on Emerging Trends in Information Technology and Engineering (ic-ETITE). https://doi.org/10.1109/ic-ETITE47903.2020.235.
17. Rojarath A et al (2021) Cost sensitive probability for weighted voting in an ensemble model for multiclass classification problems. Applied Intelligence, 51(8):3692–3707. https://doi.org/10.1007/s10489-020-02106-3.
18. Gandhi I et al (2015) Hybrid ensemble of classifiers using voting. In: 2015 International Conference on Green Computing and Internet of Things (ICGCIoT). https://doi.org/10.1109/ICGCIoT.2015.7380496.
19. Akula R et al (2019) Supervised machine learning based Ensemble model; for accurate prediction of type 2 Diabetes, presented at the 2019 SoutheastCon, Huntsville, AL, USA, pp 11–14. https://doi.org/10.1109/SoutheastCon42311.2019.9020358.
20. Kumar A et al (2019) A two stage model to predict surgical patient's lengths of stay from an electronic patient database. IEEE J Biomed Health Inform 23(2):848–856. https://doi.org/10.1109/JBHI.2018.2819646
21. Tsai PF et al (2016) Length of hospital stay prediction at the admission stage for cardiology patients using artificial neural network. J Healthc Eng 2016:7035463. https://doi.org/10.1155/2016/7035463
22. Zeng X (2022) Length of stay prediction model of indoor patients based on SPARCS data using light gradient boosting machine, Hindawi. Comput Intell Neurosci 2022:9517029. https://doi.org/10.1155/2022/9517029
23. Jain R et al (2022) Machine learning models to predict length of stay in Hospitals using SPARCS data, presented at the 2022 IEEE 10th International Conference on Healthcare Informatics (ICHI), 11–14. https://doi.org/10.1109/ICHI54592.2022.00105
24. Rahman MM et al (2022) Hospital patients' length of stay prediction: a federated learning approach. J King Saud Univ - Comput Inf Sci 34(10A):7874–7884. https://doi.org/10.1016/j.jksuci.2022.07.006.

25. Alnsour Y et al (2023) Predicting patient length of stay using artificial intelligence to assist healthcare professionals in resource planning and scheduling decisions. J Global Inf Manage 31(1) [Online]. https://orcid.org/0000-0001-7198-7197. https://doi.org/10.4018/JGIM.323059.
26. Caetano N et al. A data-driven approach to predict hospital length of stay: a Portuguese case study. In: Proceedings of the 16th international conference on enterprise information systems (ICEIS-2014). https://doi.org/10.5220/0004892204070414
27. Jaotombo F et al (2022) Machine-learning prediction for hospital length of stay using a French medico-administrative database. Journal of Market Access & Health Policy, 11(1):2149318. https://doi.org/10.1080/20016689.2022.2149318.
28. Chen L et al (2022) Six machine-learning methods for predicting hospital-stay duration for patients with sepsis: a comparative study. In: IEEE Southeast Con 2022, New Jersey, United States of America, pp 302–309. https://doi.org/10.1109/SoutheastCon48659.2022.9764052.
29. Shahid Ansari M et al (2021) Identification of predictors and model for predicting prolonged length of stay in dengue patients. Health Care Management Science 24(4):786–798. https://doi.org/10.1007/s10729-021-09571-3
30. Ebinger J et al (2021) A machine learning algorithm predicts duration of hospitalization in COVID-19 patients. Intell.-Based Med 5:100035. https://doi.org/10.1016/j.ibmed.2021.100035.
31. Li C et al (2019) Prediction of length of stay on the intensive care Unit based on least absolute shrinkage and selection operator. IEEE Access 7:108981–108987. https://doi.org/10.1109/ACCESS.2019.2934166.
32. Gutierrez JMP et al (2021) Predicting length of stay across hospital departments. IEEE Access 9:38946–38958. https://doi.org/10.1109/ACCESS.2021.3066562.
33. McCormick PJ et al (2018) Validation of the all patient refined diagnosis related group (APR-DRG) risk of mortality and severity of illness modifiers as a measure of perioperative risk. J Med Syst 42(5):81. https://doi.org/10.1007/s10916-018-0936-3.
34. Deyo et al (2009) Mapping of Charlson diagnostic categories associated with Charlson disease weights (CDW) to clinical classifications system (CCS) codes. In: ICD-9-CM to CDW mapping

BEN-RS-ANN: An Innovative Approach for Revealing Emotion from Bengali Text with Exposure to Polysemy Resolution

Taniya Seal, Anuran Bhattacharya, Shatarupa Das, Saheli Patra, Debapratim Das Dawn, Abhinandan Khan, Sanjit Kumar Setua, and Rajat Kumar Pal

Abstract Polysemy resolution with comprehension of emotion detection is a decisive aspect of linguistic processing of Bengali text. The purpose of emotion detection in natural language processing is to increase the accuracy of sentiment and emotion analysis by resolving multiple meanings of words within a given context. In emotion identification, the same word can carry different emotional nuances based on its context, making polysemy resolution essential for precise interpretation. Emotion detection from textual content plays an important role in understanding human communication and behaviour in the digital age. In this context, this paper presents ten polysemous words with three types of emotions, namely, positive, negative and neutral, where each type of emotion consists of thirty documents. Additionally, this paper proposes an innovative concept for polysemy-based feature collection with a combination of measuring the significance of words by considering both their frequency in a document and their rarity across the entire corpus. It employs a resource-scarce artificial neural network (RS-ANN) classifier for revealing emotion from Bengali text with exposure to polysemy resolution.

Keywords Bengali natural language processing · Emotion detection from Bengali text · Resource-scarce artificial neural network

T. Seal (✉) · A. Bhattacharya · S. Das · S. Patra · A. Khan · S. K. Setua · R. K. Pal
Department of Computer Science and Engineering, University of Calcutta, Acharya Prafulla Chandra Roy Shiksha Prangan, JD-2, Sector-III, Saltlake 700106, Kolkata, West Bengal, India
e-mail: taniyaseal1992@gmail.com

D. Das Dawn
Department of Computer Science and Technology, Bundwan Polytechnic, Madhupur 723129, Purulia, West Bengal, India

A. Khan
Product Development and Diversification, ARP Engineering, 147 Nilgunj Road, Kolkata 700056, West Bengal, India

© The Author(s), under exclusive license to Springer Nature Singapore Pte Ltd. 2025

N. Chaki et al. (eds.), *Applied Computing for Software and Smart Systems*, Lecture Notes in Networks and Systems 980, https://doi.org/10.1007/978-981-97-9762-2_13

1 Introduction

Disambiguation is a critical aspect of *natural language processing* (NLP) because it allows us to accurately detect sentiment and emotion and recognise named entities. Many words in NLP have multiple meanings and *polysemy* resolution aims to disambiguate the correct sense of a word in a particular context. Although there is a lot of work on this domain in other languages, there has not been much work on this domain in the Bengali language, where the disambiguation of words in Bengali language is very high. However, several research papers have been published on the disambiguation of the Bengali language, but there is still a gap due to the high-level morphological complexity of the Bengali language. Over the last half decade, many researchers have done a lot of challenging work on Bengali polysemy resolution, but due to a lack of resources, and publicly accessible resources, Machine learning based linguistic research on the Bengali language has not progressed. However, polysemy resolution is an Artificial Intelligence complete problem, that is, a problem having its solution at least as hard as the most difficult problems. In the field of artificial intelligence, Bengali polysemy resolution specially deals with Bengali polysemous words in different pieces of text and provides the exact sense or exact emotion of that word for that particular context. Polysemy resolution refers to the task of disambiguating and determining the specific meaning of a word that has multiple meanings or senses in a given context. It involves identifying the intended sense of a polysemous word based on the surrounding words, and context. The goal is to enhance natural language processing systems by improving their ability to accurately interpret and understand the nuances of language, particularly when dealing with words that have diverse meanings in different contexts. Detecting the correct senses and emotions of Bengali polysemous words from their respective contexts is a challenging task, and proposed work presents this challenging task in a novel approach.

Polysemy resolution has some special applications in NLP. Polysemy resolution is important in machine translation, where a word can have different meanings. And if the exact meaning of that polysemous word is not known then it can be mistranslated. In that case, polysemy resolution plays a special role in translating from another language to Bengali or from Bengali to another language. Besides, in the case of information retrieval or question-answering systems, a question can be answered correctly by determining the correct sense or meaning of a word by polysemy resolution. There is no doubt that currently there are no information retrieval sites in Bengali language but if there are, they are very few. In that case, if we can make the polysemy resolution in Bengali language, then we will be able to create an information retrieval site in Bengali language, which means the work of information retrieval in the Bengali language will progress. Polysemy resolution is important in speech recognition for the same reason. In that case, any device can recognise Bengali words very easily and this system is especially needed in present times. These systems are very common in other languages but not so much in Bengali, so polysemy resolution on Bengali words is very important.

The rest of the paper is organised as follows. Section 2 provides a brief overview of the various literature reviews done in the field of polysemy resolution and emotion detection. In Sect. 3, we introduce our proposed method for emotion detection from Bengali text with polysemy resolution. Section 4 describes our proposed Bengali dataset. Section 5 describes the results of our work and Sect. 6 analyses the obtained results. Section 7 presents the relevant conclusions regarding this work.

2 Literature Survey

There is very little word on Bengali polysemous resolution, comparatively. It is currently in its infancy. Its main reason is the lack of a dataset.

However, for Bengali polysemous words, Haque and Hoque [1] generated a *word sense disambiguation* (WSD) system that could find the polysemous word with actual meaning from a Bengali sentence. This work stands on lexical and syntactic information along with semantic information. They built a machine-readable dictionary as they used a knowledge based approach to build their system. And this WSD system achieved 82.40% accuracy for limited Bengali sentences.

Pandit and Naskar [2] used the *k*-NN algorithm for polysemous resolution in Bengali. First, they removed stop words from the entire training and test set. Then, they ran a Bengali stemmer on the entire training test set and resolved the POS ambiguity of the target words by applying the Stanford POS tagger. Lastly, they used a *k*-NN-based WSD approach on the test set. It achieved 71% accuracy overall.

Pal et al. [3] presented a study on polysemous resolution in Bengali. They used two sequential phases. The first phase comprised four well-known approaches, *decision tree* (DT) (68.30% accuracy), *support vector machine* (SVM) (79% accuracy), *artificial neural network* (ANN) (78.23% accuracy) and *Naïve Bayes* (NB) (82.30% accuracy). In the second stage, a combined approach is proposed on the results obtained from the preliminary experiment. The first approach produces 68.30%, 79%, 78.23% and 82.30% accurate results, respectively. The second approach produces 70.92%, 79.15%, 79.53% and 83% accuracy, respectively.

Das Dawn et al. [4] proposed a dataset for evaluating polysemous resolution techniques in Bengali. They generated a dataset of 100 polysemous words, available publicly and can be downloaded for free. In this dataset, each sense of each polysemous word contains ten paragraphs with 3–5 lines, where each word consists of 3–4 disjoint senses. They also proposed a framework for supervised target-word WSD. They used six methods and achieved the highest accuracy of 80% among all the methods.

Although emotion detection has been done earlier in other texts, the work on emotion detection in Bengali texts has not been done in such a way. The work of detecting emotions in Bengali texts has been done by Rabeya et al. [5]. To demonstrate how often individuals express their emotions in the last part of a sentence, a lexicon-based backtracking method is introduced to identify the correct sentiment of the sentence that can give 77.16% accurate results.

A new corpus called 'BEmoD' is presented by Das et al. [6] for emotion classification in Bengali texts. Although there are several datasets available for emotion classification in Bengali, this work categorised 5200 texts of 'BEmoD' into 6 basic emotions including joy, fear, anger, sadness, surprise and disgust. This work explores each different category of emotion text, and which type of emotion words people use to express a particular emotion.

Ruposh and Hoque [7] proposed a computational approach for recognising emotion from the Bengali corpus. They generate a technique that can identify the six basic emotions, such as happiness, sadness, anger, fear, surprise and disgust. They also develop a corpus consisting of 1200 emotive words. Those words are used to train SVM classifiers to detect different emotions. Their proposed system can recognise emotion with 73% accuracy.

3 Proposed Methodology

The proposed methodology integrates a comprehensive data preprocessing phase to cleanse and standardise raw data, ensuring its quality and reliability. Subsequently, feature extraction techniques are employed to derive relevant information from the dataset, enhancing the model's ability to capture key patterns. Additionally, a meticulous feature selection process is implemented to identify and retain the most influential variables, optimising the model's performance and interpretability. This multifaceted approach aims to enhance the overall robustness and efficiency of the predictive modelling process. The flowchart in Fig. 1 illustrates the workflow of proposed polysemy resolution method with emotion detection.

3.1 Data Preprocessing

Data preprocessing is a more important part of Polysemy Resolution Emotion Detection from Bengali text. These processes remove or fix incorrect, corrupted, redundant, incorrectly formatted or incomplete data within a dataset. When much data is collected from many places and combined to make a dataset, then many duplicate, wrongly formatted data exist in it. Now, if the data is left in this state, then the algorithm and classifier will not work properly, and the outcome will not be expected. However, there is no specific process that cleans all datasets in the same way. Different datasets have different cleaning methods. Thus, we perform the data cleaning operations and convert the data to a raw format for further feature selection process.

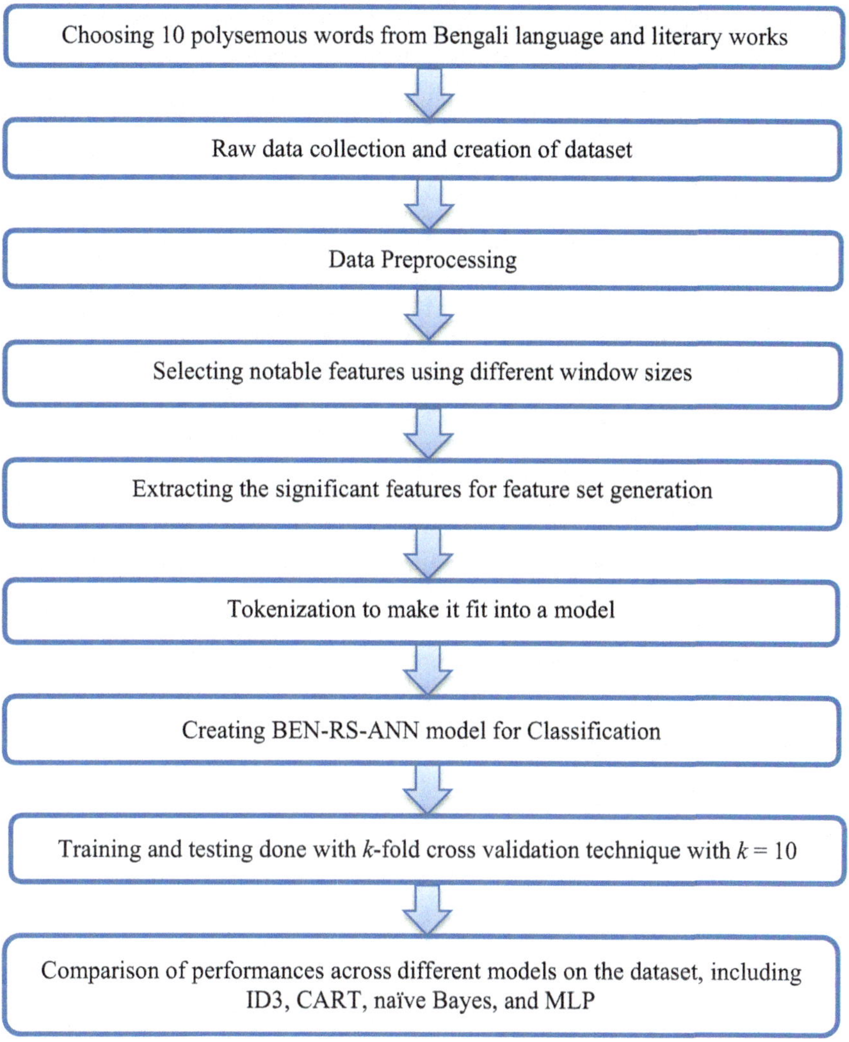

Fig. 1 Flowchart for the proposed system of emotion detection from Bengali text based on polysemy resolution

3.2 *Feature Selection*

Following the meticulous completion of the data cleaning phase, our subsequent pivotal step involves the meticulous selection of relevant features to perform the reduction in the input variables for the model by utilising only the relevant features thus reducing the problem of overfitting and enhancing generalisation. In the intricate domain of emotion detection, with a specific focus on resolving polysemy-related challenges, the nuanced significance of each word surrounding the target polysemous word becomes paramount and, in a paragraph, sometimes there may be an inherent dependency on the surrounding sentences. To methodically address this concern, we proposed a feature selection technique. In our comprehensive methodology, at first, we consider not only the immediate context but also the broader emotional context.

Specifically, we include the sentence immediately preceding and following the sentence containing the target word. This strategic inclusion captures the essence that the emotion associated with a given sentence can be influenced and potentially altered by the sentences immediately preceding and following it. By considering a final sentence from the paragraph along with one sentence before and after it, our feature selection approach accounts for the dynamic and interconnected nature of emotional expressions in text, contributing to the robustness and effectiveness of our emotion detection framework. Next, we introduce another meticulously structured approach to construct a corpus of sentences annotated with emotion labels. This process seamlessly incorporates feature selection techniques to discern and select the most relevant elements for our model.

Our chosen method involves a thoughtful modification of paragraphs, utilising the concept of window size to capture the contextual words surrounding the target polysemous word. Through this approach, we meticulously select features that contribute most effectively to the nuanced task of emotion detection. To delve into the specifics, our methodology introduces a novel approach to building context around the target word, denoted as T. Let S represents the sentence, W represents the window size and T represents the target word. The relationship to generate the context words around our target word is expressed as follows:

$$\mathrm{C\,window(T)} = \{\mathrm{S[P(T) - W], S[P(T) + W]}\} \tag{1}$$

Here, S[P(T) − W] represents the word in the sentence, S, located W positions to the left of the target word, T, S[P(T) + W] represents the word in the sentence, S, located W positions to the right of the target word, T and C window (T) denotes the set of context words meticulously selected to provide a comprehensive understanding of the emotional context surrounding the target polysemous word. This intricate process ensures that our model is trained on a rich and pertinent set of features, contributing to the robustness and effectiveness of our emotion detection framework. With the help of these obtained features, we then proceed with the feature extraction part with the idea of converting these texts to numerical representations.

3.3 Feature Extraction

The feature extraction part comprises various statistical and data analysis tools to dig out the feature set to be provided to the model for training. In our case, the feature set is the numerical representation of the text data and to obtain that a simplistic approach, maybe to incorporate the binary bag of words, can be employed, signifying the presence of words in each sentence. However, this approach lacks consideration for the importance of individual words within a sentence. In the context of emotion detection and the resolution of polysemy, the significance of each word surrounding the target polysemous word becomes crucial. To address this, we adopted the *term frequency-inverse document frequency* (TF-IDF) approach [8], which gauges the importance of each word by evaluating its frequency within a document and its rarity across the entire corpus.

Given that different senses of a polysemous word may elicit distinct emotional responses, and recognising that the surrounding words significantly influence polysemy, it becomes imperative to account for the significance and rarity of each word. The *term frequency* (TF) component measures how often a term appears in a specific document, calculated as the ratio of the number of times the term occurs to the total number of words in that document. This highlights the importance of a term within a specific context. The *inverse document frequency* (IDF) factor accounts for the rarity of a term across the entire corpus. It is calculated as the logarithm of the ratio of the total number of documents to the number of documents containing the term. This step helps mitigate the influence of commonly occurring words and emphasises the uniqueness of terms that are less frequent but more informative. The mathematical approach to computing both is as follows:

$$f_1(t, d) = \frac{\text{freq}(t, d)}{\sum \text{freq}(t', d)}, \tag{2}$$

where $f_1(t,\ d)$ represents the term frequency and inverse document frequency respectively, $\text{freq}(t, d)$ represents the number of times a word t appears in document d and $\sum \text{freq}\left(t', d\right)$ is the sum of the frequencies of all words in document d.

$$f_2(t, D) = \log\left(\frac{N}{df(t, D)}\right), \tag{3}$$

where $f_2(t, D)$ gives our inverse document frequency, N is the total number of documents in the corpus and $df(t, D)$ is the number of documents in the corpus that contain the word, t.

$$f_{12}(t, d, D) = f_1(t, d) \times f_2(t, D) \tag{4}$$

This results in a numerical score that represents the importance of the term, t in document d within the context of the entire corpus D. The TF-IDF matrix is then formed by applying this calculation to all unique terms in each document across the corpus. The final TF-IDF matrix is created from the third equation and this matrix is an $m \times n$ matrix, where m denotes the number of rows (number of documents in our case) and n denotes the number of columns (number of unique terms or features). The documents in our case are the sentences of the corpus and after creating the matrix we then proceed to test it on the classifiers and obtain the results.

3.4 Classification Model

After the feature extraction is completed, the next step is to train the obtained features. Four different popular classifiers are used, viz. ID3, CART [9], Naïve Bayes [10], MLP [11], and our proposed BEN-RS-ANN for the classification of Bengali sentence emotion based on word polysemy resolution. This model gave the most efficient result in terms of the accuracy score over the remaining four classifier models. The proposed RS-ANN model is trained using the *term frequency–inverse document frequency* (TF-IDF) matrix, which serves as a representation of the meticulously pre-processed dataset for Bengali sentence emotion classification. The training process involves minimising the categorical cross-entropy loss function and maximising accuracy through iterative adjustments of model parameters. This results in a well-optimised RS-ANN capable of discerning intricate emotional nuances in Bengali sentences.

The architecture of the RS-ANN comprises three layers: the first hidden layer, responsible for initial feature extraction, utilises neurons with the *leaky rectified linear unit* (Leaky ReLU) activation function. This function allows a small, non-zero gradient for negative inputs, preventing dead neurons and facilitating the learning of robust features from the TF-IDF matrix. The second hidden layer refines features extracted in the first layer, contributing to a hierarchical feature extraction process. This enables the model to capture complex patterns and relationships within the TF-IDF data, enhancing its ability to understand subtle emotional nuances in Bengali sentences. Deciding the number of neurons in the hidden layer is a very important part of determining the architecture of the neural network as the hidden layers play a vital role in the final output prediction. As a result of which, we have used a heuristic approach to determine the number of hidden layer units. The heuristic equation utilised is:

$$N = \sqrt{f \times c}, \tag{5}$$

where N is the number of hidden layer neurons, f is the number of features in the input data and c is the number of classes in the classification task. This heuristic aims to strike a balance between capturing the complexity of the data and avoiding overfitting.

The output layer of the RS-ANN produces the final classification based on learned features derived from the TF-IDF matrix. For emotion classification, the output layer likely has neurons corresponding to different emotion classes (e.g. *negative, neutral, positive*). The SoftMax activation function is commonly used in multi-class classification problems, converting raw output scores into probabilities, facilitating interpretation and comparison of the model's confidence in each emotion class. Categorical cross-entropy, a suitable choice for multi-class classification, measures dissimilarity between predicted probabilities and true class labels based on the TF-IDF features.

The Adam optimizer is employed to iteratively adjust model parameters during training, integrating the advantages of RMSprop and momentum. This approach provides a balanced and efficient means of updating weights in the RS-ANN based on the TF-IDF features. However, without a performance evaluation, the model is likely to get prone to overfitting issues—a reason for which we have added a performance evaluation on a validation set of the data as a proactive approach to address the issue of overfitting and hence potentially surpassing the performance of MLP. After successful training, the RS-ANN provides predictions for the emotional content of Bengali sentences, outperforming other classifiers like ID3, CART, Naïve Bayes and MLP in terms of accuracy. Figure 1 shows the architecture of our neural network model.

4 Dataset

A novel dataset in Bengali has been developed utilising news articles, social media posts and several narrative books, containing ten polysemous words with three distinct senses that related to distinct emotions. A total of 30 paragraphs were prepared on specific emotions (say, positive, negative, neutral) for each meaning of a specific polysemous word from 10 selected polysemous words. Each paragraph has at least 3 to 5 sentences. Here, 30 paragraphs each of 3 different senses of a polysemous word, i.e. a total of $30 \times 3 = 90$ paragraphs are prepared. Thus, the Bengali dataset is formed with $90 \times 10 = 900$ paragraphs of selected 10 polysemous words according to their respective senses and emotions. The help of various sources, newspapers, Facebook, blogs, Wikipedia, notable literary works of various authors, etc. have been taken in the construction of paragraphs. The nearest English translations and transliterations of the selected 10 polysemous words with 3 distinct senses and emotions are listed in Tables 1 and 2.

5 Experimental Results

We implemented the k-fold cross-validation [12] technique with the value of k selected as 10 and the window size as 5 (the selection for these is discussed in Sect. 6) and based on these our proposed model performed well in all the cases compared to other classifiers and hence was selected as the best model for our proposed work. The next two Sects. 5.1 and 5.2 highlight the details of the confusion matrix [13] and the word-wise accuracy calculation of our model.

5.1 BEN-RS-ANN Confusion Matrix

Unveiling the intricate landscape of emotion classification with the BEN-RS-ANN model, a pinnacle in *binary emotion recognition* utilising *recurrent neural networks.*

Table 1 A brief description of the dataset

	Word	Senses	
1	সিদ্ধি (Sid'dhi)	Positive	প্রাপ্তি (*Prāpti: Acquirement*)
		Negative	গঞ্জিকা (*Gañjikā: Drug*)
		Neutral	অষ্ট-সিদ্ধি (*Aṣṭa-sid'dhi: Realization*)
2	রাগ (Rāga)	Positive	অনু-রাগ (*Anu-rāga: Affection*)
		Negative	ক্রোধ (*Krōdha: Anger*)
		Neutral	সংগীতের স্বরবিন্যাস-পদ্ধতি (*Saṅgītēra sbarabin'yāsa-pad'dhati: Musical notation*)
3	হার (Hāra)	Positive	মালা (*Mālā: Garland*)
		Negative	পরাজয় (*Parājaẏa: Defeat*)
		Neutral	অনুপাত (*Anupāta: Ratio*)
4	কড়া (Kaṛā)	Positive	কঠোর (*Kaṭhōra: Rigorous*)
		Negative	চামড়ায় ঘর্ষণজনিত কাঠিন্য (*Cāmaṛāẏa gharṣaṇajanita kāṭhin'ya: Frictional hardness of the skin*)
		Neutral	কড়াই (*Kaṛā'i: Cooking pan*)
5	ফিট (Phiṭa)	Positive	উপযুক্ত (*Upayukta: Proper*)
		Negative	মূর্ছা যাওয়া (*Mūrchā yā'ōẏā:Fainting*)
		Neutral	গজের এক-তৃতীয়াংশ (*Gajēra ēka-tṛtīẏānśa: One-third of a yard*)

Table 2 A brief description of the dataset (contd…)

	Words	Senses	
6	পাকা (Pākā)	Positive	নিপুণ (*Nipuṇa: Skillful*)
		Negative	ডেঁপো (*Ḍēm̐pō: Precocious*)
		Neutral	পক্ব (*Pakba: Ripe*)
7	কালা (Kālā)	Positive	কৃষ্ণ (*Śrīkṛṣṇa: Shri Krishna*)
		Negative	বধির (*Badhira: Deaf*)
		Neutral	কালো (*Kālō: Black*)
8	মাথা (Māthā)	Positive	বোধশক্তি (*Bōdhaśakti: Intellect*)
		Negative	ব্যর্থতা (*Byarthatā: Failure*)
		Neutral	মস্তক (*Mastaka: Head*)
9	গরম (Garama)	Positive	উষ্ণ (*Uṣṇa: Warm*)
		Negative	ঔদ্ধত্য (*Aud'dhatya: Arrogance*)
		Neutral	গ্রীষ্ম (*Grīṣma: Summer*)
10	মেলা (Mēlā)	Positive	উৎসব পার্বণ ইত্যাদি উপলক্ষে আমোদ-প্রমোদের ব্যবস্থা যুক্ত অস্থায়ী হাট (*Utsaba pārbaṇa ityādi upalakṣē āmōda-pramōdēra byabasthā yukta asthāẏī hāṭa : Temporary bazaars with entertainment arrangements on the occasion of festivals, etc.*)
		Negative	অনেক (*Anēka: A lot*)
		Neutral	প্রসারিত করা (*Prasārita karā: To expand*)

Fig. 2 RS-ANN architecture

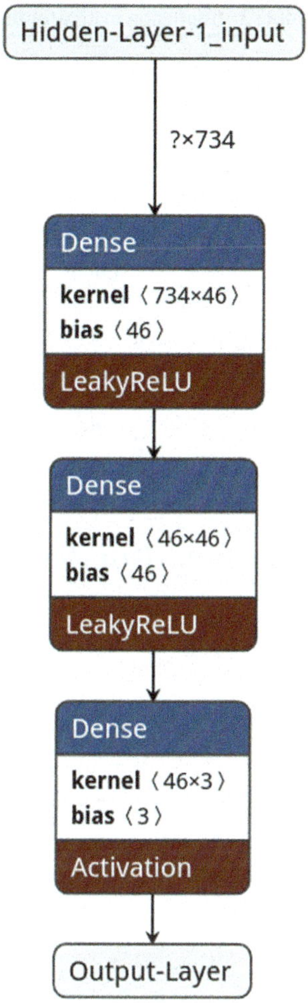

This sophisticated model harnesses the power of advanced neural networks to decipher and categorise emotions. The ensuing confusion matrix as shown in Table 3 serves as a captivating mosaic, revealing the nuanced interplay of true positives, false positives, true negatives and false negatives. Within its intricacies lies a profound exploration of BEN-RS-ANN's prowess in capturing the subtle nuances of human emotions, providing a visual narrative of its discernment capabilities.

Table 3 Confusion matrix for the dataset

		ফিট			গরম			কালা			কড়া			মাথা		
		Actual			Actual			Actual			Actual			Actual		
		√	×	–	√	×	–	√	×	–	√	×	–	√	×	–
Predicted	√	46	0	3	25	2	0	28	4	6	59	3	2	23	3	4
	×	0	49	3	7	76	2	1	64	1	7	16	2	1	90	0
	–	2	0	43	5	3	80	3	7	42	8	0	48	3	4	24
		মেলা			পাকা			রাগ			সিদ্ধি			হার		
		Actual			Actual			Actual			Actual			Actual		
		√	×	–	√	×	–	√	×	–	√	×	–	√	×	–
Predicted	√	28	2	3	53	0	0	43	3	5	44	0	0	48	6	3
	×	1	36	2	1	83	0	4	53	1	2	45	5	0	100	1
	–	0	4	35	3	8	23	4	1	43	0	4	53	2	5	43

5.2 *Word-Wise Accuracy Calculation*

The accuracy of each word for the selected model of BEN-RS-ANN is shown in Fig. 3 for a detailed score analysis of each word trained in our model. Therefore, in the proposed work of emotion detection in Bengali text with the exposure of Bengali polysemy word resolution our RS-ANN model gave an average accuracy of 90.08% after applying our heuristic method to select the hidden layers of the network and this model's accuracy is considered the best among the other 4 classifiers applied.

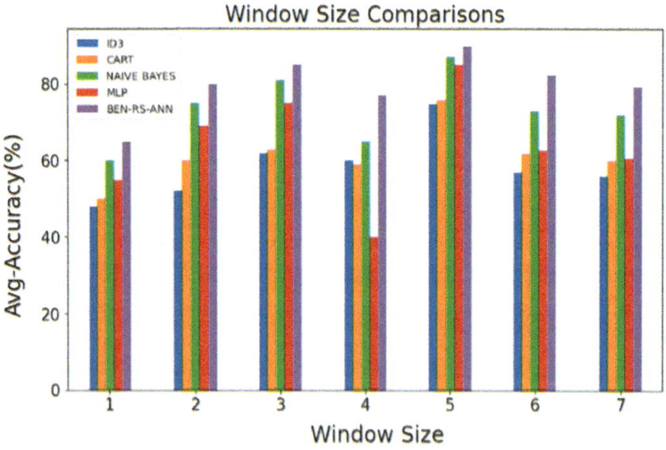

Fig. 3 Comparisons with state-of-the-art-technologies with different window size

6 Results Analysis

To start our model prediction work, we have used the k-fold cross-validation approach to assess the performance and generalisation of a machine learning model. k-fold cross-validation involves partitioning the dataset into k subsets or folds. The model is then trained on $k - 1$ folds and validated on the remaining fold. This process is repeated k times, with each fold serving as the validation set exactly once. The performance metrics are averaged over the k folds to provide a more robust evaluation of the model. Initially, we compared the performance of each classifier and our BEN-RS-ANN model with various values of k among 2, 5, 10, 15, 20, 25, 30 and 35 to select the base value for further experiment result generation and to select the most suitable window size.

6.1 Comparative Results Analysis with Respect to State-of-the-Art Technologies

Introducing the result analysis with state-of-the-art technology, we have leveraged advanced analytics, cutting-edge machine learning algorithms and sophisticated visualisation tools [14]. In this context, the graph as shown in Fig. 3 serves as a visual representation of key insights, shedding light on intricate patterns and trends extracted from our meticulously processed data. This graphical representation aims to enhance the understanding of the profound implications derived from our state-of-the-art analytical approach. In Fig. 3, the MLP classifier which is a type of *artificial neural network* (ANN) with the most basic architecture of ANN has achieved a lower accuracy score than our proposed BEN-RS-ANN model which utilises a heuristic approach to select a favourable number of hidden layer neurons making the working with less resource data more efficient.

6.2 Result Analysis with Variation of Window Sizes

Next, we generate bar graphs as shown in Fig. 4 based on the average accuracies of the 10 words obtained after training on all the models and based on their performance, we selected 5 as our window size as it achieved the best accuracy score in all cases and after 5 the accuracy scores start to decrease.

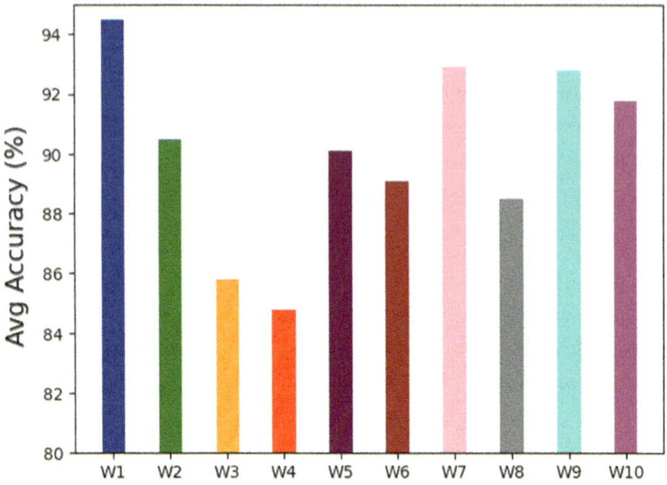

Fig. 4 Comparison of accuracy of each word

6.3 Average Accuracy with Variation of k in the k-Fold Cross-Validation

To select the best value of k, based on the performances of each model, we obtained the set of average accuracies of each and then chose the value of k as 10 with the analysis shown in Fig. 6. In analysing the line graphs for each set of values for, it was observed that set 10 exhibited a relatively stable and moderate upward trend for the average accuracy estimates. The values in set 10 showed consistent growth with moderate fluctuations, indicating a balanced pattern. Also, after 10 there is constant growth in other line plots like $k = 10$ with slight variations as a result of which $k = 10$ is selected as the base for other operations like best windows size selection and generating the performance evaluation tool.

The precision, recall and F-score of proposed classification model for each word are shown in Table 4.

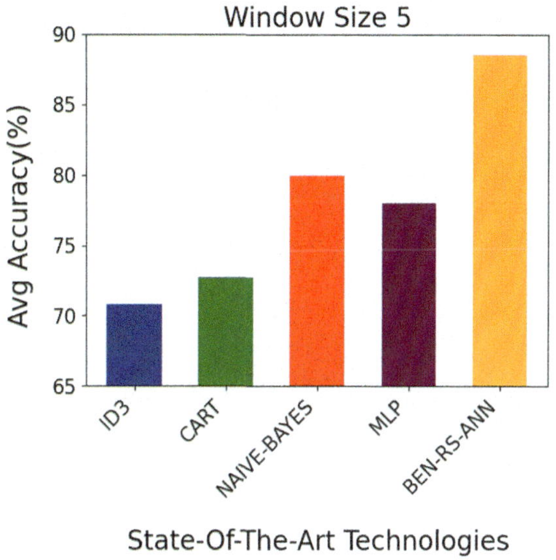

Fig. 5 Result comparison (Accuracy) with respect to state-of-the-art technologies with window size 5

Table 4 Precision, recall and F-score of BEN-RS-ANN for each word

	Precision	Recall	F-score
W1	0.9583	0.9857	0.9718
W2	0.9944	0.9945	0.9933
W3	0.8618	0.9006	0.8805
W4	0.8533	0.8983	0.8755
W5	0.9072	0.9149	0.9110
W6	0.9087	0.9124	0.9103
W7	0.9149	0.9179	0.9162
W8	0.9099	0.9132	0.9114
W9	0.9132	0.9237	0.9184
W10	0.8974	0.9400	0.9174
Average	0.9119	0.9301	0.9206

7 Conclusion

The work represents a new dataset consisting of 10 polysemy words, where each word carries 3 distinct senses associated with 3 distinct emotions (*positive*, *negative*, *neutral*) and each sense consists of 30 distinct paragraphs, which can be also used in future in any work with Bengali polysemy words. To do the work properly, remove unnecessary punctuation from the dataset. To make our emotion detection system better, we carefully choose important features and reduce unnecessary information. We focus on solving challenges related to multiple meanings of words, especially when a word has different meanings. We suggest a method where we look not only

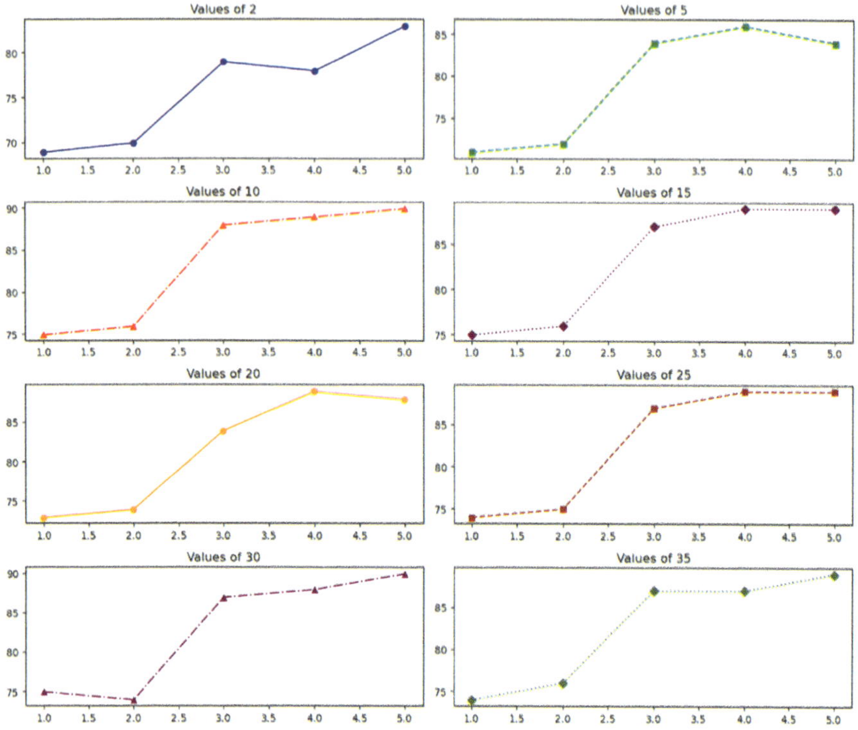

Fig. 6 Line graphs for k-value comparisons

at the sentence with the target word but also the sentences before and after it. This helps us understand the emotions better in a paragraph and improves the accuracy of our emotion detection system. Then, in the feature extraction stage, we create the $m \times n$ matrix, where m represents the number of documents and n number of unique features. Now, in the next step, it trains the obtained features for 4 different classifiers; these are ID3, CART, Naïve Bayes and MLP. It achieves 74.85%, 75.76%, 85% and 83% accuracy, respectively. Finally, we trained it with our proposed classification model, BEN-RS-ANN (that is generated for the classification of Bengali emotion on polysemy words) and received an accuracy of 88.59% after that, we even applied our heuristic method to determine the number of hidden neurons and got an improvement to 90.08% accuracy overall. In that case, proposed classification model provides more accuracy than the previously mentioned well-known models. Remember, in that case, TF-IDF in feature extraction generates a $m \times n$ sparse matrix, also this approach cannot build any semantic links between words. It does not consider the semantic meaning of words; it cannot understand the context of the words.

References

1. Haque A, Hoque MM (2016) Bangla word sense disambiguation system using dictionary based approach. ICAICT, Bangladesh
2. Pandit R, Naskar SK (2015) A memory based approach to word sense disambiguation in Bengali using *k*-NN method. In: IEEE 2nd international conference on recent trends in information systems (ReTIS). IEEE, pp 383–386
3. Pal AR, Saha D, Naskar SK, Dash NS (2021) In search of a suitable method for disambiguation of word senses in Bengali. Int J Speech Technol 24:439–454
4. Dawn DD, Khan A, Shaikh SH, Pal RH (2022) A dataset for evaluating Bengali word sense disambiguation techniques. J Ambient Intell Hum Comput:1–30
5. Rabeya T, Ferdous S, Ali HS, Chakraborty NR (2017) A survey on emotion detection: a lexicon based backtracking approach for detecting emotion from Bengali text. In: 20th International conference of computer and information technology (ICCIT). IEEE, pp 1–7
6. Das A, Iqbal MDA, Sharif O, Hoque MM (2020) BEmoD: development of Bengali emotion dataset for classifying expressions of emotion in texts. In: International conference on intelligent computing and optimization. Springer International Publishing, Cham, pp 1124–1136
7. Ruposh HA, Hoque MM (2019) A computational approach of recognising emotion from Bengali texts. In: 5th International conference on advances in electrical engineering (ICAEE). IEEE, pp 570–574
8. Luhn H (1957) A statistical approach to mechanized encoding and searching of literary information. IBM J Res Dev 1(4):309–317
9. Quinlan J (1986) Induction of decision trees. Machine Learn 1:81–106. Gordon A, Breiman L, Friedman J, Olshen R, Stone C (1984) Classification and regression trees. Biometrics 40(3):874
10. Sarkar K, Bhowmick M (2017) Sentiment polarity detection in Bengali tweets using multinomial naive Bayes and support vector machines. In 2017 IEEE Calcutta conference (CALCON), pp 31–36
11. Rosenblatt F (1958) The perceptron: a probabilistic model for information storage and organization in the brain. Psychol Rev 65(6):386
12. Anguita D, Ghelardoni L, Ghio A, Oneto L, Ridella S, Others (2012) The 'K' in K-fold cross validation. In: ESANN, pp 441–446
13. Townsend J (1971) Theoretical analysis of an alphabetic confusion matrix. Percept Psychophys 9:40–50
14. Midway S (2020) Principles of effective data visualization. Patterns 1(9)

Flood Detection in UAV Images of Urban Area Using Machine Learning and Deep Learning Techniques

Tirthankar Ghosh, Arati Paul, and Nabendu Chaki

Abstract The present work uses unmanned aerial vehicle (UAV) images for flood detection in urban areas. Two Machine Learning (ML) techniques, namely k-nearest neighbours (KNN) and random forest, are initially employed to classify flood-affected areas based on extracted features from the UAV images. Subsequently, two Deep Learning (DL) methods, UNet and DeepLabv3+, are implemented for semantic segmentation of the images to obtain precise flood maps. The evaluation metrics include accuracy, precision, recall, and F1-score. This work demonstrates the effectiveness of different ML and DL techniques for flood detection using UAV images in urban areas. The comparison reveals that UNet offers superior accuracy compared to KNN, random forest, and DeepLabv3+. The higher accuracy of UNet is attributed to its ability to capture both local and global contextual information, enabling precise delineation of flood extents in urban areas. The findings also emphasize the significance of deep learning approaches in solving similar problems.

Keywords Flood map · UAV image · Urban flood · Deep learning · Machine learning

Arati Paul, Nabendu Chaki are the authors contributed equally to this work.

T. Ghosh (✉)
Department of Computer Science and Engineering, Techno International New Town, Block—DG 1/1, Action Area 1, New Town, Rajarhat, Kolkata 700156, West Bengal, India
e-mail: tirthankarkumarghosh@gmail.com

A. Paul
RRSC-East, NRSC, Indian Space Research Organisation, BG-2, AA-1B, New Town, Kolkata 700156, West Bengal, India

N. Chaki
Department of Computer Science and Engineering, University of Calcutta, JD Block, Sector 3, Kolkata 700106, West Bengal, India

© The Author(s), under exclusive license to Springer Nature Singapore Pte Ltd. 2025
N. Chaki et al. (eds.), *Applied Computing for Software and Smart Systems*, Lecture Notes in Networks and Systems 980, https://doi.org/10.1007/978-981-97-9762-2_14

1 Introduction

Remote sensing (RS) plays an important role in the field of disaster support and management. Frequent and severe natural disasters like floods result in economic losses, social disruption, and loss of life. RS imagery provides cost-effective solutions for generating flood inundation maps that enable us to take proactive measures for swift recovery and preparedness [1]. Visual images of the flood-affected area are critically important for analysing the severity of the disaster in order to provide a quick response and recovery. However, the availability of high-resolution (HR) satellite data in visible bands is limited due to cloud cover during the flood. The emergence of small unmanned aerial vehicles (UAVs) with inexpensive sensors presents the opportunity to collect large-scale high-resolution data after the disaster and facilitates rapid data collection from hard-to-reach areas, where data collection tasks can be unsafe for humans if not impossible. However, analysing and extracting meaningful information from such large datasets remains challenging for user communities. Recent advancements in the field of machine learning (ML) and artificial intelligence (AI) have catalyzed a surge in interdisciplinary research and development endeavours. In the present work, ML/AI-based methods are applied to extract flood-inundated regions in urban areas using UAV images. Figure 1 presents the graphical abstract of the overall workflow.

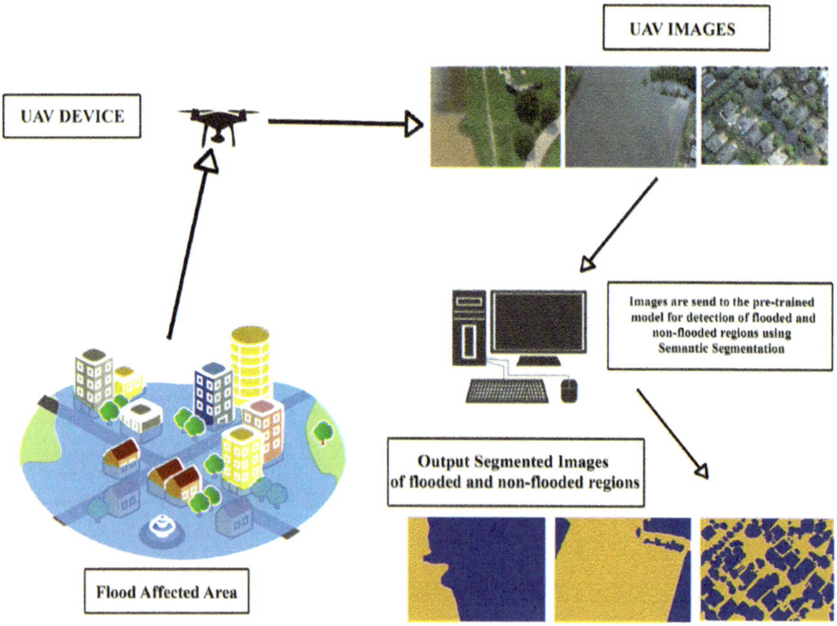

Fig. 1 Graphical abstract of the overall workflow

The primary objective is to enhance flood monitoring and to build better response capabilities in urban areas. This has been targeted with the leverage of RS imagery, specifically through the application of high-resolution UAV images. The focus is on developing appropriate flood detection and monitoring methodology using ML and deep learning (DL). In the present work, four ML and DL-based techniques are implemented for the extraction of flood-inundated regions in UAV data. Subsequently, their performances are compared using standard metrics to suggest the best-performing method for the said purpose.

2 Literature Review

Flood monitoring in urban areas using UAV images has been an active research area in recent years, and one technique that has gained attention is semantic segmentation. This study [2] provides a solution using UAV/Drones for observing the disaster occurred area. The drones over here continuously take pictures and send them to a central hub for observation and analysis of the type of disaster. A deployed deep learning architecture utilizing feature concatenation operates at the central hub for disaster classification. It's designed to intake image data captured by a UAV surveying the impacted area. The proposed model demonstrates superior efficiency, outperforming leading CNN models and closely related recent studies in terms of accuracy. A study by [3] proposed a semantic segmentation-based method for detecting and mapping flooded areas in urban environments using UAV images. Similarly, a study by [4] used a semantic segmentation-based approach to map flooded areas in an urban region using UAV images. In another study, [5] proposed a semantic segmentation-based method for detecting and mapping flooded areas in urban environments using multispectral UAV images and demonstrated the potential of the approach in mapping flooded areas in complex urban environments. Additionally, [4] developed a semantic segmentation-based method to map flooded areas in urban regions using UAV images with varying levels of image resolution using a modified UNet architecture. Another study [6] used a semantic segmentation-based approach to detect and map flooded areas in an urban environment using multi-temporal UAV images. The study utilized a CNN-based model to segment images into flood and non-flood categories and demonstrated the effectiveness of the approach in mapping changes in flooded areas over time. In the study by [7], a comparative analysis has been performed between two deep learning architectures UNet and DeepLabV3, where, according to the dataset, DeepLabV3 gives better performance with respect to the performance measurement metrics like precision, recall, IOU, and overall accuracy. Several studies have demonstrated the effectiveness of DeepLabv3+ for flood monitoring using UAV images. For instance, [8] applied DeepLabv3+ to segment flood-affected areas in urban regions using high-resolution UAV imagery. This paper [9] addresses challenges in semantic information extraction for disaster relief using UAV networks. Disrupted ground data distribution post-disaster leads to biased semantic segmentation, affecting rescue operations. The article [10] discusses the challenges in utilizing

Table 1 A summary of existing post-disaster UAV datasets

Dataset	Resolution of images
AIDER [16]	Varies in size
ISBDA [17]	Varies in size
FloodNet [6]	4000 × 3000

data-driven deep learning techniques for urban flood mapping due to limited annotation data availability and presents solutions through weakly supervised semantic segmentation (WSSS). Another study by [11] used DeepLabv3+ for flood mapping in rural areas using UAV images which highlighted the effectiveness in handling complex flood scenarios and its potential for real-time flood monitoring applications. Deep learning-based methods, such as UNet [12] and DeepLabv3+ [13], have gained popularity due to their ability to learn hierarchical representations from raw image data, capturing both local and global contextual information. Random forests [14], on the other hand, are ensemble learning methods that combine multiple decision trees to make predictions. K-nearest neighbour (KNN), a non-parametric algorithm, is another popular ML technique for image segmentation or classification [15].

3 Dataset Description

The present research focuses on extracting flooded/inundated regions in UAV images pertaining to urban areas. Some post-disaster UAV datasets are listed in Table 1. It is important to note that these datasets' availability, quality, and suitability may vary depending on the specific research questions and objectives. After careful evaluation, the FloodNet [6] dataset is found to be the most appropriate for the present work.

Flooded area extraction in UAV images using ML/AI requires a good number of images and corresponding labels. FloodNet[1,2] dataset comprises imagery captured utilizing DJI Mavic Pro quadcopters following the devastation caused by Hurricane Harvey in August 2017. This collection effort was concentrated in Ford Bend County, Texas, and adjacent areas directly affected by the hurricane. Spanning multiple flights conducted from August 30 to September 04, 2017, the dataset offers valuable insights into the aftermath of the hurricane. It provides a rich resource for studying flood dynamics and impacts in urban areas. A concise overview of the available image quantity and corresponding instances is given in Table 2 .

FloodNet dataset could be a good option for semantic segmentation. However, some of the challenges associated with UAV dataset are as follows:

[1] FloodNet dataset link1: https://drive.google.com/drive/folders/1sZZMJkbqJNbHgebKvHzcXYZHJd6ss4tH?usp=sharing.

[2] FloodNet dataset link2: https://drive.google.com/drive/folders/1g1r419bWBe4GEF-7si5DqWCjxiC8ErnY?usp=sharing.

Table 2 Number of images available under different classes

Object class	Images
Building-flooded	245
Building-non-flooded	880
Road-flooded	264
Road-non-flooded	1175
Vehicle	813
Pool	531
Tree	1885
Water	984

1. **Lack of a proper large dataset**: The lack of a proper dataset containing images with target objects (here, flooded area) and their respective masks/labels.
2. **Post-disaster data acquisition**: Post-disaster data acquisition is the most challenging task in drone imagery. Strong winds and rain can lead to wrong results and may also lead to damage to the devices.
3. **Very high-resolution data**: Unlike satellite or aerial images, UAV provides higher-resolution imagery. Therefore, similar algorithms/processes may not always yield good results for all the UAV datasets. Hence, performing suitable preprocessing operations is required to produce satisfactory output for a defined objective using UAV data.
4. **Reflection and shadow effect**: Since UAV images are acquired from a lower altitude, sometimes these images are affected by cloud/solar reflection on water, in addition to the shadows of trees, buildings and clouds. Few sample images, affected by reflection/shadow effects, are depicted in Fig. 2.

The following potential reasons are considered while choosing the FloodNet dataset in the present work:

1. **Large and diverse dataset**: The FloodNet dataset is a large and diverse dataset that includes over many labelled UAV images.
2. **High-quality labels**: The FloodNet dataset has been labelled by experts, which means that the labels are likely to be accurate and reliable.

(a) (b) (c) (d)

Fig. 2 Images showing reflection and shadow effects: (**a**) cloud reflection, (**b**) tree shadow, (**c**) solar reflection, (**d**) cloud shadow

3. **State-of-the-art performance**: The FloodNet method has demonstrated state-of-the-art performance in flood mapping using satellite images. This suggests that the dataset is suitable for training models that can achieve high levels of accuracy and generalization.
4. **Open access**: The FloodNet dataset is publicly available, which means that it is easy to access and use for research and development purposes.

4 Data Preprocessing

This section describes different data preprocessing steps involved in the present work. Although the FloodNet dataset provides many UAV images, only a portion of this dataset, which contains flooded images with labels for 'Building-flooded' and 'Road-flooded', is usable in the present work. Only 45 such labelled samples are used in the present work which are further divided into training, testing, and validation sets in 60:20:20 proportions, respectively. Training and testing samples are used in developing and tuning the ML/ DL-based models, while the validation set is used for performance evaluation. The sample dataset is shown in Fig. 3. Developing the ML/DL-based model required the image and the corresponding mask to be of equal dimensions. While verifying so, it was observed that the images are of 4532 × 3046

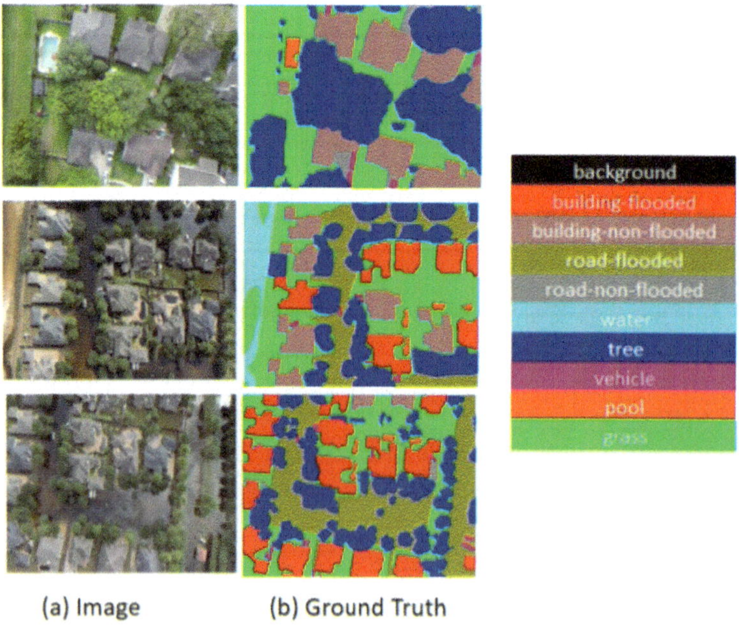

Fig. 3 Sample dataset

dimensions while their masks are 4000 × 3000. Hence, images are cropped with an approximation of the respective masks. Then, the dataset is re-scaled in the range [0,255] for easy handling. The original labelled masks consist of 8 classes. However, the present objective requires only 2 class labels, i.e., flooded and non-flooded. Therefore, the 'building flooded' and 'road-flooded' masks are renamed as 'flooded', and the rest class labels are assigned as 'non-flooded' in the labelled masks.

5 Methodology

The literature review in Sect. 2 shows that in the domain of semantic segmentation with UAV images, researchers have identified several effective ML and DL methods. Among ML approaches, KNN and Random Forest (RF) classification have demonstrated significant promise in accurately delineating semantic features in UAV images. On the other hand, DL methods such as UNet and DeepLabv3+ have emerged as powerful tools for semantic segmentation tasks.

5.1 Processing Images with ML Techniques

The ascendancy of ML methodologies has positioned them as potent instruments for image segmentation, offering automated and accurate solutions. These techniques employ various algorithms, such as RF, support vector machines (SVMs), and KNN, to tackle the segmentation task.

5.1.1 Image Segmentation Using RF Classifier

Algorithm 1: Flood Detection using Random Forest Classifier and UAV Images

Input: UAV image dataset D, Training dataset T, Number of trees in the forest N
Output: Flood detection results

Training: Train the Random Forest classifier using the training dataset T with N trees;

Testing: foreach *UAV image I in dataset D* **do**
  Apply preprocessing steps to normalize and enhance the image;
  Extract relevant features from I (e.g., color, texture, shape);
  Classify I using the trained Random Forest classifier;
  if *the majority of trees classify I as flooded* **then**
    Add I to the flood detection results;

The Algorithm 1 states the general steps for performing image segmentation using an RF classifier. The algorithm takes as input a training dataset T, an image to be segmented I, and the number of trees in the forest N.

The RF classifier[3] is trained using the training dataset T with N trees during the training phase. Once the classifier is trained, the algorithm moves to the segmentation phase. For each patch or pixel P in image I, preprocessing steps are applied to normalize and enhance the patch or pixel. Relevant features are extracted from P. Then, the RF classifier is used to classify P based on the learned decision boundaries. The corresponding class label is assigned to P in the segmented image. During the testing phase, the algorithm iterates over each UAV image I in dataset D. The values for N are considered as 10, 50, and 100.

5.1.2 Image Segmentation Using KNN Classifier

The Algorithm 2 gives the generalized processing steps for flood detection using the KNN classifier[4] in UAV images. The algorithm considers the input dataset as D, which contains all the UAV images, a training dataset T, and the number of nearest neighbours k. The goal is to classify each UAV image in D in flooded and/or not-flooded regions.

The KNN classifier is trained using the training dataset T and k nearest neighbours during the training phase. In the testing phase, each UAV image I in the dataset D undergoes preprocessing steps for normalization and enhancement. Relevant features, such as colour, texture, and shape, are then extracted from the preprocessed image. The values of k are considered as 5, 10, and 20.

Algorithm 2: Flood Detection using k-NN Classifier and UAV Images

Input: UAV image dataset D, Training dataset T, Number of nearest neighbours k
Output: Flood detection results

Training: Train the KNN classifier using the training dataset T and k nearest neighbours;

Testing: **for** *each UAV image I in the dataset D* **do**
- Apply preprocessing steps to normalize and enhance the image;
- Extract relevant features from I (e.g., colour, texture, shape);
- Classify I using the trained KNN classifier;
- **if** *the majority of neighbours are classified as flooded* **then**
  - Add I to the flood detection results;

[3] RF classifier: https://github.com/mahesh147/Random-Forest-Classifier.

[4] KNN classifier: https://github.com/vivek2319/K-Nearest-Neighbors.

5.2 Processing Images with DL Techniques

Semantic segmentation using DL techniques proved to be highly effective for flood monitoring using UAV imagery. Convolutional neural networks (CNNs) can efficiently extract features from images. Deep neural networks (DNNs), such as UNet or DeepLabv3+, can accurately detect flood-affected areas in UAV images.

5.2.1 Image Segmentation Using DeepLabv3+

The basic DeepLabv3+ algorithm is given in Algorithm 3, and the network architecture is given in Fig. 4. Since the size of the training set is very small here, a pre-trained model M[5] is considered in the present work for implementing the DeepLabv3+ algorithm. The algorithm takes as input a UAV image from dataset D and a pre-trained DeepLabv3+ model M. The measure used for loss calculation is 'dice loss', and the optimizer is Adam, with a learning rate of 0.001. A batch size of 15 images was taken, and the model was trained for 20 epochs. During the training phase, the algorithm iterates through each UAV image I in the training set T. Preprocessing steps are applied to normalize and enhance the image, ensuring optimal input for the

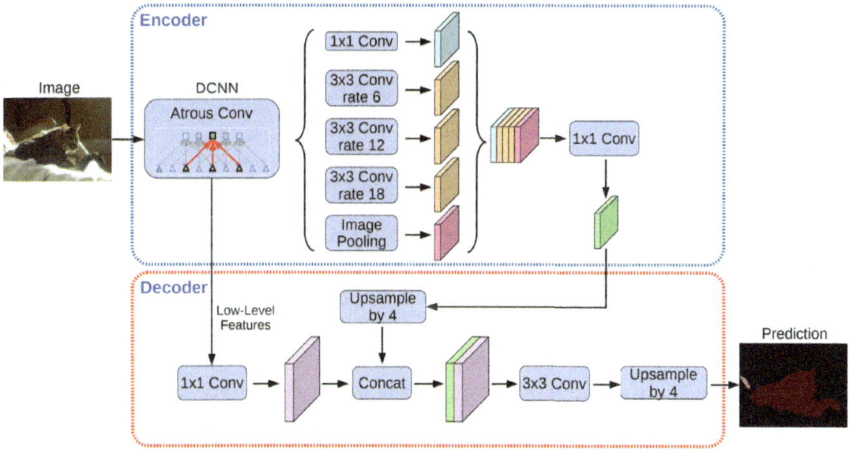

Fig. 4 DeepLabv3+ model architecture [3]

[5] DeepLabv3+: https://github.com/lattice-ai/DeepLabV3-Plus.

DeepLabv3+ model. Once the model is trained, the image I from D is fed into the model M to generate a predicted segmentation mask $\hat{y}$.

Algorithm 3: Semantic Segmentation with DeepLabv3+

Input: Training set of images $(x_i, y_i)_{i=1}^{n}$, where x_i is an image and y_i is its corresponding segmentation mask, Learning rate α, Number of epochs N

Output: Trained DeepLabv3+ model M

Initialization: Randomly initialize the DeepLabv3+ model parameters W;

Training: for $i = 1$ *to* N **do**

  for *each image-label pair* (x_i, y_i) *in the training set* **do**

    Forward pass: Compute the output of the DeepLabv3+ model M given input x_i to get the predicted segmentation mask $\hat{y}_i$;

    Compute the loss function $L(y_i, \hat{y}_i)$ between the true segmentation mask y_i and the predicted mask $\hat{y}_i$;

    Backward pass: Compute the gradient of the loss function with respect to the model parameters $\nabla_W L(y_i, \hat{y}_i)$;

    Update the model parameters using the gradient descent optimization algorithm: $W \leftarrow W - \alpha \nabla_W L(y_i, \hat{y}_i)$;

Testing: for *each test image* x **do**

  Forward pass: Compute the output of the DeepLabv3+ model M given input x to get the predicted segmentation mask $\hat{y}$;

5.2.2 Semantic Segmentation with UNet

The Algorithm 4 illustrates the methodology that is followed in the present work to detect the flooded area in UAV images using UNet. The algorithm starts with a preprocessing step where the UAV images are read and divided into smaller patches of dimension 256 × 256 as required by the network (Fig. 5). Firstly, the images and corresponding masks are cropped /resized to a dimension divisible by 256, and then they are divided into smaller patches. Subsequently, the mask values are one hot encoded. This preprocessed dataset is subsequently used for training and predicting using the UNet model.

The pre-trained UNet model[6] is loaded and assembled in the training phase with the appropriate loss function and optimizer. The measure used for loss is 'total loss', calculated by the summation of dice loss and (1 * focal loss), and the optimizer is Adam, with a learning rate of 0.001. A batch size of 15 images was taken, and the model was trained for 100 epochs using the training dataset. The trained model is applied to the validation images to estimate performance metrics. The output consists

[6] UNet: https://github.com/bnsreenu/python_for_microscopists/tree/master/228_semantic_segmentation_of_aerial_imagery_using_unet.

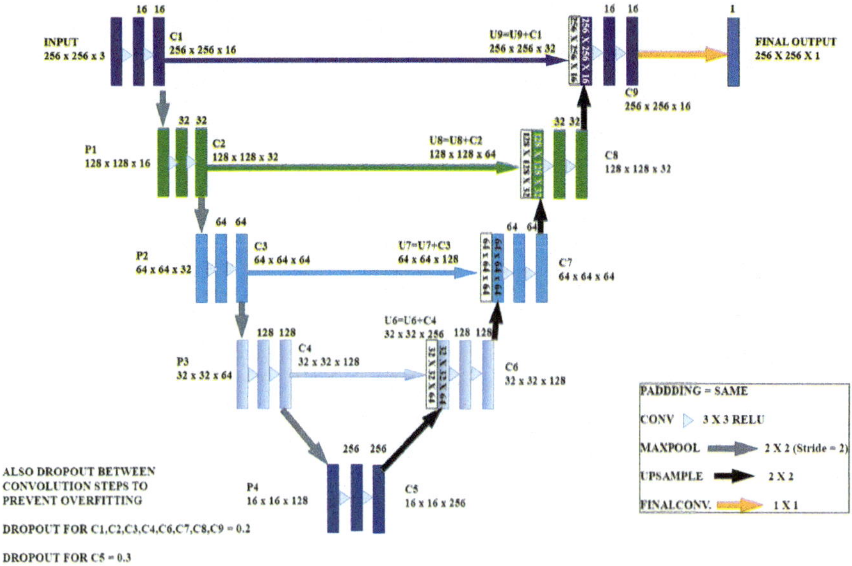

Fig. 5 UNet architecture

of a semantic segmentation map, highlighting flooded regions in images. In the post-processing step, the resulting patches are merged together to form the mask of the original dimension. Lastly, smooth blending technique is used for a better merging of the predicted images. The predicted masks representing the flooded areas are overlaid on the original UAV images to identify the affected areas.

Algorithm 4: Flood Monitoring Algorithm using UNet and UAV images

---

Input: UAV images of a flood-affected area

Output: Semantic segmentation map of the area with flooded regions highlighted

Pre-processing: Read in UAV images; Divide images into smaller sections; Apply data augmentation techniques to increase dataset size;

Training: Load pre-trained UNet model; Compile model with appropriate loss function and optimizer; Fit model to the training data using early stopping and model check pointing techniques;

Inference: Load trained model; Apply model to unseen UAV images; Generate semantic segmentation map; Threshold the map to identify flooded regions;

Post-processing: Overlay flooded regions on original UAV images; Calculate flooded area and severity; Save segmented images and flood information for future analysis;

---

6 Results

The ML (RF, KNN) and DL (DeepLabv3+, UNet) techniques, as described in the previous section, are applied to the validation dataset. The qualitative and quantitative results are presented in this section.

6.1 Quantitative Analysis

The performance of each method is evaluated using precision, recall, F1-score, and accuracy. Table 3 presents the evaluation results of different methods for flood detection.

Regarding precision, both RF and KNN achieved similar values of about 0.71. DeepLabv3+ performs significantly less with a precision of 0.67. However, UNet outperforms all other methods with a precision of 0.81, indicating its ability to classify flooded areas accurately. UNet demonstrates superior performance concerning a recall value of 0.85. RF and KNN achieve recall values of 0.82 and 0.81, respectively. DeepLabv3+ gains a recall value of 0.68, slightly lower than the other tested methods.

When considering the F1-score, UNet stands out as the best performer with an F1-score of 0.88. DeepLabv3+, RF, and KNN exhibit lower F1-scores of 0.68, 0.61, and 0.61, respectively. Regarding overall accuracy (OA), UNet achieves the highest value of 0.94, indicating its effectiveness in accurately classifying flooded areas. RF and KNN follow with an accuracy of about 0.72. DeepLabv3+ achieves a lower accuracy of 0.69. The lower performance of DeepLabv3+ may be attributed to the model complexity and fewer training samples.

6.2 Qualitative Analysis

Qualitative analysis is performed by visually comparing the predicted outputs of different methods using the validation dataset. The predicted output is displayed in Fig. 6 for three images along with their respective labels. RF and KNN predictions are quite similar for all the images. DeepLabv3+ shows rather a less performing

Table 3 Evaluation results

Method	Precision	Recall	F1-Score	Accuracy
Random Forest [14]	0.71	0.82	0.61	0.72
KNN [18]	0.71	0.81	0.61	0.72
DeepLabv3+ [13]	0.67	0.68	0.68	0.69
UNet [4]	**0.81**	**0.85**	**0.88**	**0.94**

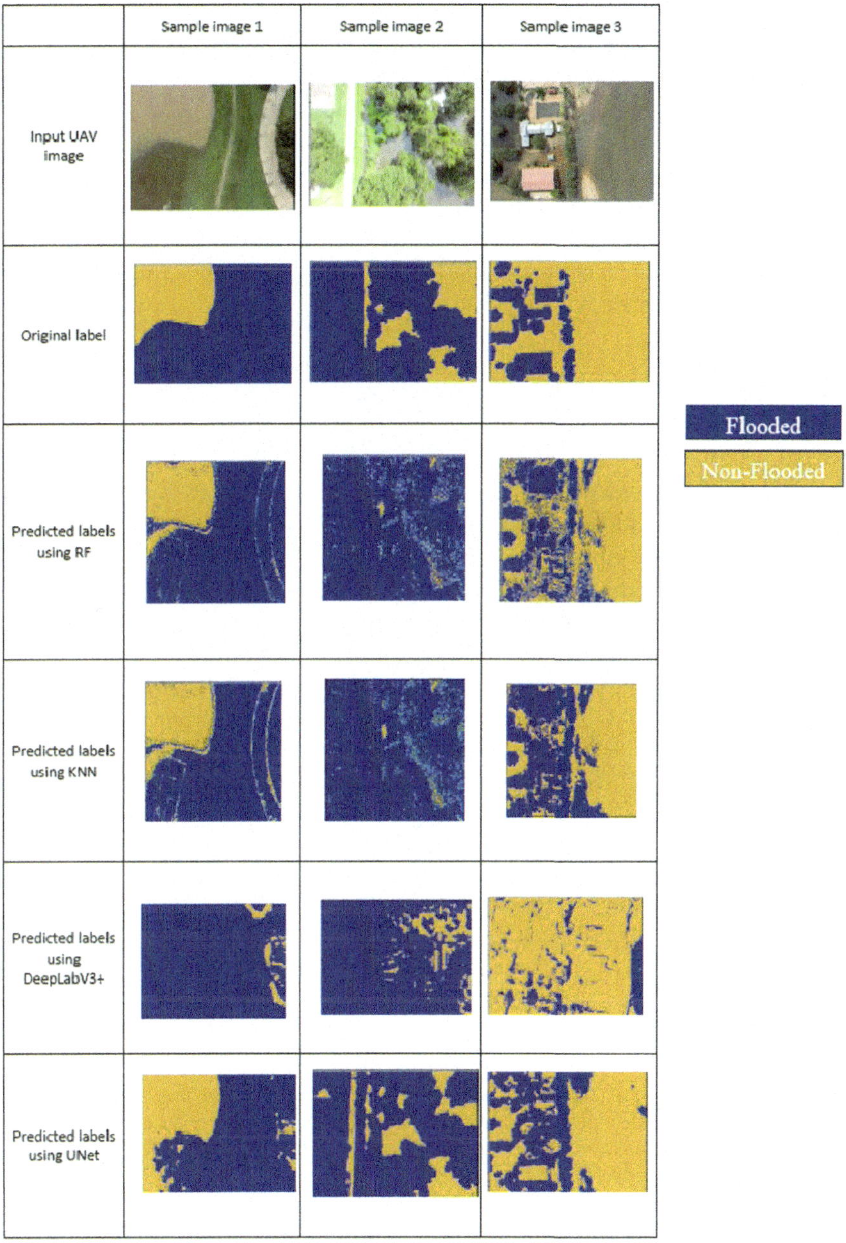

Fig. 6 Visual comparison of the predicted labels using different methods

result as a lack of large amounts of labelled training data to achieve high performance and such algorithms are often treated as black boxes, making it challenging to interpret their decisions. Upon visual inspection of the prediction results, it is evident that UNet exhibits superior performance compared to other tested algorithms. The images predicted by UNet demonstrate higher clarity, sharper edges, and better semantic segmentation, indicating a more accurate understanding of image features and context. Conversely, RF, KNN, and DeepLabv3+ produce predictions that are comparatively wrong classifications. Furthermore, RF and KNN demonstrate similar prediction results, but slight variations exist in their outputs, highlighting the differences in their underlying algorithms.

7 Conclusions

The present work aims at flood detection in urban areas using UAV images, highlighting the importance of leveraging remote sensing technologies for disaster management. Among several publicly available datasets, the FloodNet dataset has been used. The review also identified several challenges and opportunities for future research in this area. One of the significant issues is to improve on accuracy of the semantic segmentation algorithms that can work well on large and diverse datasets. Another challenge is integrating UAV images with other data sources, such as weather and sensor data. This would surely improve the accuracy of flood forecasting and, therefore, alert the stakeholders and decision-makers at right time. This study explores the application of ML and DL techniques, such as semantic segmentation, for accurate flood detection and monitoring in urban regions. This demonstrates the effectiveness of DL methods for segmenting UAV images into flooded and non-flooded categories. This article provides insight into flood management and disaster response in urban environments.

References

1. Jaisakthi S, Dhanya P, Jitesh Kumar S (2021) Detection of flooded regions from satellite images using modified unet. In: Computational intelligence in data science: 4th IFIP TC 12 international conference, ICCIDS 2021, Chennai, India. Revised Selected Papers, 18–20 Mar 2021. Springer, pp 167–174
2. Bashir MH, Ahmad M, Rizvi DR, El-Latif AAA (2024) Efficient CNN-based disaster events classification using UAV-aided images for emergency response application. Neural Comput Appl:1–14
3. Zaffaroni M, Rossi C et al (2020) Water segmentation with deep learning models for flood detection and monitoring. Int J Inf Syst Crisis Response Manag:24–27
4. Ronneberger O, Fischer P, Brox T (2015) U-net: convolutional networks for biomedical image segmentation. In: Medical image computing and computer-assisted intervention–MICCAI 2015: 18th international conference, Munich, Germany. Proceedings, Part III 18, 5–9 October 2015. Springer, pp 234–241

5. Hernandez D, Cecilia JM, Cano J-C, Calafate CT (2022) Flood detection using real-time image segmentation from unmanned aerial vehicles on edge-computing platform. Remote Sens 14(1):223
6. Rahnemoonfar M, Chowdhury T, Sarkar A, Varshney D, Yari M, Murphy RR (2021) Floodnet: a high resolution aerial imagery dataset for post flood scene understanding. IEEE Access 9:89644–89654
7. Sinha RK, Kushwaha M, Choudhary J, Singh DP, Pandey M (2024) Flood image segmentation of UAV aerial images using deep learning. In: IEEE international students' conference on electrical, electronics and computer science (SCEECS), pp 1–6. https://doi.org/10.1109/SCEECS61402.2024.10481993
8. Sharifzadeh S, Tata J, Sharifzadeh H, Tan B (2019) Farm area segmentation in satellite images using deeplabv3+ neural networks. In: Data management technologies and applications: 8th international conference, DATA 2019, Prague, Czech Republic. Revised Selected Papers 8, 26–28 July 2019. Springer, pp 115–135
9. Yu T, Yang H, Nie J, Yao Q, Liu W, Zhang J, Cheriet M (2024) Bias-compensation augmentation learning for semantic segmentation in UAV networks. IEEE Internet Things J:1–1. https://doi.org/10.1109/JIOT.2024.3373454
10. He Y, Wang J, Zhang Y, Liao C (2024) An efficient urban flood mapping framework towards disaster response driven by weakly supervised semantic segmentation with decoupled training samples. ISPRS J Photogramm Remote Sens 207:338–358
11. Thapa A, Neupane B, Horanont T (2022) Object vs pixel-based flood/drought detection in paddy fields using deep learning. In: 12th International congress on advanced applied informatics (IIAI-AAI). IEEE, pp 455–460
12. Paul A, Bhoumik S (2021) Classification of hyperspectral imagery using spectrally partitioned HyperUnet. Neural Comput Appl. https://doi.org/10.1007/s00521-021-06532-3
13. Chen LC, Zhu Y, Papandreou G, Schroff F, Adam H (2018) Encoder-decoder with atrous separable convolution for semantic image segmentation. In: Proceedings of the European conference on computer vision (ECCV), pp 801–818
14. Breiman L (2001) Random forests, vol 45. Springer, pp 5–32
15. Paul A, Tripathi D, Dutta D (2018) Application and comparison of advanced supervised classifiers in extraction of water bodies from remote sensing images. Sustain Water Resour Manag. https://doi.org/10.1007/s40899-017-0184-6
16. Kyrkou C, Theocharides T (2020) Emergencynet: efficient aerial image classification for drone-based emergency monitoring using atrous convolutional feature fusion. IEEE J Sel Top Appl Earth Obs Remote Sens 13:1687–1699. https://doi.org/10.1109/JSTARS.2020.2969809
17. Zhu X, Liang J, Hauptmann A (2020) MSNet: a multilevel instance segmentation network for natural disaster damage assessment in aerial videos
18. Mucherino A, Papajorgji PJ, Pardalos PM (2009) k-nearest neighbor classification. Springer, New York, NY, pp 83–106

Supervised Classification Approach for Precise Cell Type Identification Improves Single Cell Data Analysis

Adrija Das, Gourab Das, Amlan Chakrabarti, and Zhumur Ghosh

Abstract Single Cell RNA sequencing technology aims towards addressing the problem of cellular heterogeneity in biological systems. It unveils the transcriptomic differences at a single cell level within a cell population. Precise cell type prediction within apparently similar cells with subtle differences among them greatly influences single cell RNA sequence data analysis. The proposed machine learning based prediction model attempts to classify different cellular subtypes within apparently similar cell population. The prediction model has been trained with 68k Peripheral Blood Mononuclear Cell (PBMC) dataset. It predicted cellular subtypes on an unknown Peripheral Blood Mononuclear Cell 4k validation dataset with greater accuracy.

Keywords Cell-annotation · Cellular heterogeneity · Classification · Machine learning · Single cell RNA sequencing · Supervised method

1 Introduction

A cell is the fundamental unit of a biological system. Multicellular organisms exhibit a wide variety of unique cell types. This initiates the complex problem of cellular heterogeneity which contributes to disease etiology [1]. A deeper understanding of cellular heterogeneity within diseased system allows the identification of precise drug targets and provides a roadmap towards the development of novel therapeutic approaches [2]. Hence, characterization of individual cells is the key focus of NGS

A. Das · A. Chakrabarti
A. K. Choudhury School of Information Technology, University of Calcutta, Kolkata, India
e-mail: adrijaoly@gmail.com

A. Chakrabarti
e-mail: acakcs@caluniv.ac.in

G. Das · Z. Ghosh (✉)
Department of Biological Sciences, Bose Institute, Kolkata, India
e-mail: zhumur@jcbose.ac.in

G. Das
e-mail: gourab3gd@jcbose.ac.in

© The Author(s), under exclusive license to Springer Nature Singapore Pte Ltd. 2025

N. Chaki et al. (eds.), *Applied Computing for Software and Smart Systems*, Lecture Notes in Networks and Systems 980, https://doi.org/10.1007/978-981-97-9762-2_15

technologies. This has lead to the generation of single cell RNA-seq(scRNA-seq) data which provides a zoomed view of the system to a much larger extent relative to traditional profiling methods that assess bulk RNA populations. All Biological sciences including developmental biology [3], neurobiology [4], cancer biology [5], and autoimmune [6] and infectious diseases [7] have harnessed the scRNA-seq technology in great effect. Mapping single cells to a reference cell annotation atlas is always a critical task in scRNA-seq data analysis. Cell annotation for the validation dataset requires granular analysis of gene function for any particular cell type [8]. A crucial challenge in this domain is to construct an optimized classifier to classify closely related cell types. Defining cell identity is a difficult task for several reasons. For starters, the gene expression change is mostly gradual happening across a wide range. Secondly transcriptional changes that would enable the classification of cells could not be biologically significant in terms of their cellular activity [9]. To determine the molecular profile of individual cells, single-cell studies necessitate the application of amplification processes. Amplification, on the other hand , launches technical fluctuations that increases noise at a molecular level and complicates the process of differentiating between naturally occurring transcript level fluctuations and genuinely meaningful changes in gene expression patterns [10].Annotating cell types manually is tedious and has low reliability. To address these constraints, computational approaches for automatic annotation of cell clusters have recently evolved. The downstream analysis of scRNA-seq data initiates with dimensionality reduction and clustering technique [11, 12]. The annotation of the clusters is done in three ways i.e. marker gene based, correlation based and annotation by machine learning model based on supervised classification approach [13]. Marker gene database-based annotation [14] incorporates various cell marker databases. Reference cell markers have been cumulated from various literatures and scRNA-seq analyzed data. Correlation based methods [14]correlates gene expression values across the well-known reference and query dataset to infer similarities at either single cell or cluster level. Annotation by machine learning technique uses supervised classification upon labeled scRNA-seq dataset. Finally the trained classifier can be used to identify cell types for any query dataset. The main novelty of this paper lies in addressing a unique problem of identifying cellular heterogeneity within apparently similar cell population rather than novelty in the model itself. Such has not been addressed properly in reports of recent times [15, 16], as they mostly work towards classifying distinctly heterogeneous cell population [15, 16] which is relatively easier to solve. We have devised a supervised machine learning based approach consisting a novel feature selection methodology in order to solve this problem. This feature selection is one of the most critical steps in this problem, as precise prediction relies upon the optimal feature engineering procedure from hugely sparse scRNA-seq data. The existing tools [15, 17] do not elaborate on precise feature engineering methodology. On top of the pre-existing mechanisms that are used to classify the PBMC datasets, our approach incorporates enriched biological features constituting pathway and functional annotation of the differentially expressed set of significant genes using IPA [18] corresponding to each cluster, followed by exhaustive literature mining to identify the marker genes from within them, which can classify the clusters precisely.

2 Related Work

ScRNA-seq plays an important role in solving the problem of inter-cellular heterogeneity, rare cell type identification, cell genealogies, somatic mosaicism, microbes, and disease evolution [19]. Chen et al. [20] presents detailed procedural overview of currently available scRNA-seq technologies, as well as methods for various scRNA-seq data analyses such as quality control, alignment, expression counting, normalization, batch correction, dimensionality reduction, feature selection, cell clustering, differential expression analysis, and others. SingleCellNet [16] provides a machine learning based supervised classification of scRNA-seq data to identify cell types. The methodology of SingleCellNet constructs a cell basis binary matrix dataset derived by pairwise comparisons, feature selection has been done by template matching to find the most significant gene set pair and ultimately multi-class random forest classifier has been implemented to classify the cells. They have demonstrated that SingleCellNet is capable of identifying cells across species and measures cell identity in a way that is resilient to various scRNAseq platforms. A group of scientists had introduced a ready to use R package ELeFHAnt [17] which harmonized and annotate label of scRNA-seq data following supervised machine learning approach. EleFHAnt provides an ensemble kind of learning which is based upon logistic regression, support vector machine and random forest. It gives three unique functions in the form of label harmonization, cell type annotation and deduce relationship. The function LabelHarmonization harmonizes cell classes from various datasets which were combined into a uniform atlas, with cell labels allocated to each cell cluster in the combined dataset. One specialized function of EleFHAnt designed to infer the relations which assisted in this process via contrasting cell types or other metadata in two datasets and determining similarities between them. CellTypeAnnotation provides identification of the cells. ACTINN (Automated Cell Type Identification using Neural Networks) [21] has been trained on datasets with established cell kinds and predicts cell types for other datasets using a neural network with three hidden layers. The neural network has been designed keeping the number of the input nodes according to the genes present in the training dataset. The number of nodes in the output layer is equal to the number of cell labels in the training set. Table 1 provides a quick snapshot of available machine learning based cell annotation algorithms.

Most of the machine learning based models have been trained with dataset comprising of heterogeneous target class which in turn makes the problem easier enough to classify the cell types. Hence, none of these existing classifiers have addressed the problem of clustering cell types which are apparently similar but have subtle differences. This study aims to build a machine learning based prediction model to identify different cell types from apparently similar cell population. The proposed

Table 1 Comparison of different machine learning based cell annotation tools

Tool name	Language	Computational approach	Species	Repositories	Limitations	References
SingleCellNet	R Python	Pair transformation-feature selection-train Random Forest multiclass classifier	Mouse	Tabula muris	Lack of transparent sparsity reduction technique, heterogeneous data reduces the real life biological applications	[16]
EleFHAnt	R	Random Forest + Logistic regression + Support vector machine = Label harmonization, Cell type annotation, deduce relationship	Human	EMBL-EBI	Reason of using Random Forest and support vector machine is not clearly mentioned, heterogeneous data lacks the depth of biological implications	[17]
ACTINN	Python	Neural network having three hidden layer corresponding to 100,50 and 25 nodes. Input and output layer of neural network is number of genes and cell types of training set	Human mouse	Tabula muris Winkels et al. 10x Genomics	This method does not clearly mention how it handle the sparsity issue , no precise feature selection technique has been described	[21]

prediction model is based on multiclass Light Gradient Boosting Model (LightGBM) algorithm which has been used to classify the cells. Parameters of the LightGBM algorithm have been optimized using Optuna tool [22].

3 Material and Methodology

The training and validation dataset have been taken from 10x Genomics database (Datasets—[23]) (provided in Table 2). The cell annotations of PBMC 68k dataset have been collected from [24]. PBMC 68k dataset has been used as training dataset and PBMC 4k dataset has been used for validation purpose (Fig. 1).

Table 2 Input dataset

Dataset	#Cells	#Genes	Organ	Organism	Protocol	Application
PBMC 68k	68579	32738	Peripheral blood mononu-clear cells	Human	10x	Training dataset
PBMC 4k	4340	33694	Peripheral blood mononu-clear cells	Human	10x	Validation dataset

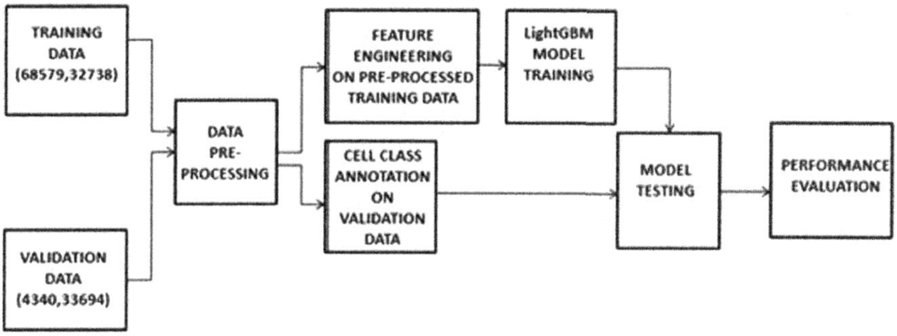

Fig. 1 Schametic diagram of the proposed methodology

3.1 Data Pre-processing

The first step of both training and validation is to pre-process the data using standard Seurat protocol [25] utilising cell-gene matrix. Seurat is an R tool for quality control, data processing, and exploration of scRNA-seq data. Data pre-processing step includes quality control, normalization, variable feature extraction. In the quality control phase, the mitochondrial gene count greater than 7.5% and 8 % were filtered out for training and validation data respectively ((a) and (b) of Fig. 2). After the basic scaling procedure, the preprocessed validation data undergoes Principal Component Analysis (PCA) for dimensional reduction followed by clustering and visualization using Uniform Manifold Approximation (UMAP) algorithm [26].

3.2 Feature Engineering On Pre-processed Training Data

ScRNA-seq data is expressed as cell-gene matrix which is a high-dimensional sparse matrix where the majority of the genes have zero values. The variability of the gene

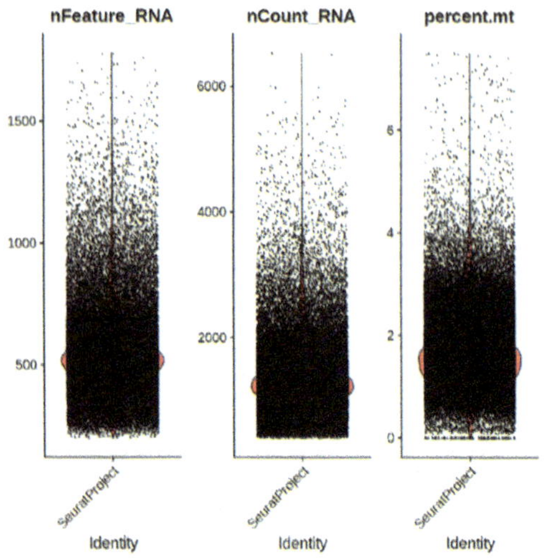

(a) PBMC 68k

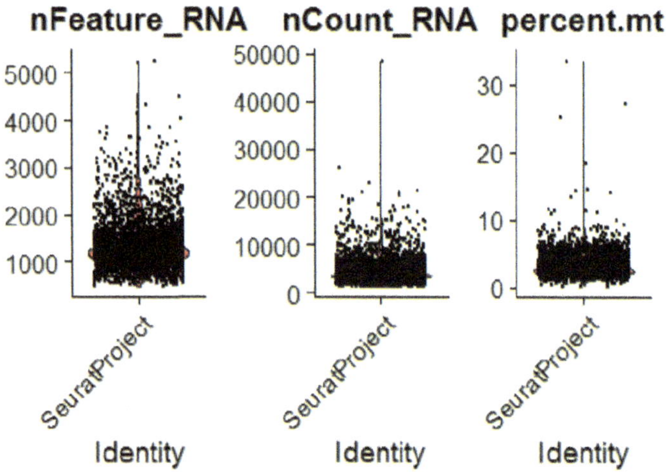

(b) PBMC 4k

Fig. 2 Quality checking for all dataset

expression across the cells serves the purpose of choosing gene as feature. Further, sparsity reduction becomes one of the most crucial challenges in scRNA-seq data analysis as it induces noises [27] and this in turn overfits the machine learning model [28]. Sparsity within the cell-gene matrix originates due to dropout event at cellular level [29] .We have seen that the number of cells in the smallest cluster

(2092 cells) of the training dataset is greater than 1% of the totals cells (680 cells) in training dataset. That's why we have addressed sparsity issue within the training dataset by considering only those genes which have expression in at least 1% of the cells.

3.3 System Model

We have utilized R based libraries such as Seurat [25], celldex [30], Limma [31], SingleR [30] to perform analysis upon scRNA-seq data and python based tools such as lightgbm [32] and optuna [22] packages have been incorporated to create machine learning model and its parameter optimization.

3.4 Cell Class Annotation

Primarily the validation dataset did not contain any class level information. Hence, we screened the upregulated DEGs (filtered with P value cut off of 0.05 and fold change cut off of 1.5) of this dataset using FindMarkers method of Seurat protocol. Further, a detailed functional annotation of the exclusive genes corresponding to individual clusters have been performed using QIAGEN's Ingenuity® Pathway Analysis [18] and exhaustive literature mining. This annotated validation dataset is further used for testing the performance of the prediction model.

3.5 Model Training and Testing

We have performed accuracy analysis on the blind test dataset using various machine learning models in order to select the best possible model. In each case LightGBM outperforms the Linear classifiers like logistic regression, non-linear classifier, decision tree classifier and random forest (data not shown). Hence, we decided to proceed with this model. LightGBM is a decision tree-based ensemble framework which integrates numerous base models [32]. It uses two algorithms for prediction, like Gradient Based One Sided Sampling (GOSS) and Exclusive Feature Bundling (EFB) to create a single optimized prediction model. GOSS permits gradient based sampling of data. Both large and small gradient sampling is used by this algorithm. When GOSS is trained with a modest gradient, the error is negligible; however, when it is trained with a big gradient, the error increases and the information gained increases accordingly. EFB algorithm combines exclusive features which reduce dimensionality and make the overall procedure faster and more efficient than other gradient boosting methodologies. All these equip LightGBM to provide better and robust prediction. Finally, our LightGBM model has been trained using preprocessed features of training data The training data (PBMC 68k data) has been split into 90:10 ratio corresponding to training and test data. Parameters of LightGBM (viz. Learning rate, reg_lambda, reg_alpha, subsample, colsample_bytree, max_depth, num_leaves, min_child_samples) have been optimized with the help of Optuna hyper parameter optimization mechanism [22] in order to get higher model accuracy. The performance of the LightGBM model on the validation dataset has been tested using various statistical parameters viz. accuracy, F1-Score, Mathew's Correlation Coefficient and Cohen's kappa.

4 Result

The training and validation datasets were filtered based on the percentage of mitochondrial gene count. Such pre-processing step yielded a set of 68443 cells and 32738 genes for the training dataset and 4169 cells and 33694 genes for the validation dataset.

4.1 Cell Class Annotation for Validation Dataset

Clustering and Visualization are the initial stages of cell level annotation of the validation dataset. The prior step of clustering is to find nearest neighbors in the dataset. The nearest neighbor finding algorithm gave eight communities within a specified dimension. Louvain clustering algorithm [33] clustered the data points into eight clusters whose visualization is provided by UMAP [26] (Fig. 3).

Next in order to resolve the closely related cell types with subtle differences, differentially expressed genes (DEGs) corresponding to each cluster were extracted

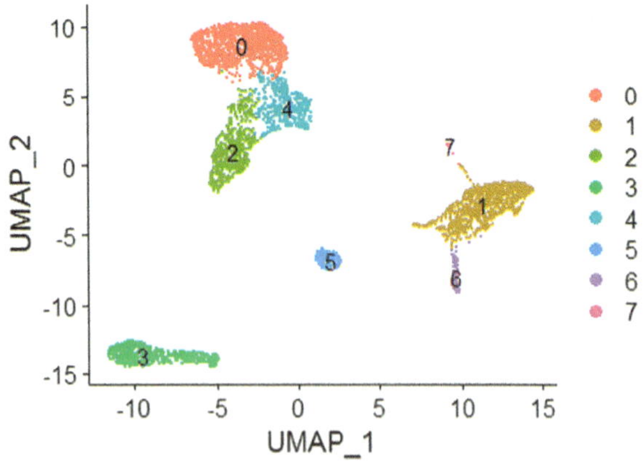

Fig. 3 UMAP of PBMC 4k data

CLUSTER NUMBER	SIGNIFICANTLY UPREGULATED DEGs	EXCLUSIVE SET OF UPREGULATED GENES
0	23	16
1	394	220
2	85	73
3	131	83
4	35	23
5	290	156
6	353	179
7	443	314

Fig. 4 Cluster specific upregulated total set of DEGs and exclusive set of DEGs

using FindMarkers method of Seurat protocol. Only significant set of up regulated DEGs were screened based on p value cut off ≤ 0.05 and fold change cut off of 1.5. Figure 4 describes the summary of DEGs and significantly upregulated genes corresponding to each cluster.

The final set of marker genes have been screened followed by detailed functional annotation of the final exclusive set of upregulated DEGs corresponding to each cluster done by using IPA followed by exhaustive literature mining. Table 3 provides the detailed functional annotation of the final set of marker genes for each cluster except cluster 0.

Cluster 0 could not be annotated properly. The highly expressed exclusive DEGs designating this cluster has direct association with T follicular helper cell [34] and some are also found in a variety of lymphoid organs, including B and T lymphocytes as well as mature dendritic cells (DC). We plan to address this discrepancy in our future work. In order to use the PBMC 4k as validation data in the prediction model, its cell level annotation is needed corroborating with the target label of the training dataset. So, it could be concluded as cluster 0—Novel cell type, cluster 1—CD14+ Monocyte, cluster 2—CD8+ Cytotoxic T, cluster 3—CD19+ B, cluster 4—CD4+/CD45RO+ Memory, cluster 5—CD56+ NK, cluster 6—CD4+/CD25 T Reg, cluster 7—Dendritic.

Table 3 Detailed information of the cluster specific pathways and marker genes

Cluster no	Cluster specific pathways	Marker genes	Functional annotation of marker genes	Designated cell type
1	1. Neutrophil extracellular trap signaling pathway 2. iNOS signaling 3. Role of IL-17A in Psoriasis 4. LXR/RXR activation 5. S100 family signaling pathway	S100A8, S100A9	These two genes are constitutively expressed by myeloid cells, including granulocytes, monocytes, osteoclasts and early-differentiation cells of the myeloid lineage [35]	CD14+ monocyte
2	1. T Cell receptor signaling 2. CTLA4 signaling in Cytotoxic T Lymphocytes 3. Immunogenic Cell Death Signaling Pathway 4. Glucocorticoid Receptor Signaling 5. Communication between Innate and Adaptive Immune Cell	CCL5, GZMK	The chemokine CCL5 is mainly expressed by T-cells and GZMK is mainly expressed in CD8+T-Cells [36, 37]	CD8+ T cell
3	1. Primary immunodeficiency signaling 2. Macrophage alternative activation signaling pathway 3. PI3K signaling in B lymphocytes 4. B cell development 5. B cell receptor signaling	IGLC3, IGLC2	These two genes are highly expressed in B-cell [38]	B-Cell
4	1. NOD1/2 signaling pathway 2. Wound healing signaling pathway 3. Pathogen induced cytokine storm signaling pathway 4. Airway pathology in chronic obstructive pulmonary disease 5. Role of pattern recognition receptors in recognition of bacteria and viruses	LTB	This has high expression in CD4+ T cells [39]	CD4+/ CD45RO+Memory
5	1. Natural killer cell signaling 2. Pathogen induced cytokine storm signaling pathway 3. Eicosanoid signaling 4. BAG2 signaling pathway	KLRF1, SPON2	KLRF1is a novel marker of bright CD56+ NK cells SPON2 had been highly expressed in CD56+ NK Cells [40, 41]	CD56+ NK
6	1. Actin nucleation by ARP-WASP complex 2. Regulation of actin-based motility by Rho 3. Integrin signaling 4. RHOGDI signaling 5. Signaling by rho family GTPases	CSF1R, FCGR3A, RHOC	These genes combinedly play a role of osteoclasts in rheumatoid arthritis signaling pathway. This pathway is also controlled by regulatory T cells [42]	CD4+/ CD25 T Reg
7	1. Activation of IRF by cytosolic pattern recognition receptors 2. Ephrin receptor signaling 3. Protein kinase A signaling 4. ILK signaling	LILRA4, IRF7	Human pDCs selectively express LILRA4 IRF7 is predominantly activated by TLR7 in pDCs, in which both TLR7 and IRF7 are expressed at extra high levels [43, 44]	Dendritic cell

4.2 Model Training

The LightGBM Model was trained using the training data and it got the accuracy as 85.52% on test data. The corresponding confusion matrix is given in Fig. 5.

4.3 Performance Evaluation

The validation dataset has been tested with the trained LightGBM. The performance on the validation dataset has been evaluated and it yields 70.5%, 75.2%, 66.3% and 64.3% as accuracy score, f1 score, Mathew's Correlation Coefficient and Cohen's kappa score respectively.

4.4 Comparison With State-of-the-Art

The efficiency of the proposed prediction model has been compared with the available pre-trained machine learning based cell annotation tool SingleCellNet [16]. The proposed model gave an enhanced accuracy of 70.5% as compared to 61.4% accuracy of SingleCellNet (Fig. 6).

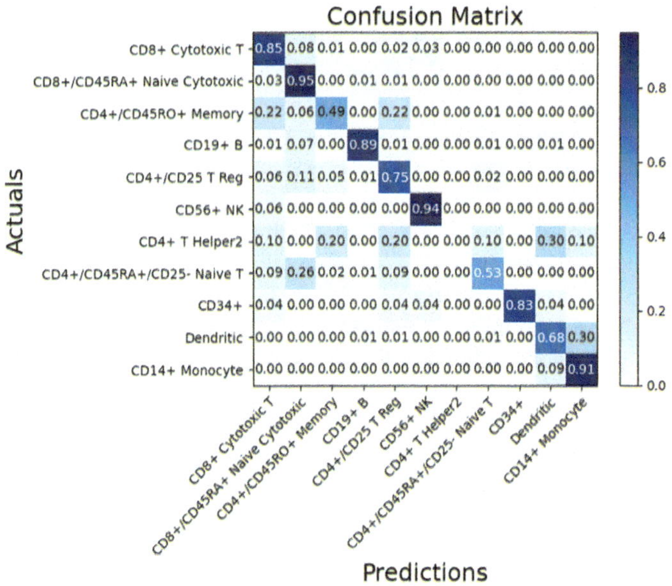

Fig. 5 Confusion matrix of the test data

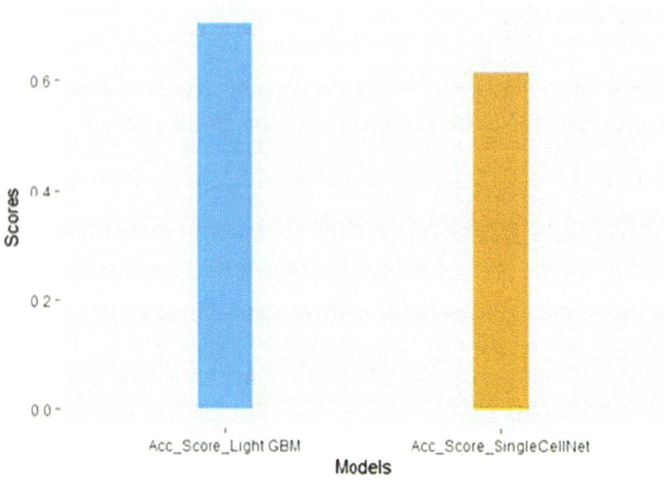

Fig. 6 Comparison between the accuracy score of LightGBM and SingleCellNet

5 Conclusion

Identifying the differences between apparently similar cell types and clustering them separately is the most crucial task in class level annotation of scRNA-seq dataset. In this work we have tried to address this problem by building a machine learning based prediction model to identify different cell types from apparently similar cell population. The proposed prediction model is based on multiclass Light Gradient Boosting Model (LightGBM) algorithm which has been used to classify the cells. Parameter optimization have been implemented on the training data (PBMC 68k data) to solve the class imbalance problem. Class levels of the validation dataset have been identified in order to test the new prediction model. The Seurat protocol gave eight clusters from the pre-processed 4169 cells of the validation dataset. Out of the eight clusters, seven were identified which were included in training dataset as well. Some of the current limitations of our work includes requirement of high level manual interventions (like exhaustive literature mining for functional annotation and pathway analysis of the marker genes) in annotating clusters. This restricts us from using larger validation dataset like PBMC 5k [45] and PBMC 6k [46] to test our prediction protocol. Further, there remained some confusion regarding identifying cluster 0 precisely. We plan to solve these issues in our future work by transferring marker annotation which automates the class annotation procedure in validation dataset using transfer learning and canonical correlation analysis [47]. This will also enable us to use larger validation dataset to test our prediction model. Overall, cell class annotation could be easier and more specific if cluster specific driver genes are identified and analyzed exhaustively. Further, in order to build a more robust classifier, cell type marker annotation of the reference dataset needs to be transferred for precise cell type identification within the unknown validation dataset.

References

1. Gupta RK, Kuznicki J (2020) Biological and medical importance of cellular heterogeneity deciphered by single-cell RNA sequencing. Cells 9(8):1751
2. Brooks M, Burness ML, Wicha MS (2015) Therapeutic implications of cellular heterogeneity and plasticity in breast cancer. Cell Stem Cell 17(3):260–271
3. Yan L, Yang M, Guo H, Yang L, Wu J, Li R, Liu P, Lian Y, Zheng X, Yan J et al (2013) Single-cell RNA-seq profiling of human preimplantation embryos and embryonic stem cells. Nat Struct Mol Biol 20(9):1131–1139
4. Mathys H, Davila-Velderrain J, Peng Z, Gao F, Mohammadi S, Young JZ, Menon M, He L, Abdurrob F, Jiang X et al (2019) Single-cell transcriptomic analysis of Alzheimer's disease. Nature 570(7761):332–337
5. Zhu S, Qing T, Zheng Y, Jin L, Shi L (2017) Advances in single-cell RNA sequencing and its applications in cancer research. Oncotarget 8(32):53763
6. Papalexi E, Satija R (2018) Single-cell RNA sequencing to explore immune cell heterogeneity. Nat Rev Immunol 18(1):35–45
7. Huang W, Wang D, Yao YF (2021) Understanding the pathogenesis of infectious diseases by single-cell RNA sequencing. Microbial Cell 8(9):208
8. Jagadeesh KA, Dey KK, Montoro DT, Mohan R, Gazal S, Engreitz JM, Xavier RJ, Price AL, Regev A (2022) Identifying disease-critical cell types and cellular processes by integrating single-cell RNA-sequencing and human genetics. Nat Genet 54(10):1479–1492
9. Kolodziejczyk AA, Kim JK, Svensson V, Marioni JC, Teichmann SA (2015) The technology and biology of single-cell RNA sequencing. Mol Cell 58(4):610–620
10. Hsiao CJ, Tung P, Blischak JD, Burnett JE, Barr KA, Dey KK, Stephens M, Gilad Y (2020) Characterizing and inferring quantitative cell cycle phase in single-cell RNA-seq data analysis. Genome Res 30(4):611–621
11. Butler A, Hoffman P, Smibert P, Papalexi E, Satija R (2018) Integrating single-cell transcriptomic data across different conditions, technologies, and species. Nat Biotechnol 36(5):411–420
12. Luecken MD, Theis FJ (2019) Current best practices in single-cell RNA-seq analysis: a tutorial. Mol Syst Biol 15(6):e8746
13. Fincher CT, Wurtzel O, de Hoog T, Kravarik KM, Reddien PW (2018) Cell type transcriptome atlas for the planarian *Schmidtea mediterranea*. Science 360(6391):eaaq1736
14. Pasquini G, Arias JER, Schäfer P, Busskamp V (2021) Automated methods for cell type annotation on scRNA-seq data. Comput Struct Biotechnol J 19:961–969
15. De Kanter JK, Lijnzaad P, Candelli T, Margaritis T, Holstege FC (2019) Chetah: a selective, hierarchical cell type identification method for single-cell RNA sequencing. Nucleic Acids Res 47(16):e95–e95
16. Tan Y, Cahan P (2019) Singlecellnet: a computational tool to classify single cell RNA-seq data across platforms and across species. Cell Syst 9(2):207–213
17. Thorner K, Zorn AM, Chaturvedi P (2021) Elefhant: a supervised machine learning approach for label harmonization and annotation of single cell RNA-seq data, pp 2021-09
18. Insights QD (2024) Qiagen digital insights: bioinformatics software. http://www.ingenuity.com/. Accessed 23 Nov 2010
19. Wang W, Gao D, Wang X (2018) Can single-cell RNA sequencing crack the mystery of cells?
20. Chen G, Ning B, Shi T (2019) Single-cell RNA-seq technologies and related computational data analysis. Front Genet 10:317
21. Ma F, Pellegrini M (2020) Actinn: automated identification of cell types in single cell RNA sequencing. Bioinformatics 36(2):533–538
22. Akiba T, Sano S, Yanase T, Ohta T, Koyama M (2019) Optuna: a next-generation hyperparameter optimization framework. In: Proceedings of the 25th ACM SIGKDD international conference on knowledge discovery and data mining, pp 2623-2631
23. 10xGenomics: Dataset (2024). https://www.10xgenomics.com/resources/datasets. Accessed 12 Dec 2019

24. Zheng GX, Terry JM, Belgrader P, Ryvkin P, Bent ZW, Wilson R, Ziraldo SB, Wheeler TD, McDermott GP, Zhu J et al (2017) Massively parallel digital transcriptional profiling of single cells. Nat Commun 8(1):14049
25. Hao Y, Hao S, Andersen-Nissen E, Mauck WM, Zheng S, Butler A, Lee MJ, Wilk AJ, Darby C, Zager M et al (2021) Integrated analysis of multimodal single-cell data. Cell 184(13):3573–3587
26. McInnes L, Healy J, Melville J (2018) Umap: uniform manifold approximation and projection for dimension reduction. Preprint at arXiv:1802.03426
27. Hou W, Ji Z, Ji H, Hicks SC (2020) A systematic evaluation of single-cell RNA-sequencing imputation methods. Genome Biol 21:1–30
28. Petrini L, Cagnetta F, Vanden-Eijnden E, Wyart M (2022) Learning sparse features can lead to overfitting in neural networks. Adv Neural Inf Process Syst 35:9403–9416
29. Kharchenko PV, Silberstein L, Scadden DT (2014) Bayesian approach to single-cell differential expression analysis. Nat Methods 11(7):740–742
30. Aran D, Looney AP, Liu L, Wu E, Fong V, Hsu A, Chak S, Naikawadi RP, Wolters PJ, Abate AR et al (2019) Reference-based analysis of lung single-cell sequencing reveals a transitional profibrotic macrophage. Nat Immunol 20(2):163–172
31. Ritchie ME, Phipson B, Wu D, Hu Y, Law CW, Shi W, Smyth GK (2015) limma powers differential expression analyses for RNA-sequencing and microarray studies. Nucleic Acids Res 43(7):e47–e47
32. Ke G, Meng Q, Finley T, Wang T, Chen W, Ma W, Ye Q, Liu TY (2017) Lightgbm: a highly efficient gradient boosting decision tree. Adv Neural Inf Process Syst 30
33. Blondel VD, Guillaume JL, Lambiotte R (2008) Lefebvre E (2008) Fast unfolding of communities in large networks. J Stat Mech: Theory Exp 10:P10008
34. Choi YS, Gullicksrud JA, Xing S, Zeng Z, Shan Q, Li F, Love PE, Peng W, Xue HH, Crotty S (2015) LEF-1 and TCF-1 orchestrate T follicular helper cell differentiation by regulating differentiation circuits upstream of Bcl6. Nat Immunol 16(9):980
35. Catalán V, Frühbeck G, Gómez-Ambrosi J (2018) Inflammatory and oxidative stress markers in skeletal muscle of obese subjects. Obesity (Elsevier):163-189
36. Aldinucci D, Colombatti A et al. (2014) The inflammatory chemokine ccl5 and cancer progression. Mediat Inflamm 2014
37. Mogilenko DA, Shpynov O, Andhey PS, Arthur L, Swain A, Esaulova E, Brioschi S, Shchukina I, Kerndl M, Bambouskova M et al (2021) Comprehensive profiling of an aging immune system reveals clonal GZMK+ CD8+ T cells as conserved hallmark of inflammaging. Immunity 54(1):99–115
38. Lee RD, Munro SA, Knutson TP, LaRue RS, Heltemes-Harris LM, Farrar MA (2021) Single-cell analysis identifies dynamic gene expression networks that govern B cell development and transformation. Nat Commun 12(1):6843
39. Zhou W, Yu M, Mao X, Pan H, Tang X, Wang J, Che N, Xie H, Ling L, Zhao Y et al (2022) Landscape of the peripheral immune response induced by local microwave ablation in patients with breast cancer. Adv Sci 9(17):2200033
40. Cao WJ, Zhang XC, Wan LY, Li QY, Mu XY, Guo AL, Zhou MJ, Shen LL, Zhang C, Fan X et al (2022) Immune dysfunctions of CD56neg NK cells are associated with HIV-1 disease progression. Front Immunol 12:811091
41. Mukherjee N, Ji N, Tan X, Chen CL, Noel OD, Rodriguez-Padron M, Lin CL, Alonzo DG, Huang TH, Svatek RS (2023) KLRF1, a novel marker of CD56bright NK cells, predicts improved survival for patients with locally advanced bladder cancer. Cancer Med 12(7):8970–8980
42. Jiang Q, Yang G, Liu Q, Wang S, Cui D (2021) Function and role of regulatory T cells in rheumatoid arthritis. Front Immunol 12:626193
43. Balan S, Saxena M, Bhardwaj N (2019) Dendritic cell subsets and locations. Int Rev Cell Mol Biol 348:1–68
44. Leylek R, Idoyaga J (2019) The versatile plasmacytoid dendritic cell: function, heterogeneity, and plasticity. Int Rev Cell Mol Biol 349:177–211

45. 10xGenomics: Pbmc 5k dataset (2024). https://www.10xgenomics.com/resources/datasets/5k-human-pbmcs-3-v3-1-chromium-controller-3-1-standard. Accessed 6 Nov 2022
46. 10xGenomics: Pbmc 6k dataset (2024). https://www.10xgenomics.com/resources/datasets/6-k-pbm-cs-from-a-healthy-donor-1-standard-1-1-0. Accessed 21 Jan 2022
47. Stuart T, Butler A, Hoffman P, Hafemeister C, Papalexi E, Mauck WM, Hao Y, Stoeckius M, Smibert P, Satija R (2019) Comprehensive integration of single-cell data. Cell 177(7):1888–1902

Quality-Driven Web Service Selection: Machine Learning Based Approach

Dewashish Kumar and Anirban Sarkar

Abstract The task of choosing a suitable web service has become increasingly difficult in today's landscape due to the growing number of services offering similar functionalities. This paper addresses the critical need for efficient web service selection based on Quality of Service parameters. A minimal set of discrete QoS characteristics were carefully chosen as inputs for several machine learning models targeted at finding and suggesting suitable web services. The performance of these models is systematically evaluated and compared. The machine learning techniques used for this work include a Support Vector Machine, Decision Tree, and Random Forest. Among these, the results show that the proposed Random Forest model consistently outperforms other techniques in terms of accuracy and effectiveness. This study contributes to the advancement of web service selection methodologies by providing insights into the comparative performance of various machine learning approaches, ultimately offering a valuable guide for decision-makers in navigating the intricate landscape of web service options.

Keywords Web service · Classifications · SVM · Decision tree · Random forest · Quality of web service

1 Introduction

In this rapidly changing world of modern software development, the increasing number of web services has become a distinguishing feature, bringing both opportunities and challenges for developers. Web services, as defined by the World Wide Web Consortium (W3C) [1], are software systems that enable compatible machine-to-machine interactions and have evolved as critical components of modern computing. These services follow established standards, such as the Simple Object Access Protocol (SOAP) [2] for communication, the Web Service Description Language (WSDL) [3]

D. Kumar (✉) · A. Sarkar
Department of Computer Science and Engineering, National Institute of Technology Durgapur, Durgapur, India
e-mail: dk.22p10208@mtech.nitdgp.ac.in

© The Author(s), under exclusive license to Springer Nature Singapore Pte Ltd. 2025

N. Chaki et al. (eds.), *Applied Computing for Software and Smart Systems*, Lecture Notes in Networks and Systems 980, https://doi.org/10.1007/978-981-97-9762-2_16

for service specification, and the Universal Description Discovery and Integration (UDDI) [4] for service directory, which together form the backbone of networked systems.

The increasing number of available web services, each with similar functionalities, has increased the difficulty of selecting the best service. Quality of Service (QoS) [5] parameters like Availability, Best Practices, Documentation, etc. have emerged as critical metrics for assessing and comparing these services. Combining all these QoS metrics to make a decision makes complicated methods necessary for effective decision-making. While traditional techniques for web service selection exist, their effectiveness in addressing the complexities of the selection process remains an area for further investigation. Finding a reliable method that can efficiently utilize the wide range of QoS characteristics to improve accuracy and lower misclassification errors is currently the challenge (Fig. 1).

The conventional techniques for choosing and categorizing web services are not as accurate or flexible as they should be given the dynamic nature of modern web service environments [5]. The problem with traditional methods is that they require combining all available QoS metrics using specific weights or formulas. Adjusting these weights and formulas manually can be complicated and challenging task. Innovative strategies that can produce the best results without being unduly complicated or resource-intensive are needed as the need for precise and effective selection mechanisms increases [6]. Use of machine learning techniques can eradicate the need for such manual tinkering and smoothly automate the process.

This paper explores various machine learning (ML) techniques such as Support Vector Machine (SVM), Decision Tree, and Random Forest, for web service classification. All of these various ML techniques are used to classify web services into categories based on their QoS metrics. These category classifications can be later used to select the better web service from available choices when needed. The rest of the work is organized as follows: Sect. 2 discusses the related works and existing literature. Section 3 gives a brief background. In Sect. 4, the proposed work is

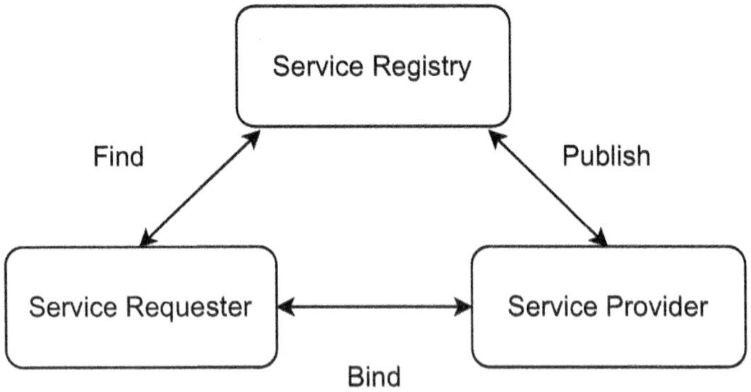

Fig. 1 Service provider, service consumer, and service registry interaction

presented in detail. The experimental details and results are given in Sect. 5. Finally, Sect. 6 provides a conclusion and some discussion on possible future work.

2 Related Work

In article by Mohammad Alrifai et al. [7], a unique strategy is presented that combines global optimization and local selection strategies. This approach addresses the challenge of selecting component services that meet end-to-end QoS standards and user preferences in service composition. Experimental assessments reveal its superiority in computing efficiency over previous methods, showing promise for real-world applications.

In another article [8], Medha Chippa et al. explore various aspects of web services, including UDDI, WSDL, SOAP, and Web Service Discovery, shedding light on the evolving landscape of web service classification. They investigate multiple classification approaches, such as Naive Bayes, SVM, KNN, PCA, and SVD, to tackle the challenges of organizing web services effectively amidst the growing number of registered services.

In another article [5], Alexandra et al. describe a hybrid system leveraging semantic technology, logic-based decision-making, and numerical calculus for classifying and promoting QoS-aware web services. The study demonstrates the use of semantic modeling and logic-based reasoning to address various requests related to online service categorization and recommendation.

In the research by Mohanty et al. [9], the evaluation and ranking of web services using Naive Bayes, Markov blanket, and Tabu search are discussed. Through experiments on a dataset derived from modern literature, they showcase the effectiveness of a Bayesian Network in classifying online services based on their quality attributes.

In article [10] by Laachemi et al., a novel technique for web service classification is proposed, combining stochastic local search (SLS) meta-heuristic for feature selection with Support Vector Machine (SVM) for classification. The study rigorously evaluates this approach on the QWS Dataset, demonstrating its effectiveness in improving web service categorization.

In the study by Yogita et al. [11], the utilization of machine learning techniques, including SVM, Naive Bayes, KNN, Bagging, and Boosting, for web service classification is reviewed. The study underscores the importance of these algorithms in enhancing the efficiency and effectiveness of service-oriented architecture and web service technology.

In another article [6] by Venkataiah et al., the application of the J48 decision tree model to estimate the quality of web-based services is discussed. Through experiments on the QWS dataset, the study highlights the significance of the web service relevance function in predicting service quality, showcasing the efficiency of the J48 classifier in this context.

3 Background and Preliminaries

3.1 Support Vector Machines (SVM)

Support Vector Machine (SVM) [12] is a machine learning algorithm designed for classification and regression tasks. It works by finding the optimal hyperplane in a high-dimensional space that best separates different classes of data. SVM is particularly effective in handling both linear and non-linear relationships between features, utilizing support vectors to define the decision boundary and maximize the margin between classes equation as:

$$f(\mathbf{x}) = \frac{1}{2}\sum_{i=1}^{N}\sum_{j=1}^{N}\alpha_i\alpha_j y_i y_j K(\mathbf{x}_i, \mathbf{x}_j) - \sum_{i=1}^{N}\alpha_i \tag{1}$$

Here, N is the number of support vectors, α_i and α_j are the Lagrange multipliers associated with the i-th and j-th support vectors, y_i and y_j are the class labels of the i-th and j-th support vectors, $K(\mathbf{x}_i, \mathbf{x}_j)$ is the kernel function representing the similarity between the i-th and j-th support vectors in the transformed feature space. The term $\sum \alpha_i$ represents the sum of all Lagrange multipliers.

The polynomial kernel is defined as

$$K(\mathbf{x}_i, \mathbf{x}_j) = (\mathbf{x}_i \cdot \mathbf{x}_j + c)^d \tag{2}$$

The constant term c influences the similarity between feature vectors x_i and x_j. The degree d of the polynomial determines the level of non-linearity introduced by the kernel.

3.2 Weighted SVM

Weighted SVM [13] is an enhancement of the standard SVM algorithm that assigns specific weights to individual data points during the training process. These weights allow for the consideration of varying importance levels for different instances in the dataset. This modification is particularly valuable in handling imbalanced class distributions or situations where certain data points hold higher significance. The weighted SVM optimizes the hyperplane, taking into account both the margin maximization and the assigned weights, resulting in a more nuanced and effective learning process equation as

$$f(\mathbf{x}) = \frac{1}{2}\sum_{i=1}^{N}\sum_{j=1}^{N}\alpha_i\alpha_j y_i y_j w_i K(\mathbf{x}_i, \mathbf{x}_j) - \sum_{i=1}^{N}\alpha_i \tag{3}$$

In the context of Support Vector Machines (SVM), several key terms play crucial roles: w_i represents the associated weight, $\mathbf{x}_i$ signifies the i-th support vector, b stands for the bias term, and c and d are constants in the polynomial kernel.

3.3 Decision Tree

A decision tree [14] is a flexible and user-friendly machine learning technique that may be used for classification as well as regression. It works by recursively splitting the dataset depending on feature criteria, resulting in a tree-like structure of decisions. At each node, the algorithm chooses the optimal feature to partition the data in order to maximize information gain (for classification) or variance reduction (for regression). The terminal nodes, or leaves, denote the ultimate choice or expected outcome. The decision tree logic can be represented using conditional statements:

$$H(S) = -\sum_{i=1}^{c} p_i \cdot \log_2(p_i) \tag{4}$$

In the context of information theory and decision trees, certain terms are fundamental: $H(S)$ denotes the entropy of the dataset S, c represents the number of classes in the dataset, p_i signifies the proportion of instances in class i in the dataset, and $\log_2$ stands for the base-2 logarithm.

3.4 Random Forest

The random forest algorithm [14] is an ensemble learning method used for classification, regression, and other tasks. It combines multiple decision trees to create a more robust and accurate model. The mathematical equation for the Random Forest algorithm involves the aggregation of predictions from individual decision trees.

Let's denote the Random Forest model as $RF(x)$, where x represents the input features. The prediction for a regression task is given by averaging the predictions of all the individual trees:

$$RF(x) = \frac{1}{N}\sum_{i=1}^{N} T_i(x) \tag{5}$$

For a classification task, the Random Forest uses a majority voting scheme, and the prediction is based on the class that receives the most votes from the individual trees:

$$RF(x) = \text{argmax}_c \left(\sum_{i=1}^{N} P_i(T_i(x) = c) \right) \quad (6)$$

Here, $argmax_c$ denotes the class c that maximizes the sum over all trees (N) in the Random Forest ensemble. P_i represents the predicted probability for the i-th tree, and $\mathbb{I}(T_i(x) = c)$ is an indicator function that evaluates to 1 if the i-th tree predicts class c for input x and 0 otherwise.

4 Proposed Work

This work aims to improve the accuracy and reliability of service selection using machine learning models. A variety of machine learning models are used in this study, such as Support Vector Machines (SVM), Weighted SVM, Decision Tree, and Random Forest. The utility of these models is assessed and compared in the classification of services into categories according to their quality so that this classification can be later used for service selection. For this work, the Quality of Web Services (QWS) dataset is used. This dataset includes a variety of QoS features that represent different aspects of services, as well as a categorical class that indicates the type of service [15]. Various QoS metrics are used as input for the machine learning models which then try to predict the class which will indicate the type or quality of the service. The flow of the proposed work is shown in the accompanying Web Service Selection Framework (Fig. 2).

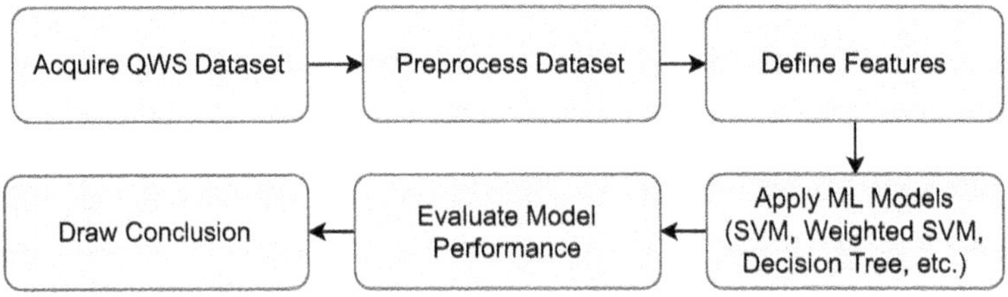

Fig. 2 Web service selection framework

Table 1 Description of QWS dataset

Service metrics	Description	Unit
Response time	Time taken by the service to respond	ms
Availability	Proportion of time the service is operational	%
Throughput	Rate at which the service processes requests	Requests/sec
Successability	Likelihood of successful completion of a request	%
Reliability	Consistency of service performance over time	%
Compliance	Adherence to predefined standards and regulations	%
Best practices	Implementation conformity to recommended practices	%
Latency	Time delay between request and response	ms
Documentation	Adequacy and clarity of service documentation	%
WsRF	Web service relevancy function	%

4.1 Quality of Web Services (QWS) Dataset Description

The Quality of Web Services (QWS) Dataset [16], which measures the Quality of Service (QoS) of genuine web services, was introduced in 2007 as part of Eyhab Al-Masri's PhD thesis work. It consists of a comprehensive collection of service metrics intended to assess the features and performance of various web services. Table 1 gives a thorough explanation of the available QoS metrics in the dataset. The online Service Crawler Engine (WSCE) was used to gather information on these web services, with the bulk coming from public web sources such as UDDI registries, search engines, and service portals. The QWS Dataset version 1.0 measures nine Quality of Service (QoS) metrics per web service for 365 web services. It also contains two extra properties. Firstly, it includes a rank of each of the web services as a percentage metric called Web Service Relevancy Function (WsRF). Secondly, it has a metric called class which classifies web services based on their overall performance into four categories Platinum (1), Gold (2), Silver (3), and Bronze (4).

4.2 Dataset Description

For the dataset used in this work, to train various ML models, a normalized form of the nine QoS metrics of (QWS) dataset is used. All the metrics are normalized between 0 and 1 to give a consistent input to the ML models. The QWS dataset categorizes web services based on QoS metrics into four classes represented as 1, 2, 3, and 4, which serve as the expected outputs for machine learning algorithms (Table 2).

4.3 SVM Algorithm

In this study, three varieties of SVM algorithm modifications are demonstrated, each precisely constructed to address particular issues. The Normal SVM serves as a benchmark, receiving a training dataset D with feature vectors $\mathbf{x}_i$ and labels y_i. After preprocessing with feature normalization, it trains on D_{train} and predicts labels for D_{test}, evaluating the accuracy. Sample Weighted SVM adjusts for data item relevance using weights during training, while Class Weighted SVM addresses uneven class distributions with generated class weights. Algorithmic issues include time complexity, which is determined by the underlying SVM training method, such as Sequential Minimal Optimization. The feasible runtime, denoted as $O(n^2)$, where n represents the number of data points, is dependent on implementation details and kernel functions.

Algorithm 1 Normal SVM

Input: Training dataset $D = \{(\mathbf{x}_i, y_i)\}$ with features $\mathbf{x}_i$ and labels y_i
Output: Accuracy, Error Rate
procedure NORMALSVM(D)
  Preprocessing:
  Normalize or scale features in $\mathbf{x}_i$
  Split D into training and testing sets
  Train Normal SVM:
  Initialize SVM model M
  Train M using the training set D_{train} with labels y_{train}
  Test Normal SVM:
  Use M to predict labels for the testing set D_{test}
  Evaluate accuracy and error rate
  Output: Accuracy, Error Rate
end procedure

Table 2 Few sample rows of dataset

Response time	Availability	Throughput	Successability	Reliability	Compliance	Best practices	Latency	Documentation	WsRF	Class
45	83	27.2	50	97.4	89	91	43.00	58	100	1
71.75	100	14.6	88	85.5	78	80	64.42	86	93	1
150.45	100	7.4	96	82.1	100	82	100.36	37	80	2
123	86	9.3	57	79.9	78	84	121.5	37	70	3
274.50	37	1.0	18	24.8	89	69	257.50	6	38	4

Algorithm 2 Weighted SVM

Input: Training dataset $D = \{(\mathbf{x}_i, y_i)\}$ with features $\mathbf{x}_i$ and labels y_i, Sample Weights w (for Sample Weighted SVM) or Class Weights (for Class Weighted SVM)
Output: Accuracy,
procedure SAMPLEWEIGHTEDSVM(D, w)
  Preprocessing:
  Normalize or scale features in $\mathbf{x}_i$
  Split D into training and testing sets
  Train SVM with Sample Weights:
  Initialize Weighted SVM model M with sample weights w_{train}
  Train M using the training set D_{train} with labels y_{train} and sample weights w_{train}
  Test SVM with Sample Weights:
  Use M to predict labels for the testing set D_{test}
  Evaluate accuracy and error rate
  Output: Accuracy, Error Rate
end procedure
procedure CLASSWEIGHTEDSVM(D)
  Preprocessing:
  Normalize or scale features in $\mathbf{x}_i$
  Split D into training and testing sets
  Calculate Class Weights:
  Compute class weights based on class distribution in the training set
  Train SVM with Class Weights:
  Initialize Weighted SVM model M with class weights
  Train M using the training set D_{train} with labels y_{train}
  Test SVM with Class Weights:
  Use M to predict labels for the testing set D_{test}
  Evaluate accuracy and error rate
  Output: Accuracy, Error Rate
end procedure

4.4 Decisison Tree Algorithm

The Decision Tree Algorithm is a flexible and interpretable model that is often employed in classification applications. After initializing with specified hyperparameters, the Decision Tree model, denoted as T, is trained using the training set D_{train} and corresponding labels y_{train}. During the testing phase, T predicts labels for the testing set D_{test}. Accuracy is one statistic used to evaluate categorization performance. Algorithmically, the time complexity of the Decision Tree model is determined by parameters such as tree depth and feature count, with values commonly ranging from $O(logn)$ to $O(nlogn)$. This is an important factor, especially when working with bigger datasets, because it affects the algorithm's efficiency.

Algorithm 3 Decision Tree Algorithm

Input: Training dataset $D = \{(\mathbf{x}_i, y_i)\}$ with features $\mathbf{x}_i$ and labels y_i
Output: Accuracy, error rate
Preprocessing:
Normalize or scale features in $\mathbf{x}_i$
Split D into training and testing sets
Train Decision Tree Model:
Initialize Decision Tree model T with hyperparameters
Train T using the training set D_{train} with labels y_{train}
Test Decision Tree Model:
Use T to predict labels for the testing set D_{test}
Evaluate the classification performance using metrics
Output: Accuracy, error rate

4.5 *Random Forest Algorithm*

The Random Forest Algorithm, a robust ensemble learning method, enhances performance by aggregating predictions from multiple decision trees, introducing diversity through random selection of attribute subsets from the training data.

Algorithm 4 Random Forest Algorithm

Input: Training dataset D, number of trees N, number of features considered at each split M, maximum depth of each tree D_{max}
Output: Accuracy, Error rate
for $i = 1$ to N **do**
    $D_i \leftarrow$ Randomly select a bootstrap sample from D
    $T_i \leftarrow$ BuildDecisionTree(D_i, M, D_{max}) ▷ Using entropy as the criterion
end for
return $\{T_1, T_2, \ldots, T_N\}$
procedure BUILDDECISIONTREE(D, M, D_{max})
    if Stopping criteria are met or D_{max} is reached **then**
        return a leaf node with the majority class in D
    end if
    $f \leftarrow$ Randomly select M features from the dataset
    $j, \theta \leftarrow$ Find the best split using entropy on features f in D
    Split the dataset into D_{left} and D_{right} using the split (j, θ)
    $T_{\text{left}} \leftarrow$ BuildDecisionTree($D_{\text{left}}, M, D_{\text{max}}$)
    $T_{\text{right}} \leftarrow$ BuildDecisionTree($D_{\text{right}}, M, D_{\text{max}}$)
    Output Accuracy, Error rate
end procedure

The algorithm's flexibility is determined by two parameters: the maximum depth of each tree (D_{max}) and the number of features examined at each split (M). Entropy is the key requirement for tree splitting, increasing the algorithm's ability to understand complex relationships within the data. The time complexity for the entire ensemble is given by the formula $O(N * D * K * log(N))$, where D is the trees' maximum depth and K is the average number of features taken into account at each split.

5 Experimental Analysis

5.1 Experimental Setup

All the experiments are conducted in a Python 3.10.12. Key libraries used in the analysis include NumPy (1.23.5), Pandas (1.5.3), Scikit-learn (0.24.2), Matplotlib (3.7.0), and Seaborn (0.13.1). The computations are carried out on an HP Ryzen machine, which provides a stable and efficient platform for the deployment of machine learning models (Figs. 3 and 4).

5.2 Results and Discussion

SVM The Support Vector Machine (SVM) with a polynomial kernel (degree=5, C=1) demonstrated a good performance in web service classification. The model had strong classification skills, especially when it came to differentiating between criteria like Documentation, Availability, and Response Time. An in-depth examination of the confusion matrix (see Fig. 5) was carried out in order to attain a better comprehension of the SVM model's performance. The confusion matrix depicts the interaction of true positive, true negative, false positive, and false negative predictions, which provides a thorough assessment of the SVM model's effectiveness in dealing with different and complex web service features.

Weighted SVM
Sample Weighted SVM: Introducing sample weights based on the 'Documentation' column, the SVM model aimed to give more importance to instances with critical documentation attributes, as shown in the confusion matrix (Fig. 3). Using this

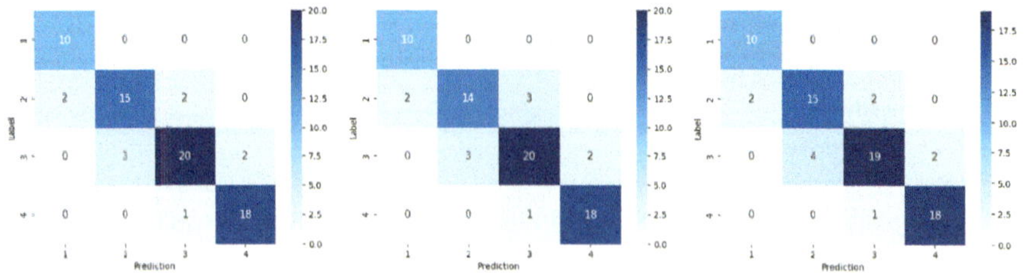

(a) (b) (c)

Fig. 3 Confusion matrix for **a** Normal SVM **b** Sample weighted **c** Class weighted

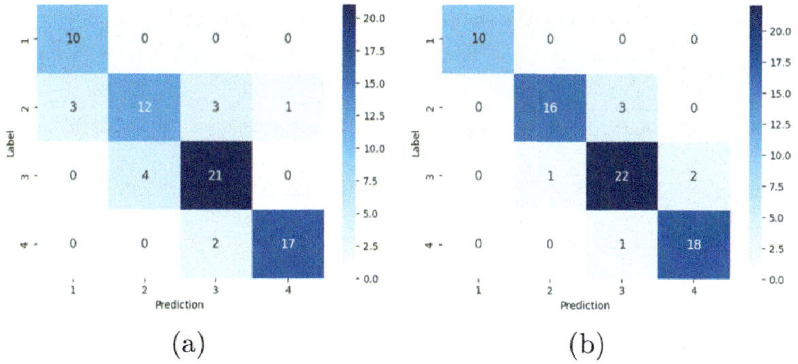

Fig. 4 Confusion matrix for **a** Decision tree **b** Random forest

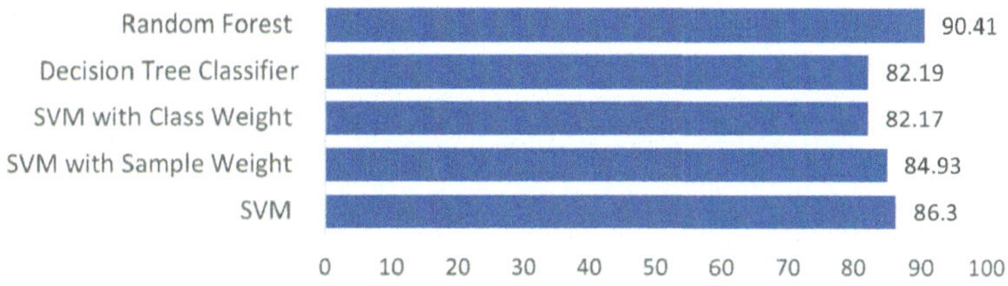

Fig. 5 Accuracy comparison of the various models

method highlights the relevance of certain characteristics in the classification process and enables the model to adjust to the importance of individual data points.

Class Weighted SVM: Investigating the influence of class weights on SVM performance, the model addressed potential imbalances in the distribution of classes. The confusion matrix (Fig. 3) highlights the effectiveness of class-weighted SVM in handling variations in the number of instances belonging to each class.

Decision Tree The Decision Tree classifier is with certain entropy and a maximum depth of 5 on the QWS dataset. This model enhances interpretability, as decisions are made based on feature thresholds. The confusion matrix (Fig. 4) offers insights into the Decision Tree's classification performance. Interpretability and ease of understanding make Decision Trees valuable for certain applications, but they may be sensitive to overfitting.

Random Forest The random forest model used in this work uses $N = 200$ number of trees. It uses entropy as the criteria for tree splitting. This Section presents an experimental investigation of the random forest method, focusing on the confusion matrix (Fig. 4). The Random Forest algorithm's ensemble learning capabilities are evaluated to determine classification performance.

Table 3 Comparison of performance metrics of different models (in %)

Model	Accuracy	Precision	Recall	F1 score
SVM	86.30	86.30	86.30	86.12
SVM with sample weight	84.93	84.93	84.93	84.67
SVM with class weight	82.17	84.93	84.93	84.71
Decision tree classifier	82.19	82.19	82.19	81.87
Random forest	**90.41**	**90.41**	**90.41**	**90.40**

5.3 Performance Metrics

In Table 3, the accuracy measure shows the proportion of properly identified cases. The precision, recall, and F1 score are also reported to get a proper understanding of the performance of the models. The SVM with a polynomial kernel stands out with a relatively high accuracy, whereas the other SVM variants exhibit lower accuracies. The accuracy that the Decision Tree Classifier attains turns out to be lowest. The Random Forest, on the other hand, stands out due to its highest accuracy as it uses ensemble strategy and can generalize well on test data.

Table 4 presents a comparison of accuracy among existing models and the best model obtained in this study, namely Random Forest. Laachemi et al. [10] implemented an SVM model achieving an accuracy of 83.52%. They also utilized a stochastic local search (SLS) meta-heuristic for feature selection with SVM, resulting in an accuracy of 84.86%. In the research by Mohanty et al. [9], Naive Bayes, Markov blanket, and Tabu search were employed, with Naive Bayes achieving the highest accuracy of 85.62%. In contrast, the Random Forest model in this study achieved an accuracy of 90.41%.

Table 4 Comparison of Exciting models with our model

Method	Accuracy (%)
SVM [10]	83.52
SVM+SLS [10]	84.86
Naive Bayes [9]	85.62
Random forest (Proposed work)	**90.41**

5.4 *Discussion*

This work focuses on accuracy as the key performance indicator when evaluating SVM, Decision Tree, and Random Forest models for web services classification. The standard SVM's classification abilities are quite impressive. With its high performance and simple analysis, the Decision Tree appears to be a good choice but a single decision tree is not able to generalize well on unseen data. The Random Forest is able to take advantage of the combined strength of multiple decision trees, as suggested by the fact that it performs better in terms of accuracy than rest of the models. Random Forest performs exceptionally well overall because of it uses an ensemble of multiple decision trees and is able to reduce overfitting and is less sensitive to noise in the dataset.

6 Conclusion

In this paper, a comprehensive approach is presented for web service classification that employs machine learning techniques. In this study, Random Forest was found to be the best performing model. The experimental results provide useful insights for researchers as well as developers, laying the groundwork for future advances in web service quality prediction. The contributions include improved model performance and insightful analyses, adding to the larger discussion about improving the reliability and applicability of machine learning models in the domain of web service classification.

In future, adding more advanced features and transformations can be considered to make the input data even more informative. It would also be interesting to investigate additional ways to explain how the model works, ensuring that users can trust and understand it better. Finally, the model's scalability can be tested on larger datasets to see how well it performs as the size increases.

References

1. Alonso G, Casati F, Kuno H, Machiraju V (2004) Web services. Springer, Berlin, Heidelberg, pp 123–149
2. Simple Object Access Protocol (SOAp) 1.1 (2000) Don Box, David Ehnebuske, Gopal Kakivaya, Andrew Layman, Noah Mendelsohn, Henrik Frystyk Nielsen, Satish Thatte, and Dave Winer
3. Christensen E, Curbera F, Meredith G, Weerawarana S et al (2001) Web services description language (wsdl) 1:1
4. van Steenderen M (2000) Universal description, discovery and integration. S Afr J Inf Manage 2(4)
5. Moraru A, Fortuna C, Fortuna B, Slavescu RR (2009) A hybrid approach to qos-aware web service classification and recommendation. In: IEEE 5th international conference on intelligent computer communication and processing, pp 343–346

6. Vaadaala V, Rao RR, Rao V (2013) Classification of web services using JForty eight. *Int J Electron Commun Comput Eng* 4(6):181–184
7. Alrifai M, Risse T (2009) Combining global optimization with local selection for efficient qos-aware service composition. In: The web conference
8. Chippa M, Priyadarshini A, Mohanty R (2019)Application of machine learning techniques to classify web services. In: IEEE international conference on intelligent techniques in control, optimization and signal processing (INCOS), pp 1–7
9. Mohanty R, Ravi V, Patra M (2012) Classification of web services using Bayesian network. J Softw Eng Appl
10. Laachemi A, Boughaci D (2016) A stochastic local search combined with support vector machine for web services classification. In: International conference on advanced aspects of software engineering (ICAASE), pp 9–16
11. Hambir Y, Sandhya P (2020) Applying various machine learning techniques for the classification of web services: a survey. J Crit Rev 7(14):1091–1103
12. Cortes C, Vapnik V (1995) Support-vector networks. Mach Learn 20:273–297
13. Huang YM, Du SX (2005) Weighted support vector machine for classification with uneven training class sizes. In: International conference on machine learning and cybernetics, vol 7, pp 4365–4369
14. Safavian SR, Landgrebe D (1991) A survey of decision tree classifier methodology. IEEE Trans Syst Man Cybern 21(3):660–674
15. Bagga P, Joshi A, Hans R (2017) Qos based web service selection and multi-criteria decision making methods. Int J Interact Multimed Artif Intell, InPress 12
16. Al-Masri E, Mahmoud QH (2007) Discovering the best web service. In: Proceedings of the 16th international conference on world wide web, pp 1257–1258

Software Engineering and Security
Machine Learning

Contextual Correlation Inference in Multi-fleet Robotic Systems

Raunak Bag, Nabendu Chaki, and Agostino Cortesi

Abstract The dependency of the Non-functional requirements (NFR) on the operating environment can lead to unexpected behaviors. This may subsequently result in threats to robotic systems after being deployed in an uncontrolled environment. This work proposes a simulation-based approach to identify the NFR correlations in environmental contexts for multi-robot systems. The simulation results are useful to infer and evaluate the different conflicts between NFRs and to study the impact of different contexts on the requirements themselves. The identified correlations would be useful for the system designers in minimizing impact and preventing problems arising from commonly overlooked issues. The adopted methodology is easily reproducible in different development scenarios.

1 Introduction

The advent of robotic systems in our lives ushered in the promise of the future. Suggesting unparalleled efficiency, robots are quickly becoming the next big thing in the brief but impactful and on some occasions revolutionary, history of technological advancements in the last half a century. Robotic systems have gained quite some ground in the last few decades, from being used in industrial assembly lines to now being trusted in some parts of the healthcare industry [1].

Robots are cyber-physical systems [2] that perceive and react to the world by making use of sensors, actuators, and control systems. As these systems evolve to handle more sophisticated functions, heterogeneous multi-robot systems [3] often organized

R. Bag (✉) · A. Cortesi
DAIS, Ca' Foscari University, Venice, Italy
e-mail: raunak.bag@unive.it

A. Cortesi
e-mail: cortesi@unive.it

N. Chaki
Department of Computer Science and Engineering, University of Calcutta, Calcutta, West Bengal, India
e-mail: nabendu@ieee.org

© The Author(s), under exclusive license to Springer Nature Singapore Pte Ltd. 2025
N. Chaki et al. (eds.), *Applied Computing for Software and Smart Systems*, Lecture Notes in Networks and Systems 980, https://doi.org/10.1007/978-981-97-9762-2_17

into fleets are witnessing wider application areas as well. It is also paramount to have guarantees on the non-functional properties like safety, of the system, especially when deployed in a shared human-occupied space. The layers of complexity of the system make the examination of non-functional properties difficult compared to a single-robot system. The presence of correlations among the non-functional requirements (NFR) if not identified in the specification stage can have a cascading negative effect [4] on the system. Existing literature reveals a deficit in the study of NFR correlations concerning environmental context for multi-fleet robotic systems.

One method to investigate the NFR correlations can be by the use of prototypes. Developing real-world prototypes though, especially for multi-fleet robotic systems, can be time-consuming and costly in some instances. Simulators provide a scalable opportunity to set the runtime environment and provide information to study non-functional properties. Our objective with this work is to explore the impact of context-sensitivity for multi-fleet robotic systems deployed often in a shared human space. The approach also attempts to expose NFR-context parameter correlations that may be present but hidden as compared to single-robot systems due to their complex nature. The proposed approach can assist system designers in being aware of potential issues arising from context and NFR correlations. We aim to build upon existing methodologies based on a simulation approach to elicit correlations among NFRs in robotic environment scenarios. The used methodology consists of the following steps briefly:

1. *Selection of robot fleet, simulator, and simulation world*: The components of the simulation have to be chosen after careful consideration of specifications and requirements elicited from the stakeholders.
2. *Task setup and simulation execution*: The tasks corresponding to the real-world targets for the robotic system have to be set up per the robotic fleet organization and capability. This is to be followed by the execution of the aforementioned tasks in the simulator and the observable NFR parameters have to be recorded.
3. *Analysis of parameter data and identifying correlations*: Finally, the experimental data has to be analyzed for correlations arising from NFR and context parameters. This analysis can be done by graphically plotting the parameters in pairs and observing the trends.

The main contribution of the work is the in-depth analysis of how to use complex simulation environments of multi-fleet robotic systems to identify correlations between non-functional requirements in different contexts, thus providing valuable information for requirements engineering [5, 6]. Our experiments provide preliminary evidence of the effectiveness of the proposed methodology.

The article has been structured as follows. Section 2 dives into the relevant literature. Section 3 briefly describes some preliminary concepts. Section 4 discusses the methodology used in this work. Section 5 discusses the results of the experiment. Finally, Sect. 6 highlights the threats to the validity of our approach, and Sect. 7 concludes.

2 Related Work

Among the many non-functional requirements usually present in robotic systems [7], security is often at the forefront. Authors in [8] highlight the limitations of Robot Operating System (ROS) 2 from a security perspective. While ROS2 improves the support for real-time multi-robot systems, it also exposes the system to a larger attack surface. The paper investigates the vulnerabilities present in SROS2 [9], the security tool of ROS2 and provides potential solutions. Some of the SROS2 vulnerabilities has already been closed while others are in the works maintained by the SROS GitHub community. The simulation setup involving the use of Open-RMF was of particular interest to our work. ROS2 has also been known to lack task management capabilities [10] crucial for multi-threaded robotic applications. The open-source nature of ROS2 allows for software architectures focusing on different aspects of native ROS2 limitations to be developed by the robotics community.

Wider deployment of robotic systems targeting multiple functionalities constitutes the use of multiple robots working in tandem, often organized into fleets. These robot fleets can be task-oriented and have heterogeneous robots or they may be function oriented involving homogeneous robots.

Heterogeneous robotic systems deployed in industrial settings such as warehouses and working in collaboration with humans can be made more efficient in their interactions with humans by the use of augmented reality bypassing the need for expensive laser scanners as found in [11]. Safety vests worn by humans can be flagged by the augmented reality system and help the robots navigate during their tasks. Path planning problem also has to be tackled for the robots to navigate within an environment. Partial knowledge about the unknown environment can be communicated among robots to improve their estimation of the surroundings and used in conjunction with petri-net models [12] of the environment by each robot to plan their next move. Task allocation and path planning for multi-robot systems are also witnessing the use of deep reinforcement learning [13] from visual information. Image processing algorithms can also assist with increased precision of robot actions such as grasping granular food materials [14]. Processing visual information does come at a cost, one which may not be feasible in some robotic systems. Use of edge-computing in robotic systems [15] involving co-design paradigms can alleviate some of the limitations that arise due to the resource-constrained nature of robotic systems.

Apart from industrial settings, multi-robot systems are also seeing an emerging application in urban mobility uses such as domestic healthcare [16], leisure, or retrieving goods. A literature review conducted in [17] identified robot designs agreed upon by user preferences based on results from three focus groups in the study. Another application of robotic systems in urban mobility can be observed in assisted living. Ambient assisted living environments can be complemented with a robot providing in-depth information on the environment dynamically from the ground. 3D depth sensors mounted with the robot can provide information on potential risks for elderly people and also those with limited mobility [18]. Smart homes can also benefit from the use of robotic systems alongside smart automation systems.

A study [19] explores the feasibility and challenges involving the integration of such systems in smart residential areas. A different study [20] investigates the degrees of autonomy achievable with the use of care-robotics for older people. A recurring gap in such studies identified in the survey [21] is the lack of consideration for the context of care. Awareness of the context of the operating environment is essential in these instances.

These studies furthermore reveal the importance of context-aware robotic systems [22], especially for those deployed in environments inhabited by humans. The reader can refer to the surveys [3, 23] on other aspects of heterogeneous and distributed robotic systems not considered in this work.

3 Preliminaries

In this section, we briefly lay out some of the concepts intended to assist the reader in better understanding the work.

3.1 Robot Operating System

The Robot Operating System (ROS)[1] was developed by Open Robotics as a bridge between the hardware and software comprising the robotic system. It has quickly become the mainstream open-source software development kit for roboticists. The ROS-based system represents its processes as a ROS-Graph consisting of an atomic-entity node. Nodes may receive and send information through message-passing. The messages include data from actuators, sensors, control systems, planners, and others [24]. The latest iteration of ROS, ROS2 [25] improves upon the limitations of ROS primarily in basic security features and uses Data Distribution Service (DDS) for communication. ROS also provides an inbuilt simulation system Gazebo.[2]

3.2 Open-RMF

Developed by Open Robotics, Open-RMF is an open-source modular software framework developed to enable multi-fleet heterogeneous robotic system interoperability with physical structures involving building management systems. The framework was initially developed for the healthcare industry but also found relevance for large public structures such as airports, offices, hotels, and more. It provides tools capable

[1] https://www.ros.org.

[2] https://gazebosim.org/.

of achieving interoperability among robots from different vendors, minimizing risks in the process. We describe some of its components briefly as follows:

1. *Traffic Editor*: The traffic editor can annotate floor plans while also generating traffic patterns involving multi-fleet robot operations. The editor can generate simulation world files that can be used even outside of RMF.
2. *RMF-Web*: RMF-Web provides an Internet interface capable of monitoring and controlling different aspects of Open-RMF deployment. Task allocation to the robot fleets is also enabled from the interface itself.

3.3 Menge Crowd Simulation

Menge [26] is a crowd simulation framework designed to simulate pedestrian movement. Menge works by decomposing the simulation problem into two sub-problems:

1. *Global Path Plan*: Generates a connection graph, defining human-reachable areas and a finite state machine, defining states and transitions for a human heading to a target.
2. *Local Collision Avoidance*: Employs Optimal Reciprocal Collision Avoidance (ORCA) [27], modeling humans as circles with modifiable radius.

4 Methodology

This section describes the methodology applied to find NFR correlations with the environment context in a robotic simulation. We build upon the methodology [28] established for single-robot systems and explore the application on multi-robot systems. The methodology can be adopted by robotic system developers working with multi-fleet systems, which will help them achieve better system design.

We assume the following necessary information is available:

- I.1 Hardware and software specification of robots in the fleet.
- I.2 Functional and non-functional requirements of the robots and fleets as gathered from system stakeholders.
- I.3 Organization information of robot fleets.
- I.4 Operational environment of robotic system's deployment.

The overall procedure can be described in three primary steps elaborated as follows:

1. *Selection of robot fleet, simulator, and simulation world*: This step describes the selection of simulation components representing their real-world counterparts for the selected operation.

(a) Select specification I.1 compliant robotic system *R* compatible with the requirements according to the information available from I.2. Preference should be given to robotic systems supporting widely adopted open-source middleware platforms like ROS2. This would ensure better compatibility among the various components of the simulation. Third-party component drivers are supported very well by open-source communities such as that of ROS2.
(b) Select simulator *S* supporting multi-fleet operations and the robots *R*. The simulator *S* should have provisions for recording and logging NFR parameters from I.2 in runtime.
(c) Select simulation environment *E* conforming to I.4 and compatible with the simulator *S*. Unavailability of the required environment *E* may be mitigated if the simulator *S* has provision for environment design. Otherwise, 3D-environment design software able to output simulator *S* compatible world file may be employed.

2. *Task setup and simulation execution*: Upon the setup of simulation components, tasks have to be assigned for simulation scenarios and then executed.

   (a) Set robot tasks *T* for the simulation scenarios according to the organization gathered in I.3. The tasks may be internally assigned among the robot fleets if a fleet manager system is available in the simulator. Otherwise, manual assignment has to be done per the functions available for the robots. Randomization of some elements of a task can elicit more nuanced cases. Simulation scenarios should be enough in numbers to provide consistent data for analysis.
   (b) Set path planner according to task information if planning is done statically. This step can be skipped if the simulator *S* is using a dynamic runtime path planner.
   (c) Identify environment constraints and the observable NFR parameters from I.2 within the simulator's capability. The NFRs to be included have to be selected by the system designer(s) conforming to I.2 taking into account the operational environment of the system.
   (d) Execute the simulation scenarios and record the observable NFR parameters.

3. *Analysis of parameter data and identifying correlations*: The analysis of experimental data can be decomposed into the following steps:

   (a) Analyze the NFR parameters in pairs for correlations among them. Graphically plotting the values for pairwise comparison should yield trends that will help in identifying NFR pairs with correlations among them.
   (b) Analyze the impact of contextual factors such as the simulation environment and its novel features on the NFR parameters recorder for the robotic system fleets. Context parameters can be plotted against NFR parameters graphically and contextual-sensitive nature can be exposed when present. Context parameters are simulation environment dependent and up to the discretion of the system designer to be selected as deemed relevant.

The methodology described above attempts to address the gap in the investigation of context-NFR correlations in multi-fleet robotic systems, as gathered from the literature in Sect. 2. The presence of these correlations may impact the system in often unexpected ways. The following section elaborates on the experimental validation of this methodology, providing evidence of its effectiveness.

5 Experimental Evaluation

This section validates the proposed methodology for multi-fleet robotic systems in simulation. The experiment simulations were run on a computer with AMD Ryzen 9 6900HS CPU with 32 GB DDR5 4800MHz RAM and AMD Radeon RX 6000 series GPU, running Ubuntu 22.04 Operating System.

The experiments were automated to the extent of the capabilities of the frameworks used. The time to conduct the experiments will vary depending on the tasks chosen for the scenarios as they depict closely the real-world execution time for the task. The setup time for the simulation consisted primarily of the module and framework installation time.

The prerequisite knowledge for the experiments as required by the methodology mentioned in Sect. 4 is as follows:

- The system specifications for the robots in the simulation (SESTO, MiR100, Magni, Hospi) were gathered.
- The functional goals G were defined according to the robots in the fleets' capabilities:

  1. *Clean*: Navigating to and cleaning a well-defined zone within the simulation environment E.
  2. *Delivery*: Delivering an item from one location in the simulation environment to another.
  3. *Patrol*: Navigating and monitoring a specified path in the simulation environment. The path can be looped as many times as specified in the task request.

- The organization information on the robot fleets was specific to the simulation environments used and was gathered according to the world demonstrations available in the simulation framework.
- Operational environments involving high interactivity were selected from the Open-RMF demos.

5.1 Experiment Steps

The proposed methodology is illustrated experimentally as follows:

1. *Selection of robot fleet, simulator and simulation world*: We have selected the following components for the simulation:
   (a) Robots SESTO, MiR100, Magni, and Hospi were selected according to their functionality to carry out the three types of functional goals G.
   (b) The Open-RMF[3] framework's simulation demos were selected for the experiments. ROS2 Humble was selected as the middleware platform and the corresponding branch of Open-RMF was installed. The simulation was run on Gazebo 11.
   (c) The simulation operating environments chosen for this experiment were
      i. Hotel[4] environment consisting of three floors with four available robots divided into three fleets according to their function.
      ii. Airport[5] environment comprising an airport terminal with several boarding gates with static as well as simulated dynamic human movements. Menge crowd simulation runs as a plugin in this environment for human motion. Ten robots divided into three fleets are available in this environment.

   Open-RMF also has Traffic Editor[6] tool to auto-generate 3D world files for simulation and can annotate 2D floor plans to do so.

2. *Task setup and simulation execution*:
   a. The robot tasks were assigned to the fleet manager of Open-RMF in the simulation from the web portal RMF-Web[7]. The fleet manager distributes the tasks among fleets and ultimately to the robots based on their functions and availability. The task elements like task type, destination, number of loops, priority, and schedule were randomized to gather as much unbiased data as possible.
   b. Open-RMF employs fleet adapters that use a bidding mechanism to assign tasks to fleets and robots with the most suitable bid. Path planning is done dynamically during runtime accounting for changing circumstances during the simulation run. Fleet adapters are also responsible for traffic decongestion for overlapping robot paths. If task priorities are assigned (0 for no preference, 1 or more for higher priority), robots gain precedence according to their task priority. For same-priority tasks, precedence is given to the task with a shorter

---

[3] https://github.com/open-rmf/rmf_demos.

[4] https://github.com/open-rmf/rmf_demos/blob/main/rmf_demos_maps/maps/hotel/hotel_L1.png.

[5] https://github.com/open-rmf/rmf_demos/blob/main/rmf_demos_maps/maps/airport_terminal/airport_terminal.png.

[6] https://osrf.github.io/ros2multirobotbook/traffic-editor.html.

[7] https://github.com/open-rmf/rmf-web.

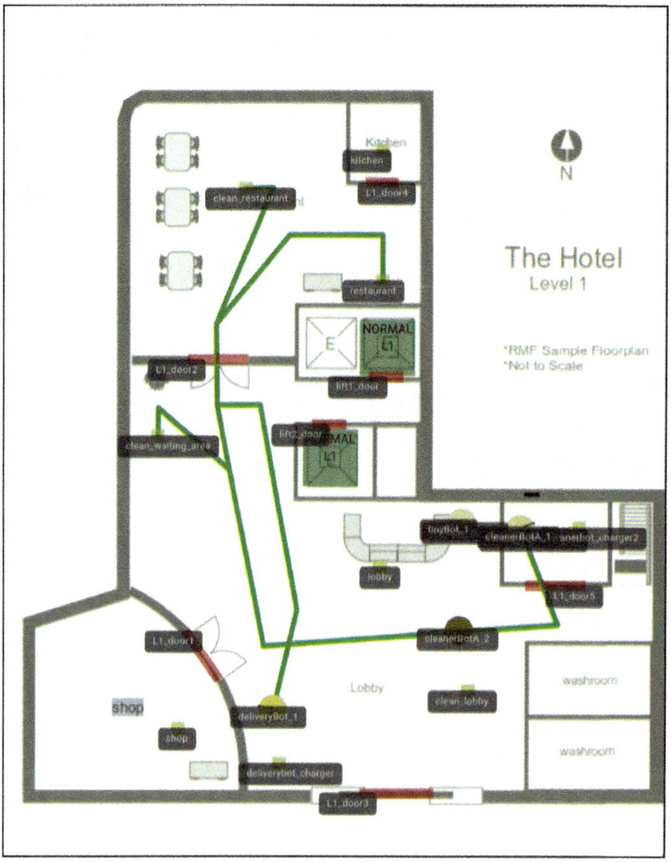

Fig. 1 Hotel ground floor map for a scenario with overlapping robot paths

next intermediate goal. An instance of a scenario in both environments with overlapping path tasks is shown in Figs. 1 and 2.

c. The identified observable NFR parameters are as follows:

   – Fleet availability: Robots available in the fleet for new tasks.
   – Task completion time: Time to complete the assigned tasks.
   – Battery retention: Battery capacity remaining after execution of the task by a robot.
   – Navigation events: Decomposition of a task into navigation and action goals.
   – Traffic resolution events: "Wait" events as a resolution step for traffic congestion during navigation.

d. The simulation was executed with 30 scenarios for each of the two environments selected. The parameter values recorded were obtained from the status

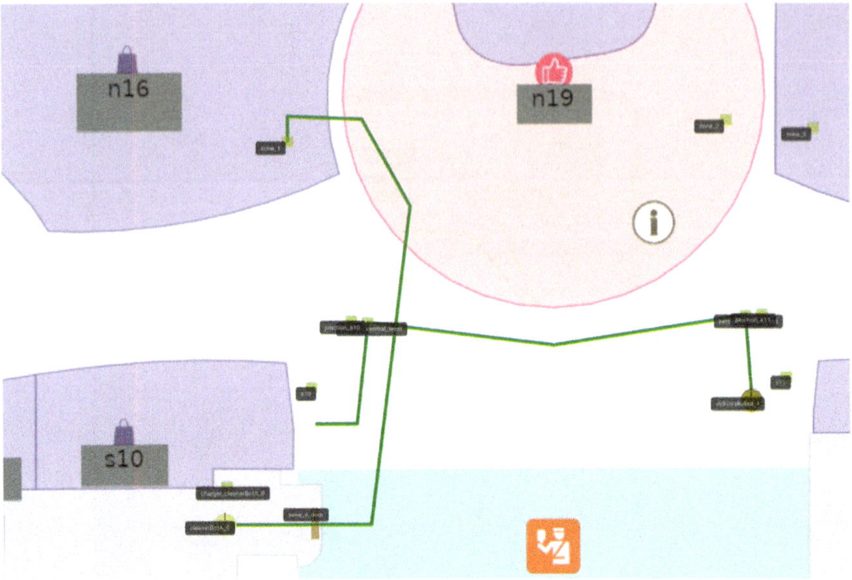

Fig. 2 Airport map for a scenario with overlapping robot paths

Table 1 Hotel environment results

No. of active robots	No. of scenarios	Average task completion time	Average navigation events	Average traffic resolution events
1	4	331.7	26.25	0
2	14	385.5	44.8	2
3	12	356.4	54.8	4

update and log from the RMF-Web panel as well as terminal command outputs. Information about tasks such as task type, destination, number of loops, and priority was also tabulated along with the observable parameter values in the previous sub-step.

3. *Analysis of parameter data and identifying correlations*: The analysis of the recorded data obtained in the previous sub-step has been done in two parts and elaborated in the following subsection.

Summaries of the recorded data from hotel and airport environments are presented in Tables 1 and 2.

5.2 Discussion

This section provides the result of the analysis made on recorded parameters from the experiment. The analysis has been done in three parts as explained next. The parameters of interest for analysis include battery retention, scenario completion time, navigation events, and traffic resolution events.

Hotel Environment The hotel simulation environment consists of three floors, accessible by a lift and sliding doors sectioning the ground floor, both of which are interactable by the robots. The environment consists of three robots operating in two fleets capable of cleaning and patrolling tasks. We compare the different parameters for the hotel environment simulation scenarios as follows:

Figure 3 compares the number of active robots, the average number of navigation events, and the average number of traffic resolution events among those. It is evident from the graph that an increase in the number of active robots follows an increase in the average number of navigation events and average traffic resolution events almost linearly. Increasing the number of active robots from 1 to 2 in the scenario also increases the average number of navigation events by 70.6% from 26.25 to 44.8. In contrast, with three active robots, the rise is 22% as compared to two active robot scenarios. The average number of traffic resolution events increases linearly in all three cases.

Next, we look at the impact of scenario execution time on the average battery retention of robots in the scenario. The observations in the graph have been ordered by the increasing complexity of the assigned tasks in the scenario. A cleaning task is considered to be more complex than a patrolling task since it takes much more time

Table 2 Airport environment results

No. of active robots	No. of scenarios	Average task completion time	Average navigation events	Average traffic resolution events
1	3	459.3	13.6	0
2	14	545.6	20.8	0.1
3	9	519.7	30.4	0.3
4	4	719.2	46.2	2.5

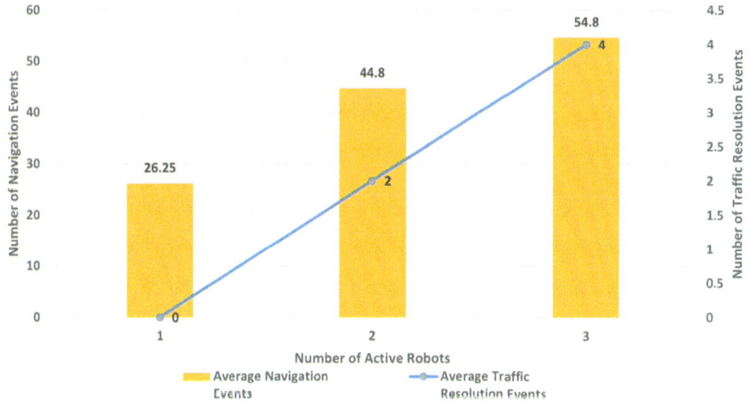

Fig. 3 Hotel scenario execution time and events

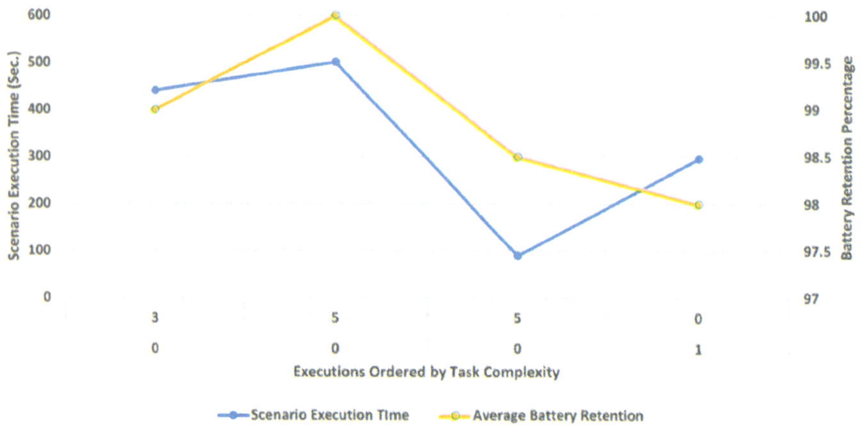

Fig. 4 Hotel environment with one active robot

to complete on average. The bottom row of the X-axis labels represents the number of cleaning tasks while the row above represents the number of patrolling tasks. Each column of the labels represents one scenario.

The graph in Fig. 4 shows the result for scenarios involving a single active robot. The scenario with only one cleaning task shows the expected trend of a decrease in average robot battery retention with an increase in the scenario execution time. We are not able to gain meaningful insight into the other cases from this graph. This may warrant further investigation in the nature of the task locations for those scenarios.

Figure 5 plots scenarios with two active robots. We can see a general expected trend of an increase in the average battery retention of robots with a decrease in the scenario execution time. Deviations such as the last observation for scenarios with one cleaning and three patrolling tasks requires further investigation into the location of the assigned tasks as the environmental context may come into play not evident from this graph.

We can see a similar expected trend in Fig. 6 for three active robot scenarios. The rise in scenario execution time results in a fall in the average battery retention percentage of the robots in the scenario.

Airport Environment The airport simulation environment consists of a large terminal area with ten robots operating in three fleets capable of cleaning, patrolling, and delivering tasks. The recorded parameters are analyzed for correlations as follows.

In Fig. 7, we try to find the relation in scenario parameters among the number of active robots, the average number of navigation events, and the subset of traffic resolution events within them. We observe a gradual increase in both the average number of navigation events and traffic resolution events in the scenarios employing one to three active robots. Further increasing the number of active robots to four results in a larger jump in the number of average navigation events while also witnessing a steep increase in the average traffic resolution events. The average number of navigation events goes up by 52% while average traffic resolution events see a jump of over seven times the number from three active robot scenarios.

Figures 8, 9, 10, and 11 look for the relation between scenario execution time and average battery retention percentage for different numbers of active robots. The observations have been ordered by the increasing task complexity scenarios. Of the three types of tasks robots in this scenario are capable of, cleaning tasks are considered to be the most complex followed by patrolling tasks next and finally delivery tasks. The complexity has been determined in terms of the task completion time for each type on average. The last row of the X-axis labels represents the number of cleaning tasks while the row above represents the number of patrolling tasks. The top row represents the number of delivery tasks. Each column of the labels represents one scenario. The graph in Fig. 8 validates the expected behavior of the simulation for a single active robot. The observations consist of a maximum of four tasks in each scenario. The decrease in the battery average battery retention is almost proportional to the increase in the scenario's execution time for each observation.

Figures 9 and 10 showcase a similar behavior, that is, the battery retention goes down as the time to execute the scenario goes up in most of the cases. The rate may not be proportional and this can be attributed to the particular objective of the tasks, primarily in the location of the tasks. The system designer may investigate the influence of the task locations in the environment of concern.

The graph for four active robots shown in Fig. 11 indicates adherence to the expected tendency of the scenario execution time increasing when the average robot battery retention decreases in the second and third observations. The first and fourth observations buck from the expected trend. While not particularly insightful directly, this may serve as a starting point for further investigation for the system designer to elicit potential correlation among the task objectives itself with scenario execution time and average battery retention of robots.

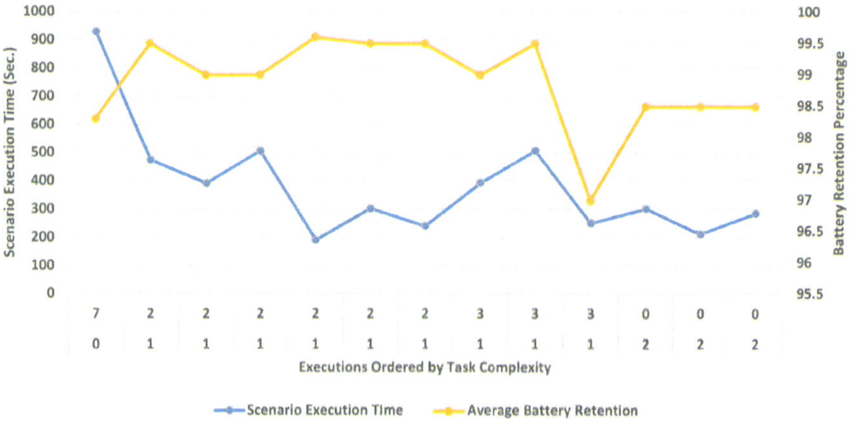

Fig. 5 Hotel environment with two active robots

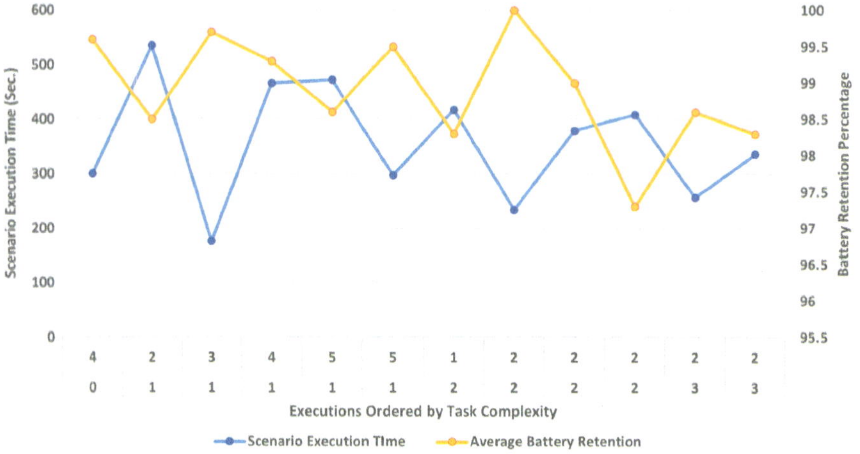

Fig. 6 Hotel environment with three active robots

Context Impact Analysis Finally, we compare the two environments used in the simulation for the experiments to discern the impact of context on the task execution scenarios.

We compare the three parameters: the number of events with traffic resolution, the average navigation events, and the average scenario execution time between the two environments in Fig. 12. The scale of the environments can be judged from the maps referred to in 4 and 5.

From the graph, on one hand, we observe that the average number of navigation events and the number of events with traffic resolution is much higher in the hotel

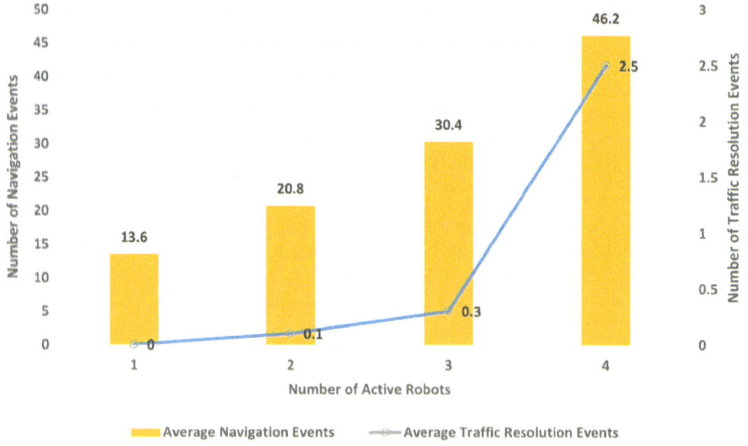

Fig. 7 Airport scenario execution time and events

environment than in the airport environment despite having less number of maximum active robots. This can be attributed to the fact that the airport map is much larger on a single level than the hotel environment and also has more open areas for the robots to navigate in. This results in the sub-task objectives represented by the navigation events being simpler for the airport environment. The number of traffic resolution events is also reduced as the robots have more open operating areas and can choose fewer overlapping paths. The hotel environment in contrast consists of many environmental elements within a small area, reducing the operating area of the robots and causing more frequent path overlaps as well as more navigation events due to the increased complexity of navigation.

On the other hand, we also observe that the average scenario execution time is quite high for the airport environment in comparison to the hotel environment. This can be explained by the scale of the airport environment. The larger environment also results in task locations that are further from the robots and thus take a longer time to complete.

Apart from the observations already made in this section from the analysis of recorder parameters, the simulation also ran into a few problems during the execution of the scenarios in the environments. Of particular interest was the occurrence of deadlock in certain rare cases. One instance of such occurrence was in the hotel environment when two robots, on different floors, were requesting access to the shared resource lift at the same time. The Open-RMF planner was not able to resolve the deadlock and the simulation had to be terminated. In another instance, a robot navigating to an objective in the airport environment got stuck due to the presence of a lamp-post in its path which the planner was not able to recognize.

The occurrence of these unexpected behaviors within the simulation nonetheless highlights the presence of factors such as a deadlock possibility and unlabeled or incorrectly labeled environment elements that the robot is unable to recognize. The

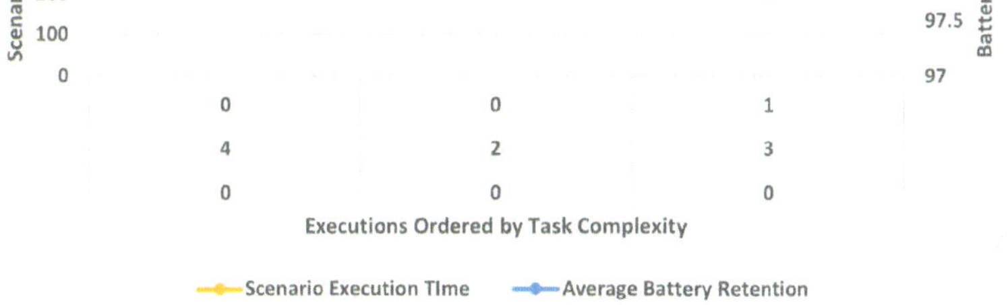

Fig. 8 Airport environment with one active robot

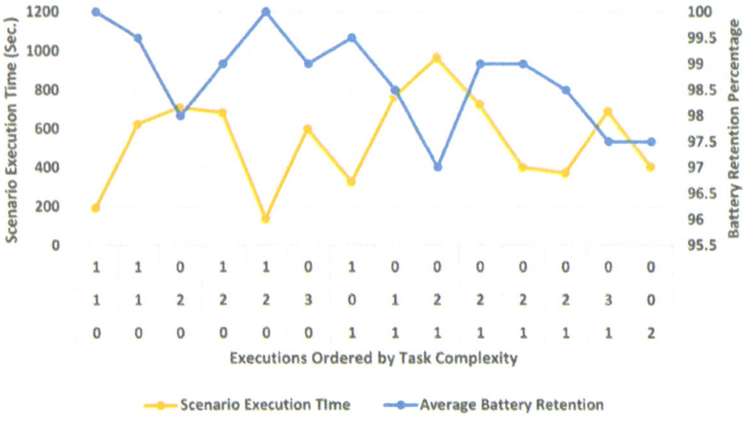

Fig. 9 Airport environment with two active robots

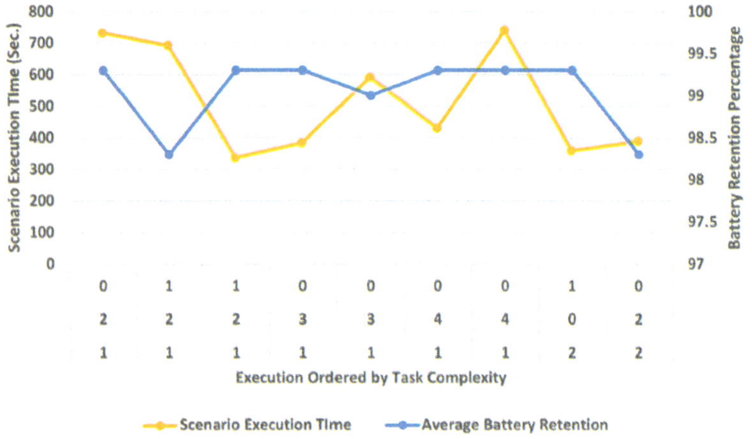

Fig. 10 Airport environment with three active robots

system developer may take note of these issues that can also impact real-world execution and improve the system design.

6 Threats to Validity

Simulation can be very effective as compared to real-world testing and verification. Robots in simulation can be charged instantly, downtime for resetting between runs is not present, and collision events do not have physical consequences. These are some of the benefits of using a simulation approach. Nonetheless, the approach and

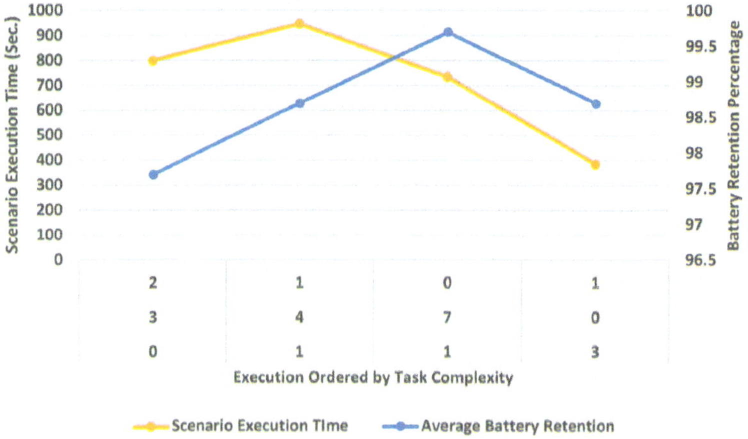

Fig. 11 Airport environment with four active robots

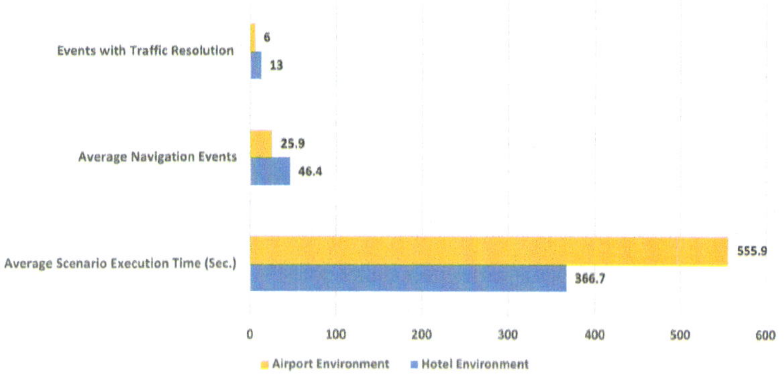

Fig. 12 Comparison of hotel and airport environment parameters

methodology adopted in this work have drawbacks too. We highlight the limitations of the proposed methodology that are threats to its validity:

1. Availability of a suitable simulator may be lacking for the system to be developed. While there has been significant improvement in the simulation tools available to a developer, there is still limited open-source support for some robot types that require very complex physics simulation. Projects like Open-RMF are aimed at tackling this problem but they still have some way to go before they are easily able to cater to different system designer's requirements.
2. The simulation environment available to a system designer may not be able to depict real-world interactions in a realistic enough way for all intents and purposes [29]. The complexity of the scenarios within the simulation is limited by a simulator's capabilities. This may hinder the application of the proposed approach for more demanding and critical systems such as in healthcare. The reliability of

simulation data also depends on the complexity of the tasks and the simulator's capabilities. Setting up a simulation may also be a time-consuming process depending on the availability of robots and virtual environment.
3. The system designer is required to have a good understanding of NFR and contextual parameters relevant to the system being designed to derive the correlations between parameters. Identification of proper parameters to be analyzed is crucial in obtaining useful information about the contextual-sensitive nature of the system. The work also assumes the expertise of the system designer in the operational environment of the system and the ability to identify relevant NFR constraints with regard to that context.

7 Conclusion

The advent of robotics has recently seen substantial research on heterogeneous and multi-fleet robotic systems. Despite that, there has been little focus on the analysis of NFR-context correlation for these systems that may negatively impact system design and performance if not handled carefully. Layers of complexity in multi-robot systems are only convoluted more by the presence of environmental contexts that can have an impact on the various NFR constraints relevant to the system. This work aims to build upon pre-existing methodologies for NFR correlation analysis and proposes a suitable methodology for multi-fleet robotic systems. The result obtained from the experiments demonstrates the validity of the proposed approach and identifies NFR-context relations for the generated scenarios.

This work may be further extended in the future to work around one of the current limitations of this approach in its closeness to real-world scenarios. The use of digital twins can considerably improve the simulation quality, especially in terms of interactivity with the simulation environment and depiction of more complex properties in the operational domain.

Acknowledgements Work partially supported by SERICS (PE00000014—CUP H73C220008 9001) under the NRRP MUR program funded by the EU—NGEU, and iNEST-Interconnected NordEst Innovation Ecosystem funded by PNRR (Mission 4.2, Investment 49 1.5) NextGeneration EU (ECS_00000043—CUP H43C22000540006).

References

1. Cooper S, Di Fava A, Vivas C, Marchionni L, Ferro F (2020) Ari: the social assistive robot and companion. In: 29th IEEE international conference on robot and human interactive communication (RO-MAN). IEEE, pp 745–751
2. Baheti R, Gill H (2011) Cyber-physical systems. Impact Control Technol 12(1):161–166
3. Rizk Y, Awad M, Tunstel EW (2019) Cooperative heterogeneous multi-robot systems: a survey. ACM Comput Surv (CSUR) 52(2):1–31

4. Mairiza D, Zowghi D (2011) Constructing a catalogue of conflicts among non-functional requirements. In: Evaluation of novel approaches to software engineering. Springer, Berlin Heidelberg, pp 31–44
5. Roy M, Bag R, Deb N, Cortesi A, Chaki R, Chaki N (2023) SCARS: suturing wounds due to conflicts between non-functional requirements in autonomous and robotic systems. Softw Pract Exp 54:759–795
6. Roy M, Das S, Deb N, Cortesi A, Chaki R, Chaki N (2023) Correlating contexts and NFR conflicts from event logs. Softw Syst Model 22(6):1987–2010
7. Brugali D (2019) Non-functional requirements in robotic systems: challenges and state of the art. In: IEEE international conference on real-time computing and robotics (RCAR), pp 743–748
8. Deng G, Xu G, Zhou Y, Zhang T, Liu Y (2022) On the (in)security of secure ros2. In: Proceedings of the 2022 ACM SIGSAC conference on computer and communications security. CCS '22, Association for Computing Machinery, New York, NY, USA, pp 739–753. https://doi.org/10.1145/3548606.3560681
9. Mayoral-Vilches V, White R, Caiazza G, Arguedas M (2022) Sros2: usable cyber security tools for ros 2. In: 2022 IEEE/RSJ international conference on intelligent robots and systems (IROS). IEEE, pp 11253–11259
10. Jo YH, Cho SY, Choi BW (2023) Towards a ros2-based software architecture for service robots. Bull Electr Eng Inf 12(5):3027–3038
11. Jost J, Kirks T, Gupta P, Lünsch D, Stenzel J (2018) Safe human-robot-interaction in highly flexible warehouses using augmented reality and heterogenous fleet management system. In: IEEE international conference on intelligence and safety for robotics (ISR), pp 256–260
12. Mahulea C, Montijano E, Kloetzer M (2020) Distributed multirobot path planning in unknown maps using Petri Net models. IFAC-PapersOnLine 53(2):2063–2068. https://www.sciencedirect.com/science/article/pii/S2405896320332675
13. Wang D, Deng H (2021) Multirobot coordination with deep reinforcement learning in complex environments. Expert Syst Appl 180:115128
14. Paul H, Qiu Z, Wang Z, Hirai S, Kawamura S (2022) A ros 2 based robotic system to pick-and-place granular food materials. In: IEEE international conference on robotics and biomimetics (ROBIO). IEEE, pp 99–104
15. Baxi A, Eisen M, Sudhakaran S, Oboril F, Murthy GS, Mageshkumar VS, Paulitsch M, Huang M (2022) Towards factory-scale edge robotic systems: challenges and research directions. IEEE Internet Things Mag 5(3):26–31
16. Ramoly N, Bouzeghoub A, Finance B (2018) A framework for service robots in smart home: an efficient solution for domestic healthcare. Irbm 39(6):413–420
17. Bärnklau K, Rötting M, Roesler E, Siebert FW (2021) Requirement analysis for personal autonomous driving robotic systems in urban mobility. In: International conference on human-computer interaction. Springer, pp 3–18
18. Gomez-Donoso F, Escalona F, Rivas FM, Cañas JM, Cazorla M et al (2019)Enhancing the ambient assisted living capabilities with a mobile robot. Comput Intell Neurosci 2019
19. Kotsopoulos S, Nawyn J (2023) Rethinking autonomous and robotic systems in residential architecture: Assessing the motivations and values of home automation. Footprint 17(1):43–66
20. Pirhonen J, Melkas H, Laitinen A, Pekkarinen S (2020) Could robots strengthen the sense of autonomy of older people residing in assisted living facilities? A future-oriented study. Ethics Inf Technol 22(2):151–162
21. Trainum K, Tunis R, Xie B, Hauser E et al (2023) Robots in assisted living facilities: scoping review. JMIR Aging 6(1):e42652
22. Salam H, Celiktutan O, Gunes H, Chetouani M (2023) Automatic context-aware inference of engagement in HMI: a survey. IEEE Trans Affect Comput:1–20
23. Zhang J, Keramat F, Yu X, Hernández DM, Queralta JP, Westerlund T (2022) Distributed robotic systems in the edge-cloud continuum with ros 2: a review on novel architectures and technology readiness. In: 2022 Seventh international conference on fog and mobile edge computing (FMEC). IEEE, pp 1–8

24. White R, Caiazza G, Christensen H, Cortesi A (2019) SROS1: using and developing secure ROS1 systems. Stud Comput Intell 778:373–405
25. Macenski S, Foote T, Gerkey B, Lalancette C, Woodall W (2022) Robot operating system 2: design, architecture, and uses in the wild. Sci Robot 7(66):eabm6074. https://www.science.org/doi/abs/10.1126/scirobotics.abm6074
26. Curtis S, Best A, Manocha D (2016) Menge: a modular framework for simulating crowd movement. Collect Dyn 1:1–40
27. Van Den Berg J, Guy SJ, Lin M, Manocha D (2010) Optimal reciprocal collision avoidance for multi-agent navigation. In: Proceedings of the IEEE international conference on robotics and automation, Anchorage (AK), USA
28. Bag R, Roy M, Cortesi A, Chaki N (2023) Eliciting context-oriented NFR constraints and conflicts in robotic systems. Innov Syst Softw Eng:1–14
29. Afzal A, Katz DS, Goues CL, Timperley CS (2020) A study on the challenges of using robotics simulators for testing. arXiv:2004.07368

Risk Assessment in Agile Software Development

Oindrilla Ghosh, Rituparna Chaki, Nabendu Chaki, and Agostino Cortesi

Abstract Risk assessment in software engineering has seen many approaches. Despite the amount of scientific literature on risk management, the failure rate of software projects after the first installation remains considerable. The reason is probably that the dynamically changing software feature has not been imbibed in the framework, which mostly deals with risk management in retrospect. This article attempts to suggest a new approach which will take these gaps into consideration.

Keywords Software development · Risk management · Agile methodologies

1 Introduction

Every product, including software products, needs to be produced within a limited period and a somewhat fixed budget. The system has certain inherent risks and components, in the duration, of the process and in their mutual interaction. In 1991 Boehm [1] in his seminal paper discussed the techniques and suggestions for risk management. Over the years there has much work devoted solely to risk management in software engineering. Yet in recent times, as in 2020, [2] it is observed that, there is a 11% probability of software projects being shelved after the first installment. A cursory examination of public case studies [3] reveals a startling picture. The sheer

O. Ghosh (✉)
Department of Computer Science, S. A. Jaipuria College, Kolkata, India
e-mail: oindrillanatasha@gmail.com

R. Chaki
A. K. Choudhury School of Information Technology, Kolkata, India
e-mail: rchaki@ieee.org

N. Chaki
Department of Computer Science and Engineering, University of Calcutta, Calcutta, India
e-mail: nabendu@ieee.org

A. Cortesi
DAIS, Ca'Foscari University, Venice, Italy
e-mail: cortesi@unive.it

© The Author(s), under exclusive license to Springer Nature Singapore Pte Ltd. 2025
N. Chaki et al. (eds.), *Applied Computing for Software and Smart Systems*, Lecture Notes in Networks and Systems 980, https://doi.org/10.1007/978-981-97-9762-2_18

loss of resources highlights the importance and relevance of risk management in software engineering.

Many small or medium firms tend to use customized tools instead of formal risk management. Authors in [4] studied the reasons behind the unsuccessful small-scale software projects when the projects are completed by third-party firms. Risk factors along with their management are discussed in [5] for large-scale projects. Thus it is apparent that, irrespective of size, software projects are prone to possible failures. There are many reasons behind it. The concept of risk management, though ideally postulated in theory, doesn't provide the required assistance that it is supposed to in real life. The expensive and unsuccessful risk management doesn't attract many users, therefore. This leads to the escalation of threats to the point where either a project is shelved mid-way or delivers an undesirable product. In either case, it doesn't complete its life cycle. Van Scoy in [6] has highlighted how risks can be turned into opportunities by effectively introducing risk management as a standard practice.

In recent times most firms use the Agile model for their software projects. The dynamic nature of the software projects and timely delivery of products makes Agile a preferred model nowadays. While Agile models are credited with having their inherent risk management strategy because of their iterative approach, the reality is not so [7]. Any risk methodology designed, therefore, must conform to the Agile model features. In this paper, we have attempted to create a framework that works in sprints and rectifies the issues in the succeeding sprints, which were raised in the current sprint. Thus, risk management will be effectively working in parallel with the actual project.

The traditional Agile approaches offer fast delivery of products overlooking certain lacunae. Our approach offers to build a support system, where each risk management will run in slightly shorter sprints than the actual sprint. This will create the option of applying corrective measures before the conclusion of a sprint. Our framework also aims to benefit future researchers of this domain by creating a database of the risk management of the entire project after completion.

This article explores the risks ailing various software projects and compares some analysis techniques while offering a framework that we feel will contribute to the field of Agile software development.

The rest of the paper is structured as follows. Section 2 contains a literature review, Sect. 3 provides a comparative analysis of the most relevant papers, and the most relevant issues emerging from them are elaborated in Sect. 4. Section 5 introduces our dynamic risk management framework and discusses the novelties with respect to the state-of-the-art literature. Section 6 concludes.

2 Literature Survey

The causes for software risk failures range from mostly technical to some level of human contribution as well. But the results can be catastrophic as discussed in [8]. Unearthing software failures have many approaches. In [9] the Root cause analysis is used to explain the causality between the causes of failure and their interrelationships. The dimensions of software engineering success have been discussed in [10].

Various tactics have been applied to combat risk management in software engineering. While [1] introduced the concept of assessing risks, there are many other popular risk tackling models as well. The methods of risk analysis are also varied. Some are qualitative and others are quantitative. From graphical to statistical to mathematical to deep learning. The approach of handling software risks are far and wide.

Many software models have been used for risk management. Hijazi [11] gives a comparative study between the common types of models for risk management. Amongst the multiple models surveyed Agile remains the most widely used according to PMI, 2017. PMI report [12] Agile is about rapid delivery of the product and hence many conventional projects and therefore risk management steps are often ignored. The authors in [13] have studied the risks faced by and the mitigation strategies employed by the Agile practitioners and found that the non-systematic implementation of the strategies to be the root cause of failure.

Agile methodology is applicable in different forms, namely, Scrum, eXtreme Programming (XP), Kanban, Dynamic systems development method (DSDM), Rational Unified Process (RUP), lean, and ScrumBan and others. Albadarneh [14] compares three Agile methods—Scrum, XP, and DSDM and concludes DSDM being more versatile with respect to risk management. In [15] it is explored how combined use of Scrum and XP leads to better output.

3 Comparative Study

The literature survey instills the belief that risk management in software engineering if carried out systematically, can yield satisfactory results. Amongst all of the relevant papers reviewed, they are presenting a brief comparative assessment of certain papers. Table 1 shows a comparison between different papers.

4 Emerging Issues

Several striking features have surfaced from the assessment of the papers.

4.1 Data

Much of the work uses data from software repositories. While this can act as a training dataset or in the initial formulation of a technique, it is not conclusive in nature. Real-time data from the software projects in place would act as better indicators for the

Table 1 Comparison

Paper	Technique	Problems
Suresh [16]	Hybrid fuzzy	High computational complexity
Hseih [17]	Fuzzy	Risk model needs validation
Lunesu [18]	Monte Carlo simulation	Construct and external validity needs exploring
Neto [19]	Recommender	Internal, construct and external validity needs exploring
Filipetto [20]	Atropos goal model	Threats to validity needs elimination and exhaustive database required
Odzaly [21]	Agile risk tool agents	Study used student data based on historical artifacts
Gorski [22]	IT tool	Data needs processing and amateur views used in the data
Lopes [23]	Framework	Data not applied to real projects
Ghazali [24]	Risk poker	Data not applied to real projects and external validity proof remaining
Chaouch [25]	Framework	Model needs to be tested against real data and quantitatively validated

construction of the configuration of any framework. Dynamic data from the risk management running in parallel can help to predict the fate of mitigation plans. Data being limited to initial requirements runs the risk of becoming obsolete since risk scope can alter in the entirety of the life cycle.

4.2 *Procedure*

Some of the frameworks are qualitative. Risk deals with management of resources and some metrics are necessary. Many of the techniques are complex and would require unnecessary computation - a weakness management may not want to invest in. In some papers, the issues flagged are software features- their correlation with risks is not highlighted or cataloged. Manual interaction needs to be judiciously allowed and if possible reduced. Many studies have construct or internal and external validity threats, which is not favorable.

4.3 *Machine Learning Algorithms*

A sufficient database is needed for recommenders or bots to become reliable. The algorithms should be varied to select a better performer among themselves

4.4 *Optimization of Resources*

Classification of risk management is required- whether size or organization matters or the features of the project are the deciding factors. This will help in the prudent use of resources, a lucrative feature that the management would probably proceed with.

5 Dynamic Risks Reassessing Within Agile

We propose a framework that will address the issues raised in the previous section. The framework is based on the scrutiny of the papers discussed above and is compliant with the Scrum guidelines. The spirit of Agile methodology has been imbibed in the proposed framework.

5.1 *Risk Modeling*

As a first step, a pre-planning for the risk management of the software project takes place. Initially, three basic performance metrics are chosen for assessing the effects of the risk management framework; these are time limit for delivery T_l, cost overhead C_o and permissible tolerance risk level PT_l. The basic steps for software development are [1] are requirement analysis, design, coding, testing, and deployment. Our framework breaks up the entire development project into these 5 modules. The risk management cycle of identification, analysis, planning, and monitoring are performed for each module and the feedback is used to modify the module operation. Figures 1 and 2 depict the block diagram and an enhanced view of each block.

We define a risk register Risk_reg { probability P, impact I, vulnerability V}. The risk register has three attributes—the probability of each risk, the impact of each risk, the third column of consisting the vulnerable components and the scope of their influence will be present. The risk register is populated with individual risks as we discover them. The risks are classified into three overheads—time limit for delivery T_l (for example, time taken to correct errors in programs, time taken to implement each step of the module, the lines of code, the time taken to execute the code, etc.), cost overhead C_o (for example, cost of tool and talent, cost to repair

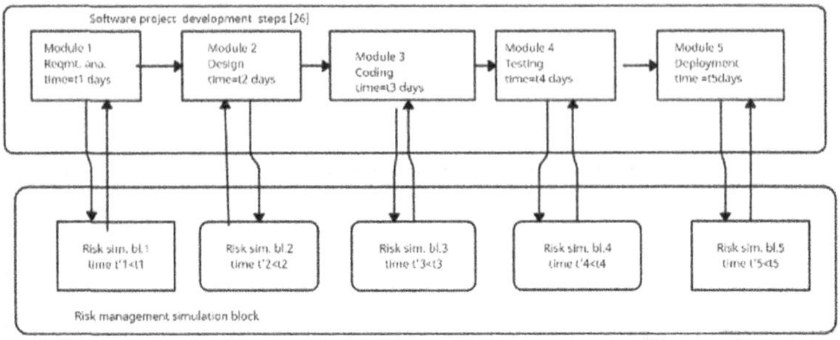

Fig. 1 Block diagram of the framework

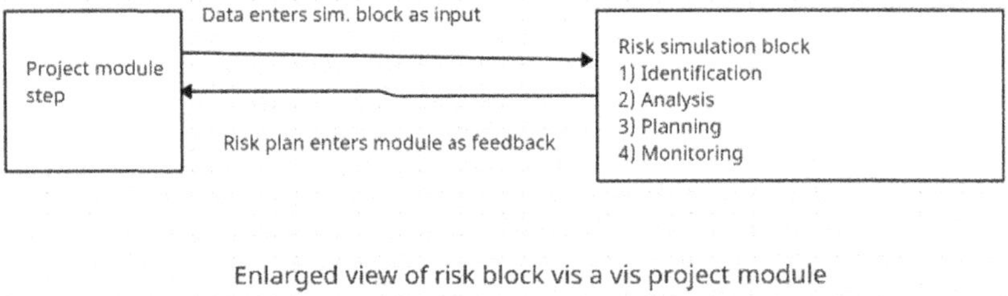

Fig. 2 Enlarged diagram for module and risk interaction

defects, budget overshoot due to time limit cross, amount of memory required for the project, maintenance costs, etc.) and permissible tolerance risk level PT_l (a maximum threshold assigned against each risk empirically). The software metrics are mainly considered and risk issues since their performance decides the viability and longevity of the product.

The risk matrix is then formed from Risk_reg as in the following manner. The probability is classified into 5 sub-divisions—p1, p2, p3, p4, and p5 on an increasing scale of 1 to 5. The impact is similarly grouped into 5 categories—i1, i2, i3, i4 and i5. Based on the product values of probability and impact each risk is plotted in the risk grid. The risk exposure can be classified into 5 categories—re1, re2, re3, re4, and re5 along similar scales. Each level is assigned a corresponding color. The risks are now prioritized in the risk matrix in the descending order of risk exposure (product of impact and probability).

For each module, its risk block (simulation block whose output is based on recommender) runs in tandem but is completed in less time. The output of the risk block acts as feedback and an indicator of corrective measures needed to be adopted in case the simulated results are not satisfactory.

5.2 Risk Table Management

For each sprint of each module the risk table is reassessed and risk matrix redrawn. The generation of the table depends on the reconfiguration of the project data. Once the risk from the table is within the tolerance level the risk is treated as a global risk and used in the feedback.

It is expected that the framework should be lightweight if it has to be Agile-compliant. Even if other models are used, risk assessment and mitigation is a parallel process running in tandem with the project and can't be computationally enormous.

There needs to be a memory-based repository of how risks were handled, artifacts, tools used, and the mitigation strategy. This can act as indicator for similar projects implemented in future. After completion of the project the table of risks for each sprint and mitigation plan should be saved in the database for future use.

5.3 Example

As an example, a project from [26] based on Android studio, JAVA, and SQLite3 which aims to build an Android application to manage contacts is considered. The project is divided into 5 tasks, like our 5 modules. The first task is about understanding Android workings, life cycle, and its purpose. We have used the online AI model generator [27]. It has been populated with Android risks sourced from the user stories on the internet as shown in Table 2. A model was built with 79% text semantic and 55% text similarity. The top five recommended risks are used to create the following table (risk register) and the risk matrix. Vulnerability column has not been considered for the requirement analysis stage. The risk matrix template used has been shared in Fig. 3. The risks from the risk register are fed back into the first task module. Since this is the requirement phase, the performance enhancement issue doesn't directly arise. But it can be added to the list of requirements considered.

Impact Probability	i1=1	i2=2	i3=3	i4=4	i5=5
p1=1	1	2	3	4	5
p2=2	2	4	6	8	10
p3=3	3	6	9	12	15
p4=4	4	8	12	16	20
p5=5	5	10	15	20	25

re5= 21 to 25	re4=16 to 20	re3=11 to 15	re2=6 to 10	re1=1 to 5

Fig. 3 Risk matrix

Table 2 Risk register for task 1

Risk	Probability	Impact
SMS trojan viruses	3	5
App crashes	2	3
Syncing error	2	2
Easy access passwords	3	3
Google now problems	2	1

There remain two major issues. Firstly, the data is not available in the dynamic format as was desired. Therefore the effect of feedback on the implementation of the suggestions in the module remains elusive. Secondly, the probability and impact have been assigned empirically, creating a serious internal validity issue. In the next section we propose an assumption which will address this issue.

5.4 Assumptions

The probabilities for each risk have been assigned intuitively and it is assumed that they remain constant throughout the execution of the project. This doesn't happen a lot in reality. We propose the use of a Dynamic Bayesian network for assessing the risk probabilities. Especially dynamic Bayesian network allows the probabilities, which are inter-related and temporal, to be assessed over a time span. Our network allows us to explore the various effects and their repercussions on the set of risk variables over a time span. As an illustrative example, let us consider the second task of the project [26]. In this phase the main setup for the project- getting the build Gradle ready is the primary function. Gradle, with all of its advantages, still has some flaws. We are considering a few of them in the list below [28]. The labels in the list as depicted in Table 3 will be used in future. The basic assumptions are listed. These serve mainly for the illustrative purposes:

1. The six risks mentioned above have been considered.
2. The time intervals considered are 5.
3. The probability assignments have been randomized to consider the scope of the analysis.
4. The parent-child dependencies have similarly been sporadic.
5. The dynamic model has been constructed on [29].

The outputs are shown in the Figs. 4, 5, and 6.

Table 3 Risk register for task 2

Risk number	Risk
R1	System file encoding
R2	Environment variable tracking
R3	Line endings
R4	Symbolic links
R5	Java version tracking
R6	Avoid changing inputs external to your build
R7	Suggestions for authoring your build
R8	Achieving stable task inputs
R9	Custom and third-party tasks

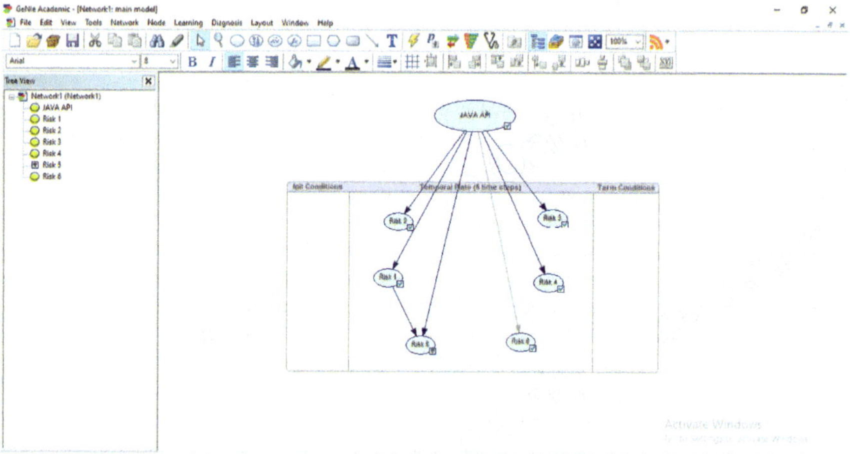

Fig. 4 Dynamic Bayesian network for task 2 risks

5.5 *Advantages*

Our intended approach is dynamic in nature. The automation of risk management has been suggested along with a proposed creation of a risk database.

5.6 *Limitations*

There are three major problems with this implementation. Firstly, in order to suggest the mitigation strategies a preliminary survey of similar projects is needed, since a global repository is currently unavailable. The trained model had very little data to be trained on. The authors wanted to use live projects as a test case but had to

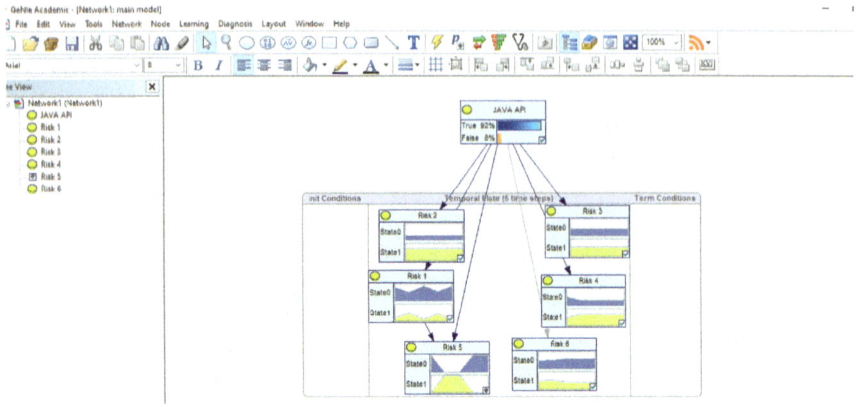

Fig. 5 Bar chart of the updated probabilities after analysis

settle for a static approach. Thirdly, the actual indicator of the risk inter-relations and their probabilities based on system output would have made the dynamic Bayesian network analysis more robust. Also our objective of updating the risk register and risk matrix and creating global risks is elusive.

5.7 Comparison with Literature Review

The high computational and time complexity of [16] has been addressed in our proposed work Framework is lower in time complexity, specially in each sprint-wise operation. The lightweight feature of Agile has been maintained.

In [17, 18, 20] there are issues of internal, construct and external validity. Our proposed framework operates on the output data and therefore is validated at every sprint. We intend to use accepted measurement scales and quantitative approaches and build a generic model which can be adapted to any system. The database contribution from using our project might help build an exhaustive risk management repository in the future.

The data, we intend to use, for the framework will be from real-time projects and processed accordingly unlike [21–25]. We plan to minimize manual interaction in order to minimize the number of errors. The effects that bias or inexperience has will simultaneously be alleviated.

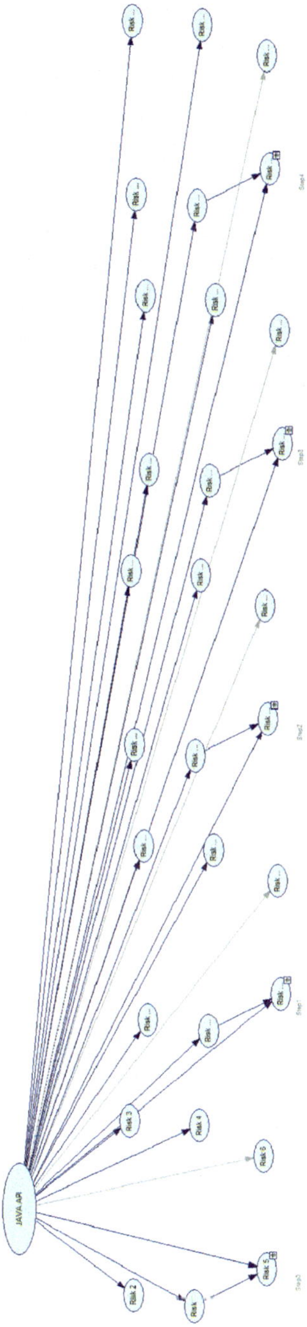

Fig. 6 Unrolled view of the total analysis

6 Conclusion and Future Work

Risk management is a continuous process throughout the SDLC as issues arise at any pint and needs to dealt with in that phase, not just in the requirement analysis. Our framework is dynamic and runs in tandem with the project and also provides a database for future similar projects.

The suggested framework can be used to analyze dynamic case study data and assess course mitigation plans. We hope to introduce further automation of the framework and take the help of machine learning in subsequent study to improve the framework.

It will be a priority, in the continuation of the work, to provide dynamic empirical evidence of the validity of the proposed model, even if an evaluation of the effectiveness in terms of reducing the percentage of failures can only be made following the adoption of this approach by a large number of software companies, and this is beyond our capabilities.

In [30] a study of using deep learning in risk management is present. We hope to use deep learning in their proposal to train the data set running in parallel with the project. Moreover we aim to integrate within our proposed framework some previous works on risks associated with conflicts among requirements [31–33], and to specialize the framework itself on specific software typologies, in particular to increase robustness of the robotics software life cycle [34].

Acknowledgements Work partially supported by SERICS (PE00000014—CUP H73C2200089-001) under the NRRP MUR program funded by the EU—NGEU, and iNEST—Interconnected NordEst Innovation Ecosystem funded by PNRR (Mission 4.2, Investment 49 1.5) NextGeneration EU (ECS_00000043—CUP H43C22000540006).

References

1. Boehm BW (1991) Software risk management: principles and practices. IEEE Softw 8(1):32–41
2. Tavares BG et al (2020) A risk management tool for Agile software development. J Comput Inf Syst 61(6):561–570
3. Failed projects. https://www.cigniti.com/blog/37-software-failures-inadequate-software-testing/
4. Jørgensen M (2014) Failure factors of small software projects at a global outsourcing marketplace. J Syst Softw 92(6):157–169
5. Pasha M, Qaiser G, Pasha U (2017) A critical analysis of software risk management techniques in large scale systems. IEEE Access 6:12412–12424
6. Van Scoy RL (1992) Software development risk: opportunity, not problem. CMU/SEI-92-TR-30 AD-A258 743, pp 1–22
7. Nyjford J (2008) Towards integrating Agile development and risk management. DSV Report Series 08-008, ISSN: 1101-8526
8. Charette RN (2005) Why software fails. IEEE Spectr 42(9):42–49
9. Lehtinen T et al (2014) Perceived causes of software project failures—an analysis of their relationships. Inf Softw Technol 56(6):623–643

10. Ralph P, Kelly P (2014) The dimensions of software engineering success. In: ICSE 2014: Proceedings of the 36th international conference on software engineering (ICSE 2014) proceedings of the 36th international conference on software engineering, Hyderabad, India, pp 24–35
11. Hijazi H, Khdour TJ (2012) A review of risk management in different software development methodologies. Int J Comput Appl 45(7):8–12
12. PMI report (2017) Success rates rise-transforming the high cost of low performance. PMI's pulse of the profession, 1–32
13. Hammad M, Inayat I, Zahid M (2019) Risk management in Agile software development: a survey. In: International conference on frontiers of information technology (FIT). IEEE, Islamabad, pp 162–164
14. Albadarneh A, Albadarneh I, Qusef A (2015) Risk management in Agile software development: a comparative study. In: IEEE Jordan conference on applied electrical engineering and computing technologies (AEECT). IEEE, Jordan, pp 1–6
15. Zuiderveld NR (2003) eXtreme programming and SCRUM: a comparative analysis of Agile methods. Int J Comput Appl. https://api.semanticscholar.org/CorpusID:12962093
16. Suresh K, Dillibabu R (2020) A novel fuzzy mechanism for risk assessment in software projects. Soft Comput 24:1683–1705
17. Hsieh M, Hsu Y, Lin C (2018) Risk assessment in new software development projects at the front end: a fuzzy logic approach. J Ambient Intell Hum Comput 9:295–305
18. Lunesu MI et al (2021) Assessing the risk of software development in Agile methodologies using simulation. IEEE Access 9:134240–134258
19. Neto A et al (2023) Towards a recommender system-based process for managing risks in scrum projects (SAC '23). In: Proceedings of the 38th ACM/SIGAPP symposium on applied computing. Association for Computing Machinery, New York NY, United States, Tallinn Estonia, pp 1051–1059
20. Filippetto AS, Lima R, Barbosa JL (2021) Victoria: a risk prediction model for software project management based on similarity analysis of context histories. Inf Softw Technol 131:106497
21. Odzaly E, Greer D, Stewart D (2017) Agile risk management using software agents. J Ambient Intell Humaniz Comput. https://doi.org/10.1007/s12652-017-0488-2
22. Górski J, Łukasiewicz K (2012) Assessment of risks introduced to safety critical software by Agile practices—a software engineer's perspective. Comput Sci 13(4):165–182
23. Lopes S et al (2021) A risk management framework for scrum projects. In: Proceedings of the 23rd international conference on enterprise information systems (ICEIS 2021). SCITEPRESS, pp 30–40
24. Ghazali S et al (2018) A risk poker based testing model for Scrum. Int J Comput Syst Sci Eng 33(3):169–185
25. Chaoucha S, Mejrib A, Ghannouchia SA (2019) A framework for risk management in Scrum development process. Proc Comput Sci 164:187–192
26. Web development program. https://www.crio.do/
27. Synthetic data platform for developers. https://www.gretel.ai/
28. Gradle problems. https://gradle.org/
29. Genie modler. https://www.bayesfusion.com/genie/
30. Aziz S, Dowling M (2018) Machine learning and AI for risk management-disrupting finance. Palgrave Studies in Digital Business and Enabling Technologies, Palgrave Pivot, Cham. ISBN: 9780198520115
31. Roy M et al (2023) SCARS: suturing wounds due to conflicts between non-functional requirements in autonomous and robotic systems. Softw Pract Exp. https://doi.org/10.1002/spe.3297
32. Roy M et al (2023) Correlating contexts and NFR conflicts from event logs. Softw Syst Model 22(6):1987–2010. https://doi.org/10.1007/S10270-023-01087-4
33. Das S et al (2024) Minimising conflicts among run-time non-functional requirements within DevOps. Syst Eng 27(1):177–198. https://doi.org/10.1002/SYS.21715
34. Bag R et al (2023) Eliciting context-oriented NFR constraints and conflicts in robotic systems. Innov Syst Softw Eng. https://doi.org/10.1007/s11334-023-00545-y

Adaptive Cyber Defence: Leveraging GANs for Simulating and Mitigating Advanced Network Attacks in IoT Environments

P. Kiran Rao, Subarna Chatterjee, P. Suman Prakash, and K. Seshadri Ramana

Abstract The progressive dissemination of the Internet of Things (IoT) and Wireless Sensor Networks (WSNs) has ushered in a new era of connectivity, with vast applications spanning from medicare to smart city infrastructure. However, this expansion has been paralleled by a corresponding increase in the sophistication and variety of cyber threats targeting these networks. Traditional cyber security measures, designed for a less dynamic threat landscape, are proving increasingly insufficient in protecting against the innovative and varied attack methods now in commonplace. This study introduces an innovative application of Generative Adversarial Networks (GANs) to address this challenge, presenting a novel framework for the simulation and mitigation of advanced network attacks, particularly focusing on Distributed Denial of Service (DDoS) and spoofing attacks which pose significant threats in IoT environments. Generative Adversarial Networks (GANs), comprising two neural networks-the generator and the discriminator compete in a game-theoretic scenario, facilitating a deep understanding of attack patterns through the generation of realistic, synthetic cyber-attack scenarios. This research exploits GANs to bridge the gap between the static

P. K. Rao (✉)
Assistant Professor, Department of CSE (Data Science), Rajeev Gandhi Memorial College of Engineering and Technology, Nandyal 518501, Andra Pradesh, India
e-mail: kiranraocse@gmail.com

S. Chatterjee
Department of CSE, MS Ramaiah University of Applied Science, Bengaluru, Karnataka, India
e-mail: subarna.cs.et@msruas.ac.in

P. S. Prakash
Department of CAI, G Pullaiah College of Engineering and Technology, Kurnool, Andra Pradesh, India

K. S. Ramana
Department of CAI, Ravindra College of Engineering for Women, Kurnoo 518001, Andra Pradesh, India

© The Author(s), under exclusive license to Springer Nature Singapore Pte Ltd. 2025
N. Chaki et al. (eds.), *Applied Computing for Software and Smart Systems*, Lecture Notes in Networks and Systems 980, https://doi.org/10.1007/978-981-97-9762-2_19

nature of traditional security protocols and the dynamic, evolving landscape of cyber threats. By training on a comprehensive dataset of known attacks and normal network activities, our proposed model, the Dynamic Adaptive Threat Simulation GAN (DATS-GAN), is capable of producing varied and realistic attack scenarios. These simulations serve a dual purpose: they not only enhance the detection capabilities and responsiveness of current security systems but also provide a basis for the development of new, adaptive security mechanisms capable of dynamically responding to the ever-changing cyber threat landscape. The effectiveness of DATS-GAN is demonstrated through extensive empirical analysis, highlighting significant improvements in the detection precision and reaction times of security frameworks within WSNs. Moreover, the generated synthetic attack scenarios provide a valuable resource for training machine learning models, leading to the advancement of adaptive security solutions that maintain a high readiness level against emerging cyber threats. The outcomes of this research hold substantial promise for the cyber security domain, showcasing the potential of GANs to revolutionize network defenses against sophisticated cyber threats in IoT and WSN environments.

Keywords IoT security · Generative Adversarial Networks (GANs) · Network attack simulation · AI in cyber defence · Cybersecurity in WSN · DDoS attack mitigation · Adaptive security mechanisms

1 Introduction

The Internet of Things (IoT) and Wireless Sensor Networks (WSNs) have ushered in a new technological paradigm, profoundly altering our interaction with the world around us. These innovations have catalyzed advancements across a broad spectrum of applications, from revolutionizing healthcare delivery and industrial automation to enabling the smart city infrastructures of the future and promoting sustainability initiatives. However, the swift proliferation of IoT and WSN technologies has also surfaced significant cyber security challenges. Deployed often in vulnerable environments, these networks are exposed to a wide array of cyber threats, magnified by their inherent limited computational capabilities which complicate the implementation of robust security protocols. Figure 1 illustrates the diverse and pervasive nature of these threats. Among the most prevalent and damaging are Distributed Denial of Service (DDoS) and spoofing attacks, both characterized by their high frequency and potential for substantial impact. These vulnerabilities underscore an urgent need for advanced security solutions capable of safeguarding these expanding networks. The landscape of cyber threats, as delineated in Fig. 2, presents a dynamic and increasingly sophisticated challenge. In response to this evolving threat environment, the integration of Generative Adversarial Networks (GANs) into cyber security strategies represents a pioneering approach. A GAN comprises two neural networks in a competitive setting: a generator, which aims to create data indistinguishable from genuine samples, and a discriminator, tasked with distinguishing between real and

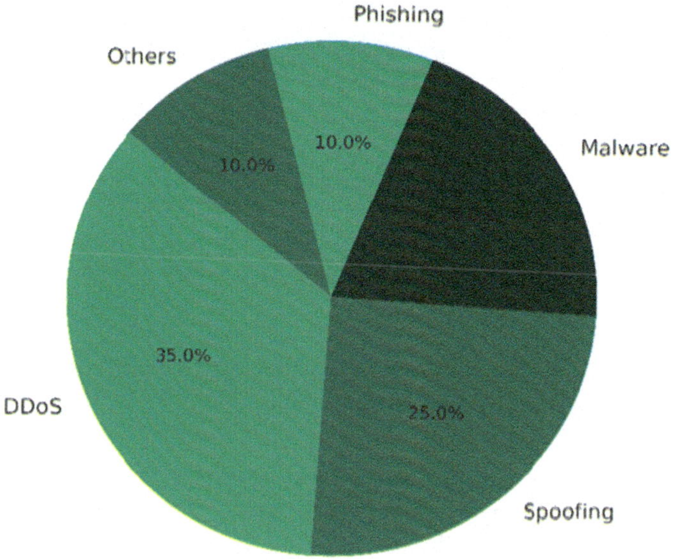

Fig. 1 Distribution of various types of cyber-attacks targeting WSNs and IoT networks

synthetic data. This adversarial framework is particularly pertinent to cyber security, enabling the generation of realistic attack simulations that enhance the preparedness of security systems and foster the development of adaptive, resilient defense mechanisms. This article specifically targets the mitigation of DDoS and spoofing attacks, which pose significant threats to the integrity and functionality of IoT and WSN ecosystems. Through leveraging GANs for the generation of intricate and realistic attack scenarios, we aim to address the disconnect illustrated in Fig. 2. Our goal is to synthesize sophisticated security measures that remain effective amidst the rapidly changing cyber threat landscape. This novel application of GANs is anticipated to make a substantial contribution to cybersecurity, particularly within the realms of IoT and WSNs, by improving network defenses against the increasingly complex and sophisticated cyber threats of the digital era. Beyond technical advancements, this study endeavors to highlight the imperative for innovative defense strategies in ensuring the security and resilience of IoT and WSN infrastructures against the backdrop of their growing integration into our daily lives and critical systems.

In summary, this research not only seeks to enhance the cyber security framework for IoT and WSNs through the application of GANs but also aims to underscore the broader implications of these technologies for the future of secure digital ecosystems. By advancing our understanding and capabilities in this area, we contribute to the foundational security infrastructure necessary to support the continued growth and evolution of IoT and WSN technologies.

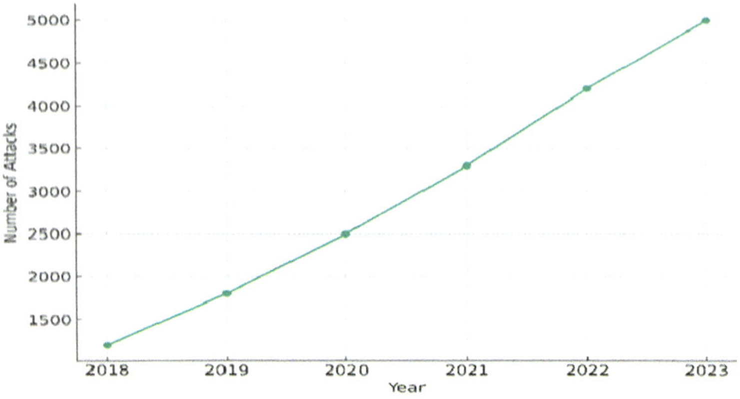

Fig. 2 Increasing timeline of cyber attacks

2 Related Works

The rapid evolution of digital infrastructure, notably within the Internet of Things (IoT) and Wireless Sensor Networks (WSN), has necessitated advanced cyber security measures. In this context, Generative Adversarial Networks (GANs) have emerged as a pivotal technology, offering innovative solutions to complex security challenges. A seminal contribution by Ferrag et al. [1] introduced a GAN-based framework specifically engineered to uncover and replicate sophisticated cyber threats targeting IoT systems. This breakthrough not only enhances detection mechanisms but also facilitates the creation of realistic simulations of cyber-attacks, significantly advancing IoT security paradigms. Exploring the versatility of GANs, Debicha et al. [2] demonstrated their applicability in blockchain environments, particularly for threat detection within Ethereum networks. This research underscores the adaptability of GANs, showcasing their potential to transcend traditional cyber security applications and extend into blockchain technology, a testament to the flexibility and innovative capacity of GANs in safeguarding digital transactions and interactions.

Further expanding the technological frontier, explorations into the compatibility of GANs with 4G and forthcoming 5G-enabled IoT networks [3] have illuminated the technology's capacity to adapt and thrive within next-generation network standards. This body of work highlights the critical role of GANs in driving the evolution toward more connected, intelligent, and secure IoT infrastructures, underscoring their strategic importance in the digital era. The research by Vitorino et al. [4] delves into the operational intricacies of GANs, showcasing their effectiveness in enhancing intrusion detection and classification within the IoT domain. This investigation reveals the strategic deployment of GANs in strengthening the security fabric of IoT ecosystems, thereby mitigating vulnerabilities and enhancing resilience against cyber threats.

In a similar vein, Mathur and Babu [5] introduced the application of Conditional GANs (C-GANs) in anomaly detection within fog-assisted IoT environments. This innovative approach illustrates the diverse applicability of GANs across varied cyber security contexts, further cementing their role as a versatile and robust solution in modern cyber defense strategies. The narrative of GANs in cyber security is further enriched by studies focusing on specific applications and advancements. The integration of Generative Edge Intelligence in the security of IoT-assisted smart grids [6] emphasizes the synergy between edge computing and GANs, highlighting a novel approach to enhancing security postures in critical infrastructure. Comprehensive analyses, such as the one conducted by Chenou et al. [7], provide a detailed overview of GANs' role across the cybersecurity landscape, elucidating the synergistic potential of these networks in fostering effective threat analysis and robust defensive responses.

The development of DA-GAN by Hoang et al. [8] aimed at augmenting cyber threat detection efficiency, and the creation of malicious packets through cyber-creative GANs by Fang et al. [9] mark significant strides in the practical application and understanding of GANs in cybersecurity. These studies not only demonstrate the innovative application of GANs in creating and countering cyber threats but also highlight the ongoing evolution of GAN technology in response to the dynamic nature of digital security challenges. Peppes et al. [10] provide a critical comparative analysis of GAN architectures, offering invaluable insights essential for selecting the most effective models in cybersecurity simulations. This research, alongside broader discussions on the potential and challenges of GANs within cybersecurity by Chyou et al. [11] and subsequent studies [12–27], underscores the multifaceted applications of GANs in enhancing digital security measures. From bolstering the robustness of intrusion detection systems to assessing and improving resilience against adversarial examples, the collective body of work reflects a deep and nuanced understanding of the strategic importance of GANs in navigating the complex landscape of cybersecurity.

In conclusion, the expansive array of research contributions elucidates the profound impact of Generative Adversarial Networks in redefining cybersecurity paradigms. By harnessing the innovative potential of GANs, the field is poised to address the multifarious and evolving cyber threats that characterize the digital age, ensuring the integrity and resilience of IoT and WSN environments against sophisticated adversaries (Table 1).

3 Methodology

3.1 Data Acquisition

The basis underlying foundation of our empirical analysis largely depends on the CICIDS2017 dataset [14] which is a consolidated repository designed by the Cana-

Table 1 Summary of studies on GANs application in cybersecurity

Study	Focus area	Main findings/contributions	Year	Limitation
1	IoT	GAN framework for intricate cyber threats detection in IoT	2023	May not effectively simulate unforeseen, novel cyber threats
2	Blockchain	Use of GANs for threat hunting in Ethereum	2021	Limited scalability and adaptability to evolving blockchain technologies
3	6G-enabled IoT	Exploration of GANs adaptability in 6G-enabled IoT networks	2022	6G network dynamics may outpace GANs' ability to simulate realistic threats
4	IoT	Enhancement of intrusion detection and classification in IoT with GANs	2023	Detection models may not generalize to new, complex attacks not represented in training data
5	Fog-assisted IoT	C-GANs for anomaly detection in fog-assisted IoT environment	2022	Challenges in integrating C-GANs across diverse and distributed fog computing architectures
6	IoT smart grids	Generative Edge Intelligence for securing IoT-assisted smart grids	2021	Smart grids' unique security requirements may not be fully met by current GAN approaches
7	Cybersecurity	Broad role of GANs in cybersecurity and threat analysis	2019	Overreliance on GAN-generated data could introduce bias into threat analysis
8	Cybersecurity	DA-GAN for improving cyber threat detection efficiency	2021	Real-world applicability may vary significantly across different cybersecurity contexts
9	Cyber threat creation	Creation of malicious packets using cyber-creative GANs	2022	Ethical considerations and potential for misuse by malicious actors
10	GAN architectures	Comparative analysis of GAN architectures for cybersecurity	2016	Not all GAN architectures may be effective or efficient for specific cybersecurity application
11	GAN overview	Overarching view of GANs' potential and challenges in cybersecurity	2023	Broad focus may overlook the nuanced, practical challenges of deploying GANs in real-world scenarios
12	Intrusion detection	Robustness of intrusion detection systems against adversarial samples	2008	Adversarial samples generated by newer, sophisticated attacks may not be detected
13	AI in cyber defense	Impact and risks of generative AI technology on cyber defense	2021	Potential for AI to be exploited by attackers, turning defensive tools into vulnerabilities
14	Deep learning IDS	Deep learning-based intrusion detection with adversaries	2018	Model robustness may decrease when confronted with highly sophisticated adversarial techniques
15	Network security	Crafting adversarial examples accounting for domain-specific constraints	2020	Adversarial examples may not account for all possible real-world operational constraints
16	Unspecified	Untargeted adversarial attack via expanding the semantic gap	2019	Techniques may not be directly transferable or as effective across different domains
17	Network IDS defense	An ensemble defense mechanism against adversarial attacks	2021	Defense mechanism effectiveness may diminish over time as attack methodologies evolve
18	IoT security	Robust deep learning-based schemes for intrusion detection in IoT	2021	IoT-specific attack vectors may require constant updating of detection models

(continued)

Table 1 (continued)

Study	Focus area	Main findings/contributions	Year	Limitation
19	Anomaly detection	Universal method to perform adversarial attack on network anomaly detection	2020	Potential for generating false positives, complicating legitimate network traffic analysis
20	IDS evaluation	Evaluating network IDS in adversarial environment	2019	Evaluation metrics may not fully capture the dynamic nature of adversarial attacks
21	IDSs adversarial training	Evaluation of adversarial training on neural networks in IDSs	2020	Training methods may not fully prevent sophisticated evasion techniques
22	IDS resistance	Investigating IDS resistance against adversaries using min-max optimization	2020	Optimizations may not be universally effective against all forms of adversarial attacks
23	Adversarial DoS	Analyzing classifiers in adversarial DoS contexts	2019	Specific focus on DoS may neglect other critical security threats and attack vectors
24	Adversarial examples	Adversarial example generation by perturbing selected pixels	2022	Focused pixel perturbation methods may not apply to non-visual data or attacks
25	Traffic classification	Target adversarial sample generation for traffic classification model	2022	Effectiveness may be limited by the complexity and variability of network traffic patterns
26	DNN robustness	Enhancing DNN robustness with spectral normalization	2021	Implementation complexity and computational overhead may limit practical applicability
27	Orthogonal regularization	Adversarial training with orthogonal regularization	2020	May not address all forms of adversarial vulnerabilities within deep neural networks

dian Institute for Cyber security. This particular dataset has been enriched with the amalgamation of a wide array of network behaviors that ranged between the benign and malignant traffic patterns. There are specifically attacks such as the Distributed Denial of Service (DDoS) [14] that have a special meaning for the Internet of Things (IoT) environment. The data were gathered in a rigorous way, under efforts to create positioning in controlled testbed conditions, with an emphasis on authentic replication of network interchange and intrusion tactics. This assembly provides a comprehensive and diverse corpus to train our Generative Adversarial Networks (GANs) imperative simulating the cyber threats in the IoT infrastructures.

3.2 *Data Preprocessing*

The preprocessing phase was started with a strong data cleansing, intended to excise such extraneous or duplicative elements that would be potentially detrimental to the integrity of the model training phase. After this, normalization was done whereby datasets were transformed to the same scale—which becomes a very important task since dumbbell traffic parameters such as packet size and transmission rates varied greatly. The next phase of feature selection is characterized by thoroughness to retain significant features that portray network dynamics and cyber-attack patterns. This leads to accurate cyber-attack emulation in IoT environments. Further, the dataset underwent comprehensive verification of labeling which niched attack typologies from "normal" traffic flows. This preprocessing regimen instrumental at refining efficacy of the GANs, ensuring synthesizing the realistic and multifarious cyber-attack scenarios reflecting the complex and emerging landscape of cyber threats in IoT networks.

3.3 Dynamic Adaptive Threat Simulation GAN (DATS-GAN)

Our research introduces an innovative Generative Adversarial Network (GAN) architecture designed to fortify cybersecurity mechanisms in Internet of Things (IoT) environments by generating synthetic yet convincingly realistic cyber-attack scenarios. The architecture is composed of two primary networks: the Dynamic Generator Network (DG-Net) and the Evolving Discriminator Network (ED-Net), which are enhanced through Conditional Scenario Modeling (CSM) and refined via an iterative process known as Adversarial Training and Feedback.

3.4 Dynamic Generator Network (DG-Net)

As a main component of our GAN model, DG-Net is meticulously architected with layers of Long Short-Term Memory (LSTM) alongside a feature injection module. Each component within DG-Net is initialized with randomly assigned weights to initiate the training process. The crux of DG-Net's functionality lies in its ability to craft attack scenarios that are indistinguishable from those observed in real-world cyber-attacks. This capability is facilitated by processing a conditional input vector, c, meticulously crafted based on the insights provided by CSM. This vector, c, encapsulates the specifics of the attack type and contextual details necessary for simulation, offering a tailored framework for scenario generation. During the training phase, DG-Net innovatively merges noise vectors, z, sampled from a distribution $p_z(z)$, with the conditional vector, c, executing the operation through the function $G(z, c; \theta_g)$, where θ_g symbolizes the parameters governing DG-Net.

The training of DG-Net is orchestrated to minimize a specifically defined loss function L_G, aimed at aligning the distribution of generated attack scenarios with that of real-world data. This objective is quantitatively defined as follows:

$$L_G = -\sum [y \log G(z, c; \theta_g) + (1 - y) \log(1 - G(z, c; \theta_g))], \quad (1)$$

where y denotes the ground truth labels, distinguishing real data from synthetic ones.

3.5 Evolving Discriminator Network (ED-Net)

In parallel, ED-Net's purpose is to discern between authentic attack scenarios and those synthesized by DG-Net. It is characterized by its convolutional layers, each initiated with randomly determined weights. ED-Net undergoes a rigorous training regimen, utilizing a blend of genuine attack samples and synthetic data generated by DG-Net. Its ultimate goal is to accurately classify these inputs as real or synthetic, a task guided by the Binary Cross-Entropy loss function L_D, expressed as:

$$L_D = -\sum [y \log D(x; \theta_d) + (1 - y) \log(1 - D(x; \theta_d))], \tag{2}$$

where θ_d are the parameters of ED-Net and x represents the input data.

3.6 Conditional Scenario Modeling (CSM)

CSM serves as a pivotal component within the IoT environment, leveraging sensor data, network traffic, and system logs. It is enriched with the latest threat intelligence, identifying known attack patterns and vulnerabilities. The resultant conditional input vectors c, utilized by DG-Net, ensure the generated attack scenarios are both relevant and timely, reflecting the dynamically evolving threat landscape.

3.7 Adversarial Training and Feedback

This stage embodies the iterative process of refining both DG-Net and ED-Net. The discriminator's performance metrics inform the fine-tuning of DG-Net's parameters (θ_g) using gradient ascent, aiming to enhance the realism of generated attack scenarios while preventing issues like mode collapse. The objective functions driving the adversarial training are formulated to minimize the gap between the discriminator's output and the ground truth labels, detailed for the discriminator and generator respectively as follows:

For the discriminator (D):

$$L_D = -\sum [y \log D(x) + (1 - y) \log(1 - D(G(z, c)))] \tag{3}$$

For the generator (G):

$$L_G = -\sum \log D(G(z, c)) \quad (4)$$

4 Result Analysis

In this section of the paper, a detailed statistical analysis pertaining to the performance of various machine learning (ML) and deep learning (DL) algorithms when learned on the original and the synthetic datasets derived from the CICIDS2017 dataset are presented. The original CICIDS2017 dataset of diverse simulated cyber attacks and a synthetic dataset generation through the DATS-GAN was specifically used in developing a DATS-GAN framework to generate a synthetic simulation of the realistic cyber-attack scenarios. The selected key ML and DL algorithms eligible for use in cyber-attack detection including Support Vector Machine (SVM), Random Forest, Deep Neural Networks (DNNs), and Convolutional Neural Networks (CNNs). Figure 3 shows the distribution of different types of cyber-attacks comparing how frequent the attacks are expressed in the original dataset against that in the synthetic dataset generated by DATS-GAN. The high degree of similarity evidenced by the close alignment of the two lines of attack patterns is indicative that the attack patterns contained in the synthetic data set are mimetic of those contained in the original data. Figure 4 shows the frequency of cyber-attacks over time, again comparing the original dataset with the synthetic data produced by DATS-GAN. The similarity in the temporal pattern of the attacks suggests that DATS-GAN effectively captures the temporal dynamics of cyber-attack scenarios. Table 3 shows a detailed comparison of some machine learning and deep learning algorithms, namely Random Forest, Support Vector Machine (SVM), Deep Neural Networks (DNNs) and Convolutional Neural Networks (CNNs) with the respective CICIDS2017 dataset. Below the performance metrics-Accuracy, Precision, Recall, and F1-Score-stand as

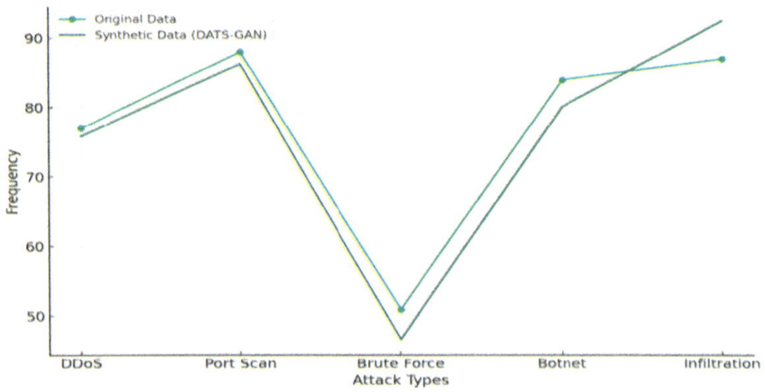

Fig. 3 Enter caption

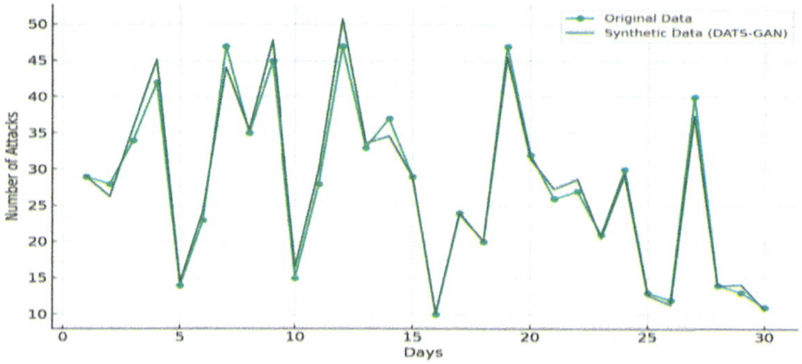

Fig. 4 Enter caption

Table 2 Performance metrics of various algorithms

Algorithm	Accuracy	Precision	Recall	F1-Score
Random Forest	0.77	0.84	0.78	0.76
SVM	0.75	0.81	0.74	0.73
DNN	0.80	0.85	0.82	0.81
CNN	0.82	0.88	0.83	0.85

Table 3 Updated performance metrics of various algorithms

Algorithm	Accuracy	Precision	Recall	F1-Score
Random Forest	0.91	0.88	0.93	0.87
SVM	0.89	0.85	0.90	0.86
DNN	0.92	0.90	0.91	0.90
CNN	0.94	0.93	0.92	0.93

a portrayal of the effectiveness of each algorithm when engaged in identification and categorization tasks of cyber-attack pattern existent within the specific dataset. For example, Random Forest shows a well-rounded performance in all the metric aspects, while CNNs get credits for being the best in the business when it comes to properly identifying attack scenarios as recorded with high precision and F1-Score. While SVMs are robust, on the other hand, they offer relatively lower performance, which may suggest potential problems for this algorithm to manage the diverse and complex attack types of the dataset (Table 2).

In Table 3, the same set of algorithms are assessed, but this time with training on synthetic data generated by DATS-GAN. A notable observation is the overall enhancement in the performance metrics for each algorithm. The synthetic dataset appears to provide a more comprehensive and varied set of scenarios, thereby enhancing the algorithms' ability to detect and classify cyber-attacks effectively. This improvement is particularly evident in the case of DNNs and CNNs, where the richness and complexity of the synthetic data significantly bolster their detection capabilities, as reflected in the heightened accuracy and F1-Scores. These results suggest that the synthetic data, with its diverse and realistic attack simulations, is highly effective in training robust cyber defence models.

5 Conclusion

In this paper, the Dynamic Adaptive Threat Simulation Generative Adversarial Network (DATS-GAN) framework is proposed for simulating and analyzing cyber-attacks in IoT environment virtually. We have been able to prove that with extensive experimentation and analysis on the CICIDS2017 dataset, DATS-GAN has the potential for the generation of realistic and diversified cyber-attack scenarios which bear close resemblance with the real-world data's threat scenario in practice. Moreover, high topology-specific evolving discriminator network (ED-Net) as well as a dynamic generator network (DG-Net) were integrated within the DATS-GAN model in order to generate the high-fidelity synthetic attack scenarios. Statistical analysis and visual representations of the results of the validity experiment showed a significant similarity between the synthetic and original data. The similarity is not about quantity alone but also in quality as exhibited by type and temporal profile of attacks.

Furthermore, DATS-GAN significantly improved the performance of different machine learning and deep learning models. The models trained with data generated from DATS-GAN led to distinct performance improvements in the key performance metrics including the accuracy, precision, recall, and F1-score relative to the training scenario where original dataset was used. This improvement points toward the potential that synthetic data generated by advanced AI techniques can bolster cybersecurity defence mechanisms. The implementation of this study is ranging wide. With more dynamism and adaptability to simulate cyber threats, DATS-GAN offers significant contributions to the cybersecurity field particularly in training and preparing for emerging threats. This framework will not help only in designing strategies to offer a potent defence but also it lends a very good guiding tool for further research in AI based cyber security solutions. The DATS-GAN framework is indeed a new state-of-the-art mechanism in the usage of GANs for cyber security. It opens up new possibilities of cutting-edge proactive analysis and mitigation of the cyber threats against complex and fast evolving domains like IoT. Further[1] research in the future can refine this model and test it in other contexts but also integrate it into real-time cybersecurity systems to enhance defence capabilities.

Acknowledgements **Disclosure of Interests.** It is now necessary to declare any competing interests or to specifically state that the authors have no competing interests. Please place the statement with a bold run-in heading in small font size beneath the (optional) acknowledgments, for example: The authors have no competing interests to declare that are relevant to the content of this

[1] If EquinOCS, our proceedings submission system, is used, then the disclaimer can be provided directly in the system.

article. Or: Author A has received research grants from Company W. Author B has received a speaker honorarium from Company X and owns stock in Company Y. Author C is a member of committee Z.

References

1. Ferrag M et al (2023) Generative adversarial networks-driven cyber threat intelligence detection framework for securing internet of things. In: 19th International conference on distributed computing in smart systems and the internet of things (DCOSS-IoT), pp 196–200. https://doi.org/10.1109/DCOSS-IoT58021.2023.00042
2. Debicha I et al (2021) Detect and reject for transferability of black-box adversarial attacks against network intrusion detection systems (2021). Preprint at arxiv:2112.12095, https://doi.org/10.1007/978-981-16-8059-5_20
3. Janković A, Mayer R (2022) An empirical evaluation of adversarial examples defences, combinations and robustness scores. In: Proceedings of the 2022 ACM on international workshop on security and privacy analytics. https://doi.org/10.1145/3510548.3519370
4. Judah M, Sierchio J, Planer M (2023) Detection of adversarial attacks on machine learning systems. https://doi.org/10.1117/12.2664015
5. Merzouk MA et al (2022) Investigating the practicality of adversarial evasion attacks on network intrusion detection. Ann Telecommun 77:763–775. https://doi.org/10.1007/s12243-022-00910-1
6. Debicha I et al (2021) Adversarial training for deep learning-based intrusion detection systems. Preprint at arXiv:2104.09852
7. Chenou J, Hsieh G (2019) Increasing the robustness of deep learning with colored noise augmentation
8. Pan Z, Mishra P (2021) Fast approximate spectral normalization for robust deep neural networks. Preprint at arXiv:2103.13815
9. Fang H et al (2022) Tracing the origin of adversarial attack for forensic investigation and deterrence. Preprint at arXiv:2301.01218
10. Zacharia T et al (2016) Evaluation of voice parameters in people with head and neck cancers: an investigational study. Middle East J Cancer 7:193–197
11. Chyou CC et al (2023) Unsupervised adversarial detection without extra model: training loss should change. arXiv:2308.03243
12. Costa SC et al (2008) Short-term cepstral analysis applied to vocal fold edema detection. https://doi.org/10.5220/0001066901100115
13. Maarouf R, Sattar D, Matrawy A (2021) Evaluating resilience of encrypted traffic classification against adversarial evasion attacks. In: IEEE symposium on computers and communications (ISCC), pp 1–6. https://doi.org/10.1109/ISCC53001.2021.9631407
14. Wang Z (2018) Deep learning-based intrusion detection with adversaries. IEEE Access 6:38367–38384. https://doi.org/10.1109/ACCESS.2018.2854599
15. Teuffenbach M et al (2020) Subverting network intrusion detection: crafting adversarial examples accounting for domain-specific constraints. https://doi.org/10.1007/978-3-030-57321-8_17
16. Wu A et al (2019) Untargeted adversarial attack via expanding the semantic gap. In: IEEE international conference on multimedia and expo (ICME), pp 514–519. https://doi.org/10.1109/ICME.2019.00095
17. Wang J et al (2021) Def-IDS: an ensemble defense mechanism against adversarial attacks for deep learning-based network intrusion detection. In: International conference on computer communications and networks (ICCCN), pp 1–9. https://doi.org/10.1109/ICCCN52240.2021.9522215

18. Fu X et al (2021) The robust deep learning-based schemes for intrusion detection in Internet of Things environments. Ann Telecommun 76:273–285. https://doi.org/10.1007/s12243-021-00854-y
19. Yang C et al (2020) U-ASG: a universal method to perform adversarial attack on autoencoder based network anomaly detection systems. In: IEEE INFOCOM 2020—IEEE conference on computer communications workshops (INFOCOM WKSHPS), pp 68–73. https://doi.org/10.1109/INFOCOMWKSHPS50562.2020.9162699
20. Peng Y et al (2019) Evaluating deep learning based network intrusion detection system in adversarial environment. In: IEEE 9th international conference on electronics information and emergency communication (ICEIEC), pp 61–66. https://doi.org/10.1109/ICEIEC.2019.8784514
21. Abou Khamis R et al (2020) Evaluation of adversarial training on different types of neural networks in deep learning-based IDSs. In: International symposium on networks, computers and communications (ISNCC), pp 1–6. https://doi.org/10.1109/ISNCC49221.2020.9297344
22. Abou Khamis R et al (2020) Investigating resistance of deep learning-based IDS against adversaries using min-max optimization. In: ICC 2020—2020 IEEE international conference on communications (ICC), pp 1–7. https://doi.org/10.1109/ICC40277.2020.9149117
23. Martins N et al (2019) Analyzing the footprint of classifiers in adversarial denial of service contexts. https://doi.org/10.1007/978-3-030-30244-3_22
24. Tomoki K, Masaomi K (2022) A method for adversarial example generation by perturbing selected pixels. In: 2022 Asia-pacific signal and information processing association annual summit and conference (APSIPA ASC), pp 1109–1114. https://doi.org/10.23919/APSIPAASC55919.2022.9980119
25. He Z, Fu S (2022) Target adversarial sample generation for malicious traffic classification model. In: Proceedings of the 2022 4th international conference on robotics, intelligent control and artificial intelligence. https://doi.org/10.1145/3584376.3584606
26. Pan Z et al (2021) Accelerating spectral normalization for enhancing robustness of deep neural networks. In: IEEE computer society annual symposium on VLSI (ISVLSI), pp 260–265. https://doi.org/10.1109/ISVLSI51109.2021.00055
27. Chang E et al (2020) Adversarial training with orthogonal regularization. In: 28th Signal processing and communications applications conference (SIU), pp 1–4. https://doi.org/10.1109/SIU49456.2020.9302247

Author Index

A
Aayush Patel, 17
Abhinandan Khan, 207
Adrija Das, 241
Agnish Paul, 49
Amlan Chakrabarti, 241
Anindita Mohanta, 129
Anirban Sarkar, 257
Anuran Bhattacharya, 207
Arati Paul, 225
Asif Iqbal Middya, 49

B
Bag, Raunak, 275

C
Chayan Halder, 161
Cortesi, Agostino, 275, 295

D
Debapratim Dawn Das, 207
Debotosh Bhattacharjee, 29, 187
Deepak Gohil, 97
Devadatta Sinha, 63
Devarsh Patel, 17
Dewashish Kumar, 257
Dharmik Jethva, 97
Dhruv Shah, 17
Dipten Kumar, 29

G
Gourab Das, 241
Grabowski, Daniel, 115

H
Het Patel, 17

K
Kaushik Roy, 161, 173
Kiran Rao, P., 309
Krish Patel, 97

M
Moumita Ghosh, 3
Mridul Ghosh, 173
Mrinal Kanti Bhowmik, 129, 145

N
Nabendu Chaki, 225, 275, 295
Nandini Basumallick, 29
Niharika Nath, 129
Novarun Deb, 17, 97

O
Oindrilla Ghosh, 295

P
Payel Rakshit, 161
Priya Saha, 145

R
Rajat Kumar Pal, 63, 207

© The Editor(s) (if applicable) and The Author(s), under exclusive license
to Springer Nature Singapore Pte Ltd. 2025

N. Chaki et al. (eds.), *Applied Computing for Software and Smart Systems*, Lecture Notes
in Networks and Systems 980, https://doi.org/10.1007/978-981-97-9762-2

Rakesh Paul, 173
Rejoy Chakraborty, 161
Rituparna Chaki, 295

S
Saeed, Khalid, 115
Saheli Patra, 207
Sanjit Kumar Setua, 207
Sankhayan Choudhury, 3
Santanu Das, 145
Sarbani Roy, 49
Sarthak Tailor, 17
Seshadri Ramana, K., 309
Shatarupa Das, 207
Sk Md Obaidullah, 173
Somnath Ghosh, 187
Soumen Roy, 63
Sourav Dey Roy, 129, 145
Subarna Chatterjee, 309
Suman Prakash, P., 309
Susobhan Ghosh, 173

T
Taniya Seal, 207
Tirthankar Ghosh, 225

U
Ujjwal Singh, 97
Utpal Roy, 63

V
Ved Vaghela, 97

Z
Zhumur Ghosh, 241

The manufacturer's authorised representative in the EU is Springer Nature Customer Service Centre GmbH, Europaplatz 3, 69115 Heidelberg, Germany. If you have any concerns regarding our products, please contact ProductSafety@springernature.com

Printed and bound by CPI Group (UK) Ltd, Croydon, CR0 4YY
01/10/2026
02225959-0003